U.S.NRC

United States Nuclear Regulatory Commission

Protecting People and the Environment

NUREG-1536
Revision 1

I0488194

Standard Review Plan for Spent Fuel Dry Storage Systems at a General License Facility

Final Report

Manuscript Completed: July 2010
Date Published: July 2010

Office of Nuclear Material Safety and Safeguards

ABSTRACT

The Standard Review Plan (SRP) for dry storage systems (DSS) provides guidance to the U.S. Nuclear Regulatory Commission (NRC) staff in the Division of Spent Fuel Storage and Transportation (SFST) for reviewing applications for a Certificate of Compliance (CoC) of a dry storage system (DSS) for use at a general license facility. This SRP is intended for use by the NRC staff. Its objectives are to:

- provide a basis that promotes a consistent regulatory review of an application for a DSS;

- promote quality and uniformity of these reviews across each technical discipline;

- present a basis for the review scope;

- identify acceptable approaches to meeting regulatory requirements; and

- develop an approach for review of each review procedure section of each chapter to assist the staff in prioritization of its review.

Title 10 of the *U.S. Code of Federal Regulations* (CFR) Part 72 (10 CFR 72), Subpart B, specifies the information needed in a license application for the independent storage of spent nuclear fuel for a site specific application. Subparts A specifies the information needed in an application for a CoC for use at a general license facility. Regulatory Guide 3.61, *Standard Format and Content for a Topical Safety Analysis Report for a Spent Fuel Dry Storage Cask*, contains an outline of the information required by the staff. This SRP is divided into 14 chapters with appendices that reflect the standard application format. Regulatory requirements, staff positions, industry codes and standards, acceptance criteria, and other information are discussed. However, the format used herein has evolved and, in some instances, superseded Regulatory Guide 3.61 to better reflect current staff practice.

In conjunction with the SRP, the SFST developed several Interim Staff Guidance (ISG) documents. An ISG addresses emergent review issues in a timely manner by staff and applicants. These ISGs were developed to address changes in requirements, reflect lessons learned and evolving technology, and document detailed technical positions. Current ISGs are available on the NRC website. Although Revision 1 of this SRP was revised to incorporate the applicable ISGs listed in Appendix C, other ISGs will continue to be developed as needed. This SRP will be revised periodically to reflect current guidance to the staff.

The review procedures sections of each chapter of this SRP have been prioritized to assist the NRC staff in its review in an effort to increase efficiency. The method used to prioritize the Review Procedures sections is documented in Appendix B. The priority of each review procedure is shown in the applicable section of each chapter.

Public Comments were solicited on this document and all public comments received are documented in Appendix D along with their resolution and associated changes to this SRP.

Comments, errors or omissions, and suggestions for improvement should be sent to the Director, Division of Spent Fuel Storage and Transportation, U.S. Nuclear Regulatory Commission, Washington, DC 20555-0001.

TABLE OF CONTENTS

LIST OF FIGURES

LIST OF TABLES

ACRONYMS AND ABBREVIATIONS

ACI	American Concrete Institute
ADE	annual dose equivalent
AISC	American Institute of Steel Construction
ALARA	as low as is reasonably achievable
ANL	Argonne National Laboratory
ANS	American Nuclear Society
ASCE	American Society of Civil Engineers
ANSI	American National Standards Institute
API	American Petroleum Institute
APSR	axial power shaping rod
ASD	allowable stress design
ASME	American Society of Mechanical Engineers
ASTM	American Society for Testing and Materials
AWWA	American Water Works Association
AWS	American Welding Society
B&PV	boiler and pressure vessel
BPRA	burnable poison rod assembly
BR	breathing rate
BWR	boiling-water reactor
CDE	committed dose equivalent
CEA	control element assembly
CEDE	committed effective dose equivalent
CFD	computational fluid dynamics
CFR	U.S. Code of Federal Regulations
CoC	Certificate of Compliance
CSFM	Commercial Spent Fuel Management Program
DBA	design-basis accident
DBE	design-basis event
DCF	dose conversion factor
DSS	dry storage system
DDE	deep dose equivalent
DE	design earthquake

DLF	design load factor
DOE	U.S. Department of Energy
EPA	Environmental Protection Agency
EPRI	Electric Power Research Institute
FR	Federal Registry
g	gram
Gr	Grashof
GTCC	greater than Class C
Gy	Gray
Gz	Graetz
HAC	hypothetical accident condition
HAZ	heat affected zone
HTGR	high-temperature gas-cooled reactor
H/U	hydrogen-to-uranium
IBC	International Building Code
ICBO	International Conference of Building Officials
ICC	International Code Council
ICRP	International Commission on Radiological Protection
INEL	Idaho National Engineering Laboratory
ISFSI	Independent Spent Fuel Storage Installation
ISG	Interim Staff Guidance
LANL	Los Alamos National Laboratory
LCO	limiting condition of operations
LDE	lens dose equivalent
LLNL	Lawrence Livermore National Laboratory
LRFD	load resistance factor design
LT	leak testing
LWR	light water reactor
mJ	milliJoule
mm	millimeter
MNOP	maximum normal operating pressure

MPa	megapascal
ms	millisecond
MT	magnetic particle examination
N	Newton
NDE	nondestructive examination
NDT	nil-ductility transition
NEI	Nuclear Energy Institute
NFPA	National Fire Protection Association
NOAA	National Oceanic and Atmospheric Administration
NRC	United States Nuclear Regulatory Commission
NRPB	National Radiation Protection Board
NRR	Office of Nuclear Reactor Regulation
OBE	operating-basis earthquake
OFA	optimized fuel assembly
ORNL	Oak Ridge National Laboratory
PNL	Pacific Northwest Laboratory
PT	liquid (dye) penetrant examination
PWHT	preheat and post-weld heat treatment
PWR	pressurized-water reactor
QA	Quality Assurance
QAPD	Quality Assurance Program Description
QC	quality control
RAI	request for additional information
RC	reinforced concrete
RCCA	rod cluster control assembly
RG	Regulatory Guide
RSICC	Radiation Safety Information Computational Center
RT	radiographic examination
SAR	Safety Analysis Report
SDE	shallow (skin) dose equivalent
SEM	scanning electron microscopy
SER	Safety Evaluation Report

SFST	Division of Spent Fuel Storage and Transportation
SI	système international (d'unités) (International System of Units)
SNF	spent nuclear fuel
SNT	American Society for Nondestructive Testing
SRP	Standard Review Plan
SSC	structures, systems, and components
SSE	safe shutdown earthquake
Sv	Sievert
TEDE	total effective dose equivalent
TEM	transmission electron microscopy
TODE	total organ dose equivalent
TS	Technical Specification
TSAR	Topical Safety Analysis Report
UBC	Uniform Building Code
UK	United Kingdom
UT	ultrasonic examination
VT	visual examination

UNITS

Btu/hr.ft.°F	British thermal units per hour-foot-degree Fahrenheit
°C	degrees Centigrade
Ci/cm^3	Curies per cubic centimeters
Ci/s	Curies per second
cm^3/s	cubic centimeters per second
°F	degrees Fahrenheit
ft	feet
ft/s	feet per second
ft^3	cubic feet
ft^3/s	cubic feet per second
g/cm^3	grams per cubic centimeters
GWd/MTU	GigaWatt days per Metric Ton Uranium
in.	inches
K	Kelvin
kg	kilogram
kgf/cm^2	kilograms force per square centimeters
kPa	kiloPascal
ksi	thousand pounds per square inch
kW	kilowatts
lb	pounds
m	meters
m^2	square meters
m^3	cubic meters
m^3/s	cubic meters per second
m/s	meters per second
mCi	millicuries (one-thousandth of a curie)
MeV	million electron volts
mg	milligram (one-thousandth of a gram)
mm	millimeters (one-thousandth of a meter)
MPa	MegaPascal (million Pascals)
mrem	millirem (one-thousandth of a rem)
mSv	millisievert (one-thousandth of a sievert)
MWd/MTU	MegaWatt days per Metric Ton Uranium
pCi/m^3	picocurie (one-trillionth of a curie)/cubic meter
PM^{10}	particulate matter (less than 10 microns in diameter)
ppm	parts per million

psi	pounds per square inch
s	second
Sv	sievert
μCi	microcurie (one-millionth of a curie)
μCi/cm^2	microcurie per square centimeter
W/m.K	Watts per meter - Kelvin

GLOSSARY

The following terms are defined here by the staff for the purpose of this document.

Acceptance Test. Tests conducted by the applicant to ensure that material or component produced in a given production run is in compliance with the material or design requirements of the application. Acceptance tests are also used to ensure that the process is operating in a satisfactory manner by using statistical data for selected measurable parameters.

Accident-Level. A term used to include both design-basis accidents and design-basis natural phenomenon events and conditions.

Areal Density. Mass per unit area, usually expressed in grams per square centimeters (g/cm^2). In this document, this term is used to describe the distribution of neutron absorber content in a material.

Adequate Margin. In the design of structures, systems, and components, the margin for safety is achieved by satisfying the acceptance criteria of the codes and standards for the specified design criteria loads, and the design basis (performance requirements). The reviewer must judge if the calculated design bases values require any margins with respect to the acceptance criteria of the codes and standards. This may depend on the uncertainties associated with the calculation of predicted design bases values (stress, displacements, etc.) used as reference for the performance of the structures.

As Low As is Reasonably Achievable (ALARA). Making every reasonable effort to maintain exposures to radiation as far below the dose limits in 10 CFR Part 20 as is practical and consistent with the purpose for which the licensed activity is undertaken taking into account the state of technology, the economics of improvements in relation to state of technology, the economics of improvements in relation to benefits to the public health and safety, other societal and socioeconomic considerations, and in relation to utilization of nuclear energy and licensed materials in the public interest (10 CFR 20.1003). Per 10 CFR 72.3, ALARA means as low as reasonably achievable taking into account the state of technology, and the economics of improvement in relation to: (1) benefits to the public health and safety, (2) other societal and socioeconomic considerations, and (3) the utilization of atomic energy in the public interest.

Benchmarking. Establishment of the bias of a computer code for a particular application by comparison of the calculated results with the measured results of relevant representative experiments. For purposes of criticality analyses, benchmarking is the process of establishing the bias of the calculational method, which includes aspects such as the computer code, cross sections set, analyst's technique, and analysis assumptions.

Bias. ANSI/ANS-8.1 defines bias as "a measure of the systematic differences between calculational method results and experimental data" and uncertainty in the bias as "a measure of both the accuracy and the precision of the calculations and the uncertainty of the experimental data." See NUREG/CR-6361 for further discussion of bias. Bias defined as the average of the differences between results and measurements may be acceptable, provided that one adequately considers the variation in the differences.

Burnable Poison Rod Assembly (BPRA). An assembly of poison rods used to absorb neutrons created in the nuclear reactor to control the power produced in the associated fuel assembly during the early core life. The BPRs are inserted into the fuel assemblies through the upper end fittings of the assembly and held in place against lift forces in the core by a retainer mechanism. BPRs within the spent fuel assembly envelope may be approved for storage in a dry storage system as part of the spent fuel assembly.

Burnup. The measure of the thermal power produced in a specific amount of nuclear fuel through fission, usually expressed in units of MWd/MTU (megawatt days per metric ton of uranium). For the purpose of assessing the allowable contents, the maximum burnup(s) of the fuel should be specified in terms of the average burnup of the entire fuel assembly (i.e. assembly average). For the purpose of assessing fuel cladding integrity in the materials review, the rod with the highest burnup within the fuel assembly should be specified in terms of peak rod average burnup.

Calculational Method. The calculational procedures – mathematical equations, approximations, assumptions, and associated numerical parameters (e.g., cross sections) – that yield the calculated results (ANSI/ANS-8.1-1998).

Canister. In a dry storage system for spent nuclear fuel, a metal cylinder that is sealed at both ends and may be used to perform the function of confinement. Typically, a separate overpack performs the radiological shielding and physical protection function.

Canning. To store damaged or consolidated spent nuclear fuel or nuclear fuel debris in a separate container and confine it in such a way that degradation of the fuel during storage will not pose operational safety problems with respect to its removal from storage [10 CFR 72.122(h)(1)].

Cask. In a dry storage system using the cask design for spent nuclear fuel, a passive stand-alone component that performs the functions of confinement, radiological shielding, decay heat removal, and physical protection of spent fuel during normal, off-normal, and accident-level conditions (NUREG-1571).

Certificate of Compliance. The certificate issued by the NRC that approves the design of a spent nuclear fuel storage cask in accordance with the provisions of Subpart L of 10 CFR 72 (10 CFR 72.3).

Code. A generic reference to a national or "consensus" code, standard, and specification, or specifically to the ASME Boiler and Pressure Vessel Code (ASME B&PV Code).

Committed Dose Equivalent (H_T,50). The dose equivalent to organs or tissues of reference (T) that will be received from an intake of radioactive material by an individual during the 50-year period following the intake (10 CFR 20.1003).

Confinement. The ability to prevent the release of radioactive substances into the environment (NUREG-1571).

Confinement System. Those systems, including ventilation, that act as barriers between areas containing radioactive substances and the environment (10 CFR 72.3).

Confirmatory Calculations. Calculations made by the reviewer to determine whether the cask design and specifications meet the requirements of the Code of Federal Regulations. These calculations do not replace the design calculations and are not intended to endorse the applicant's calculations.

Construction. Includes materials, design, fabrication, installation, examination, testing, inspection, and certification as required in the manufacture and installation of components.

Control Element Assembly (CEA) – An assembly of neutron poison elements used to control the reactor power during operations, if needed, and to provide shutdown capability. This component is designed for operations within the fuel assembly envelope, and when stored with spent fuel, fits within that envelope.

Controlled Area. For an independent spent fuel storage installation (ISFSI), that area immediately surrounding the ISFSI for which the licensee exercises authority over its use and within which ISFSI operations are performed (10 CFR 72.3). For a nuclear power plant, that area outside of a restricted area but inside the site boundary to which access can be limited by the licensee for any reason (10 CFR 20.1003).

Criticality. A measurement of the state of a fission system.

Curie. The basic unit of radioactivity. A curie is equal to 37 billion (3.7 X 10^{10}) disintegrations per second.

Damaged Fuel. Spent nuclear fuel is considered damaged for storage purposes if it cannot fulfill its regulatory or design function. Specific conditions that define damaged fuel are provided in Section 8.4.17.2 of this SRP. Section 8.6, Supplemental Information for Methods for Classifying Fuel, provides methods for classifying spent nuclear fuel as damaged.

Damaged-Fuel Can. A metal enclosure that is sized to confine one damaged spent fuel assembly. A fuel can for damaged spent fuel with damaged spent-fuel assembly contents must satisfy fuel-specific and system-related functions for undamaged SNF required by the applicable regulations.

Degradation. Any change in the properties of a material that adversely affects the behavior of that material; adverse alteration (ASTM C1174-97).

Design Bases. The information that identifies the specific functions to be performed by a structure, system, or component (e.g., spent fuel storage cask) and the specific values or ranges of values chosen for controlling parameters as reference bounds for design.

Design Earthquake. The design earthquake ground motion for a site where a cask system may be used that is determined in accordance with 72.102 or 72.103.

Design Event (I, II, III, or IV). Conditions and events as defined and used for an independent spent fuel storage installation in ANSI/ANS 57.9.

Double Contingency Principle. A design principle requiring that at least two unlikely, independent, and concurrent or sequential changes in conditions essential to nuclear criticality safety must occur before a criticality accident is possible (10 CFR 72.124(a)).

Exclusion Area. At a nuclear reactor site, the area surrounding the reactor in which the reactor licensee has the authority to determine all activities including exclusion or removal of personnel and property from the area. This area may be traversed by a highway, railroad, or waterway provided these are not so close to the facility as to interfere with normal operations of the facility, and provided appropriate and effective arrangements are made to control traffic on the highway, railroad, or waterway, in case of emergency, to protect the public health and safety. Residence within the exclusion area shall normally be prohibited. In any event, residents shall be subject to ready removal in case of necessity. Activities unrelated to operation of the reactor may be permitted in an exclusion area under appropriate limitations, provided that no significant hazards to the public health and safety will result (10 CFR 50.2).

Gray (Gy). The SI unit of absorbed dose. 1 Gy is equal to 100 rad.

Hard Receiving Surface. For a horizontal or vertical drop, need not be an unyielding surface; rather, the receiving surface may be modeled as a reinforced concrete pad on engineered fill.

High Burnup Fuel. Spent nuclear fuel with burnups (see "Burnup") generally exceeding 45 GWd/MTU.

Hoop Stress. The tensile stress in the cladding wall in the circumferential orientation.

Important Confinement Features. See "important to safety."

Important to Safety, "Important to Nuclear Safety," or "Structures, Systems, and Components Important to Safety." Those features of a dry storage system that have one or more of the following functions: (1) maintain the conditions required to store spent nuclear fuel safely; (2) prevent damage to the spent nuclear fuel cask during handling or storage; or (3) provide reasonable assurance that spent nuclear fuel can be received, handled, containerized, stored, and retrieved without undue risk to the health and safety of the public. ANSI/ANS 57.9 uses the term "important confinement features"; however, NRC does not find this term acceptable. Per Regulatory Guide 3.60, *Design of an Independent Spent Fuel Storage Installation (Dry Storage)*, "important to safety" should be substituted for "important confinement features" in the standard.

Interim Staff Guidance (ISG). Supplemental information that clarifies important aspects of regulatory requirements. An ISG provides NRC review guidance to NRC Staff in a timely manner until standard review plans are revised accordingly.

Low Burnup Fuel. Spent nuclear fuel with burnups (see "Burnup") generally less than 45 GWd/MTU.

Margin of Safety, or MofS. This term may be defined, through a factor of safety, f.s = capacity/demand, as MofS = F.S.(capacity/demand)-1 (with minimum acceptable MofS$\geq$ 0.0)."

Misloading. The placement in a cask of spent nuclear fuel in a configuration not supported by the cask's design basis or technical specifications. Also, the placement in a cask of spent nuclear fuel with characteristics that do not meet the characteristics of the cask's allowable contents.

Monitoring. Testing and data collection to determine the status of a dry storage system and to verify the continued efficacy of the system on the basis of measurements of specified parameters including temperature, radiation, and functionality and/or characteristics of components of the system. With respect to radiation, per 10 CFR 20.1003, monitoring means the measurement of radiation levels, concentrations, surface area concentrations or quantities of radioactive material, and the use of the results of these measurements to evaluate potential exposures and doses.

Neutron Absorber. Also known as "poison." Materials that have high neutron absorption cross section and are used to absorb neutrons to make a fission system less reactive. They are used to ensure subcriticality during normal/offnormal/accident-level conditions in containers of fissile materials.

Nondestructive Examination (NDE). Testing, examination, and/or inspection of a component that does not affect the functionality and performance of the component. NDE can be broadly divided into three categories: visual, surface, and volumetric examinations. Additional information may be found in the ASME B&PV Code, Section V, *Nondestructive Examination*, Appendix A.

NDE-related terms in order of increasing severity:

Discontinuity: An interruption in the normal physical structure of a material. Discontinuities may be unintentional (such as those formed inadvertently during the fabrication process) or intentional (such as a drilled hole).

Indication: Sign of a discontinuity observed when using an NDE method.

Flaw: An imperfection in an item or material which may or may not be harmful.

Defect: A flaw that, due to its size, shape, orientation, location, or other properties, is rejectable to the applicable construction code. Defects may be detrimental to the intended service of a component and the component must be repaired or replaced.

Common NDE examination methods include:

LT   leak testing
MT   magnetic particle examination
PT   liquid penetrant examination
RT   radiographic examination
UT   ultrasonic examination
VT   visual examination

Non-Fuel Hardware. Hardware that is not an integral part of a fuel assembly. Burnable Poison Rod Assembly (BPRA), Control Element Assembly (CEA), Thimble Plug Assembly (TPA), etc. are typical non-fuel hardware.

Normal Events and Conditions. Conditions that are intended operations, planned events, and environmental conditions, that are known or reasonably expected to occur with high frequency during storage operations The maximum level of an event or condition it that expected to routinely occur. The cask system is expected to remain fully functional and to experience no temporary or permanent degradation from normal operations, events and conditions. Specific normal conditions to be addressed are evaluated for each dry storage system and are documented in a safety analysis report for that system.

Normal Means. The ability to move a fuel assembly and its contents by the use of a crane and grapple used to move undamaged assemblies at the point of cask loading. The addition of special tooling or modifications to the assembly to make the assembly suitable for lifting by crane and grapple does not preclude the assembly as being considered moveable by normal means.

Off-Normal Events or Conditions. The maximum level of an event or condition that although not occurring regularly can be expected to occur with moderate frequency and for which there is a corresponding maximum specified resistance, limit of response, or requirement for a given level of continuing capability. "Off-Normal" events and conditions are similar to "Design Event II" of ANSI/ANS 57.9. An independent spent fuel storage installation structure, system, or component is expected to experience off-normal events and conditions without permanent deformation or degradation of capability to perform its full function (although operations may be suspended or curtailed during off-normal conditions) over the full license period.

Preferential Loading. A non-uniform loading configuration of spent fuel assemblies within a dry storage system, that is typically specified by assigning a fuel zone designation to each basket cell, and specifying limiting nuclear and physical parameters of SNF assemblies that can be loaded into each zone. Preferential loading is often used as a means to optimize allowable SNF parameters (e.g. burnup, cooling time, decay heat), while satisfying the shielding, criticality, and thermal performance objectives of the cask system.

Qualification Test. A test, or series of tests, that is conducted at least once for a given manufacturing process and set of material specifications to demonstrate the quality and durability of the component such as neutron absorber product over its licensed service life.

Rad. The unit of absorbed dose. 1 rad is equal to the absorption of 100 ergs per gram.

Ready Retrieval. The ability to move a canister containing spent fuel to either a transportation package or to a location where the spent fuel can be removed. Ready retrieval also means maintaining the ability to handle individual or canned spent fuel assemblies by the use of normal means

Real Individual. A person who is not a nuclear worker and who is at or beyond the controlled area of an independent spent fuel storage installation, a nuclear power plant, or other nuclear facility. For example, a real individual may be anyone living, working, or recreating close to the facility for a significant portion of the year.

Reasonable Assurance. NRC staff base their decisions on the adequacy of a dry storage system design to protect public health and safety on a variety of factors including: technical evaluations, test and operational data, compliance with NRC requirements, and insights from operational safety events.

Recovery. The capability to return the stored radioactive material to a safe condition after an accident event without endangering public health and safety. This generally means ensuring that any potential release of radioactive materials to the environment or radiation exposures is not in excess of the limits in 10 CFR Part 20 during post-accident recovery operations.

Rem. The special unit of any of the quantities expressed as dose equivalent. The dose equivalent in rems is equal to the absorbed dose in rads multiplied by the quality factor (1 rem = 0.01 sievert) (10 CFR 20.1004).

Restricted Area. An area to which access is limited by the licensee for the purpose of protecting individuals against undue risks from exposure to radiation and radioactive materials. Restricted area does not include areas used as residential quarters, but separate rooms in a residential building may be set apart as a restricted area (10 CFR 20.1003).

Retrievability. In accordance with 10 CFR 72.122(l), storage systems must be designed to allow ready retrieval of spent fuel, high-level radioactive waste, and reactor-related GTCC waste for further processing or disposal._

Safety Analysis Report (SAR). In the context of this standard review plan, the report submitted to the NRC staff by a certificate applicant to present information related to the design of a dry storage system. This document provides the justification and analyses to demonstrate that the design meets the requirements and acceptance criteria.

Safety Evaluation Report (SER). In the context of this standard review plan, the report prepared by the NRC staff to present findings and recommendations relating to the acceptability of an applicant's safety analysis and other required documents submitted as part of a certificate application. The SER also identifies the bases for those recommendations and the recommended technical specifications ("operating controls and limits" or "conditions of use").

Safety Functions. The functions that dry storage system structures, systems, and components important to safety are designed to maintain include:

- Protection against environmental conditions,
- Content Temperature Control,
- Radiation Shielding,
- Confinement,
- Sub-criticality control,
- Retrievability.

Sievert (Sv). The SI unit of any of the quantities expressed as dose equivalent. 1 Sv equals 100 rem. The dose equivalent in sieverts equals the absorbed dose in grays multiplied by the quality factor (10 CFR 20.1004).

Spent Nuclear Fuel, (SNF). Nuclear fuel that has been withdrawn from a nuclear reactor following irradiation, has undergone at least one year's decay since being used as a source of energy in a power reactor, and has not been chemically separated into its constituent elements by reprocessing. Spent fuel includes the special nuclear material, byproduct material, source material, and other radioactive materials associated with fuel assemblies (10 CFR 72.3).

Subcritical. The state at which the number of fission neutrons decreases with time and the effective neutron multiplication factor (k_{eff}) is less than unity.

Supplemental Shielding. At an independent spent fuel storage installation, an engineered radiation shield (principally neutron and gamma radiation) such as an earthen berm or concrete wall. Supplemental shielding shall be deemed as component(s) important to safety and be specified in the Technical Specifications as a condition for use of the system as designed, if credited in the shielding analyses for meeting 72.104(a) or 72.106(b) requirements.

Thimble Plug Assembly (TPA) – An assembly of short rods used to restrict the flow of coolant through a fuel assembly by being inserted into the assembly's guide tubes. This component is designed for operations within the fuel assembly envelope, and when stored with spent fuel, fits within that envelope.

Total Effective Dose Equivalent (TEDE). The sum of the deep-dose equivalent for external exposures and the committed effective dose equivalent for internal exposures (10 CFR 20.1003).

Unrestricted Area. An area to which access is neither limited nor controlled by the licensee (10 CFR 20.1003).

Validation. Demonstration of the validity of a computer code for use in a general area of application by comparison of the code's calculational results with the measured results from a variety of experiments spanning the area of intended applications.

Volume Percent. The percent of a mole of the material that is present in a volume equal to the standard volume for the material as a gas; the volume occupied by one mole of the material as a gas at standard conditions for gases (760 mm Hg [760 torr] pressure and 0°C [32°F] temperature).

INTRODUCTION

This document is a Standard Review Plan (SRP). It is intended to provide guidance to the NRC staff conducting the safety review of an application for a spent fuel dry storage system (DSS) for facilities storing spent fuel under the general license authorized by 10 CFR 72.210. A general license authorizes a nuclear power plant licensee to store spent nuclear fuel (SNF) in NRC-approved casks at a site that is licensed to operate a power reactor under 10 CFR Part 50.

This SRP was developed to promote a consistent regulatory review of an application for a DSS, present a basis for the review scope, and identify acceptable approaches to meeting regulatory requirements.

This introduction provides an overview of the DSS and the Safety Analysis Report (SAR) review process, and assists the project manager in the coordination of the review effort. It is also designed to help individual technical reviewers understand how their specific review should be coordinated and integrated with other disciplines to produce a complete Safety Evaluation Report (SER).

This SRP may be revised and updated as the need arises to clarify the content, correct errors, or incorporate modifications approved by the Director of the Division of Spent Fuel Storage and Transportation (SFST). Comments, suggestions for improvement, and notices of errors or omissions will be considered by and should be sent to the Director, Division of Spent Fuel Storage and Transportation, Office of Nuclear Material Safety and Safeguards, U.S. Nuclear Regulatory Commission, Washington, DC 20555-0001.

Use of Dry Storage Systems

In accordance with the requirements set forth in 10 CFR 72.212, a DSS may be used to store SNF in an independent spent fuel storage installation (ISFSI) under a general license. At present, any holder of an active reactor operating license under Title 10, Part 50, of the *U.S. Code of Federal Regulations* (10 CFR Part 50) has the authority to construct and operate an ISFSI using NRC-approved cask designs under the provisions of the general license.

The DSS safety review is primarily based on the information provided by an applicant, or cask vendor, in a SAR. Section 72.230 of 10 CFR Part 72 requires inclusion of a SAR in each application for approval of SNF cask storage design. Before submitting a SAR, an applicant should have designed the DSS considering as-low-as-is-reasonably-achievable (ALARA) principles for radiation protection and analyzed it in sufficient detail to conclude that it can be properly fabricated and safely operated without endangering the health and safety of the public. The SAR is the principal document in which the applicant provides the information on the design and operational features and their associated technical bases. The reviewers need to understand the design and operational features and their technical bases, including but not limited to the selection of materials and geometries, mathematical models and equations used, computer models and calculated results in order to be able to draw conclusions that the storage cask is acceptable for use.

Technical Review Oversight

Cask designers are responsible for the safety of the cask design, and the cask users are responsible for safely operating the cask system at Part 50 reactor sites and complying with appropriate safety regulations. The mission of the regulator is to license and regulate the use of each DSS and ensure adequate protection of public health and safety. The value of the NRC review team is its independent expertise in identifying and resolving potential design or operational deficiencies; potential analytical errors; significant uncertainties in novel design approaches; or other non-compliance problems. If otherwise left unchecked by the designer, user and regulator, these issues could potentially lead to the unsafe or non-compliant use of the DSS.

Several considerations may influence the depth and rigor that is needed for a reasonable assurance determination of both safety and compliance. These include the novelty of the design (as compared to existing designs); safety margins; operational experience; defense-in-depth, and the relative risks that have been identified for normal operations and potential accident conditions. Consideration should also be given to the design parameters and methodology approved in the SAR and their possible use in subsequent 10CFR 72.48(c) changes to the design or procedures by the licensee or certificate holder. Any aspect of the design or procedures that the NRC determines should not be changed by either the certificate holder or general licensee, without prior NRC approval, must be placed in the CoC conditions or in the attached technical specifications.

As described further below, each review procedure is prioritized using a graded approach that factored in many of these considerations for a typical review. The prioritization was developed with the expertise of NRC reviewers within each discipline, who have several years of regulatory experience with the current fleet of certified spent fuel storage cask designs. These priorities are intended to serve as a guidepost to the depth and rigor that is expected for a typical review; but should not be treated as absolutes for every case. It is the responsibility of the individual reviewer to assess the design and determine the ultimate rigor needed to make a safety determination, with reasonable assurance, in each review area. In other words, reviewers should consistently apply these review procedures for each case, but may need to adjust the scope of review in some areas based on safety margins, operational experience, defense-in-depth considerations, design novelty, or other issues that are unique to each proposed design.

Review Process

The purpose of the staff review is to evaluate the proposed cask design, contents and operations, and provide regulatory confirmation of reasonable assurance of safe design and construction of the cask.

The reviews are performed by project management and technical review staff with expertise in the technical discipline areas described in the review plan. Due to the complexity of the technical information in the application, coordination among the different disciplines is important to ensure a consistent, uniform, and quality review. As described in the flow charts of each chapter, technical issues can overlap between the disciplines and many rely on input from other areas.

This SRP is guidance meant to be used in unison with the current ISGs. ISGs provide guidance concerning specific, important issues that either are not currently addressed in the SRP or need clarification beyond that in the present SRP text and may delineate specific review procedures. For this reason, the staff should be familiar with ISGs that may supersede this guidance and these new ISGs should be used together with this SRP in the review of a DSS application. ISGs may be discontinued if they are fully incorporated into all applicable regulatory guidance documents. Appendix C lists the ISGs from 1 to 22, and identifies which ones have been incorporated in this revision of the SRP.

The staff may consult the SERs of previous CoC amendments, if reviewing an amendment to a currently approved design, as well as the SERs for approved systems of similar design to understand past NRC determinations regarding analyses affecting or similar to those in the application under review.

For amendments, the staff should review the entire amendment to ensure that all the licensing changes have been identified by the applicant. Amendments may range from minor changes in the design, contents, or operations, to adding new major component designs such as storage overpacks, transfer casks, and canisters.. Some amendments such as content and design changes, are founded upon the design and methodologies previously reviewed by NRC for that system. Evaluation of amendment changes to a DSS are often based on the performance of the contents, canister, and overpacks as an integrated system. As a result, portions of previously approved components, contents, or methodologies in the SAR may be re-examined to ensure that the new system under the amendment proposal meets Part 72 requirements. During the audit review of an amendment, the staff may occasionally find errors or other safety questions that affect part of the previously approved design. The staff may need to review that part of the SAR and ask questions to assure the design remains safe and compliant with applicable regulations. The questions should be limited to understanding and resolving the specific technical issue, and should consider past precedents, regulatory guidance, and risk significance, as appropriate. The staff should also consider other processes (e.g. inspections, enforcement actions, generic issue program, etc..) to resolve these potential type of safety questions with a previously approved design..

In case the reviewer finds that the information provided in the SAR is not properly justified, the reviewer may develop and then forward to the applicant questions requesting clarification of technical issues via a Request for Additional Information (RAI). The applicant's response to the RAI should be reviewed for accuracy as well as the need to update the applicant's SAR. The RAI process is repeated as necessary, consistent with NRC's in-office instructions, until the application is deemed technically acceptable, or until the application review is terminated by the NRC or withdrawn by the applicant.

Once the technical review is complete, a draft SER is written that summarizes the results of the review and the cognizant NRC Project Manager approves the SER. If the NRC intends to approve the application, the staff prepares *Federal Register* notices for a direct final rule and a companion proposed rule. The rulemaking notices identify the ADAMS numbers for the draft CoC, TSs and SER. During the rulemaking process, stakeholders and members of the public are allowed to comment on the draft CoC, TSs and SER. After addressing and responding to any public comments, the NRC staff modifies the proposed CoC, TS and preliminary SER, if necessary, and issues the Final CoC, TS, and SER. The rulemaking adds the CoC, or in the case of an amendment to an existing CoC, the CoC amendment, to the list of approved cask designs in 10 CFR 72.214.

Safety Evaluation Report and Content

The results of a SAR review are documented in an SER. The final determination of the organization of an SER is determined by the review project manager, but the SER typically is organized in the same manner as this SRP and contains the following information:

- A general description of the system, operational features, and SNF specifications.

- A summary of the approach used by the applicant to demonstrate compliance with the regulations, and a description of the reviews that the staff performed to confirm compliance.

- Comparison of systems, components, analyses, data, or other information important in the review analysis to the acceptance criteria, in addition to, conclusions regarding the acceptability, suitability, or appropriateness that this information provides reasonable assurance the acceptance criteria has been met.

- Summary of aspects of the review that were selected or emphasized; matters that were modified by the applicant: aspects of the cask's design that deviates from the criteria stated in the SRP; and the bases for any deviations from the SRP.

- Summary statements for evaluation findings at the end of each chapter.

Content of SRP

Each chapter of the SRP is organized into the following sections:

- Review Objective
- Areas of Review
- Regulatory Requirements
- Acceptance Criteria
- Review Procedures
- Evaluation Findings

Review Objective. This section provides the purpose and scope of the review and establishes the major review objectives for the chapter. The reviewer should obtain reasonable assurance during the review that the objectives are met. It also discusses the information needed or coordination expected from reviewers of other SAR chapters to complete the subject technical review.

Areas of Review. This section describes the systems, components, analyses, data, or other information and their sequence in the discussion of acceptance criteria and review procedures sections of each chapter.

Regulatory Requirements. This section summarizes the regulatory requirements from 10 CFR Part 72 pertaining to the given SAR section. This list is not all inclusive (e.g., some parts of the regulations, such as 10 CFR Part 20, are assumed to apply to all chapters of the SAR). 10 CFR Part 72 sections applicable to a DSS are listed in 10 CFR 72.13(d). In addition, 10 CFR 72.13(c) is important to the applicant to ensure that the general licensee does not violate those conditions. The reviewer should read the complete language of the current version of 10 CFR Part 72 to determine the proper set of regulations for the section being reviewed.

Acceptance Criteria. This section addresses the design criteria and in some cases specific analytical methods that NRC staff reviewers have found to be acceptable for meeting regulatory requirements, specified in 10 CFR Part 72, that apply to the given SAR chapter. The acceptance criteria are organized in accordance with the review areas established in Section 2 of the specific chapter and identify the type and level of information that should be in the application.

These acceptance criteria typically set forth the solutions and approaches that staff reviewers have previously determined to be acceptable in addressing a specific safety concern or design area that is important to safety. These solutions and approaches are discussed in the SRP so that staff reviewers can implement consistent and well-understood positions as similar safety issues arise in future cases. These solutions and approaches are acceptable to the staff, but they are not the only possible solutions and approaches.

Substantial staff time and effort has gone into developing these acceptance criteria. Consequently, a corresponding amount of time and effort may be required to review and accept new or different solutions and approaches. Thus, applicants proposing solutions and approaches to new safety issues or analytical techniques other than those described in the SRP may experience longer review times and more extensive staff questioning in these areas. An alternative for the applicant is to propose new methods on a generic basis, apart from a specific license application. Such an alternative proposal could consist of a submittal of a Topical Safety Analysis Report (TSAR). This type of application could form the basis for either a change in the staff interpretation of the regulatory requirements or support a request for rulemaking to change the requirements themselves.

Review Procedures.

This section presents a general approach that reviewers typically follow to establish reasonable assurance that the applicable acceptance criteria have been met. As an aid to the reviewer, this section may also provide information on what has been found acceptable in past reviews. Standards that have been found acceptable in specific licensing reviews, or are desirable, but not specifically identified in existing regulatory documents, are identified in this section. Since many of the reviews are interdisciplinary, the reviewer should coordinate with other reviewers, as necessary, for identification of issues in other SAR chapters.

Each review procedure has been assigned a HIGH, MEDIUM or LOW priority, following application of the prioritization process described in Appendix B. These priorities are intended to provide guidance to the reviewer regarding the relative level of effort typically applied in implementing each procedure. As previously discussed, unique aspects of an application may result in an adjustment to the scope of review in a specific technical area. Specifically, the following can be used as general guidance on the implications of the priorities for the staff review:

> **HIGH** priority means the NRC staff review should ensure all items in the applicant's submittal are complete and correct as specified in the review procedure. This represents the most comprehensive review where many of the analytical methods, assumptions, and supporting references are evaluated. The reviewer may need to perform independent confirmatory analysis to validate the results of the safety analysis calculations. It is expected a reviewer would spend approximately 60 percent of his or her review time focused on the high priority review procedures.

> **MEDIUM** priority means the NRC staff should review the applicant's submittal for completeness and correctness in key areas. This represents a review in which key analytical methods, key assumptions, and key supporting references are checked and

evaluated. It is expected a reviewer would spend approximately 30 percent of his or her review time focused on the medium priority review procedures.

LOW priority means the NRC staff review should ensure that the applicant's submittal contains all of the requested information. A limited review of selected portions of the application for correctness would be performed. Given its relative significance, the reviewer should generally consider the applicant's analysis to be complete and accurate and forego independent confirmation, unless there is a reason to believe otherwise. However, if a problem is detected, the reviewer must thoroughly evaluate and resolve the issue. It is expected a reviewer would spend approximately 10 percent of his or her review time focused on the low priority areas.

The prioritized review procedures are intended to ensure that staff focus most of their effort on the areas considered to have the greatest impact on safety and compliance with regulatory limits. While some issues could possibly escape detection and resolution through this audit review, they would be of lower regulatory significance. It is important to remember that the priority designations were developed on a generic basis and may need to be adjusted depending upon the characteristics of specific applications. It is the responsibility of the individual reviewer to assess the design and determine the ultimate rigor needed to make a safety determination, with reasonable assurance, in each review area.

Finally it should be noted that a low or medium priority review procedure does not mean an application is exempted from any associated regulatory requirement, design requirement, or safety analyses that is expected within the review objectives and acceptance criteria in this SRP.

Evaluation Findings. This section provides example summary statements for evaluation findings to be incorporated into the SER for each area of review. The evaluation findings are prepared by the reviewer based on the satisfaction of the regulatory requirements. The findings are published in the SER.

GENERAL INFORMATION EVALUATION

1.1   Review Objective

The purpose of reviewing the general description of the Spent Fuel dry storage system (DSS) is to ensure that the applicant has provided a non-proprietary description, or overview, that is adequate to familiarize reviewers and other interested parties with the pertinent features of the system.

1.2   Areas of Review

The general description should be reviewed by all reviewers, regardless of their specific review assignments, to obtain a basic understanding of the DSS, its components, and the protections afforded for the health and safety of the public. Because much of the information relevant to this initial aspect of the DSS review is presented in more detail in other chapters of this SRP, this chapter focuses on familiarization with the DSS and consistency of the DSS general description with the remaining chapters of the safety analysis report (SAR). The SAR should be reviewed for adequacy of the DSS and DSS support system descriptions and drawings. Areas of review addressed in this chapter include the following:

> *DSS Description and Operational Features*
> *Drawings*
> *DSS Contents*
> *Qualifications of the Applicant*
> *Quality Assurance*
> *Consideration of 10 CFR Part 71 Requirements Regarding Transportation*

1.3   Regulatory Requirements

This section presents a summary matrix of the portions of U.S. Code of Federal Regulations (CFR) Part 72, "Licensing Requirements for the Independent Storage of Spent Nuclear Fuel, High-Level Radioactive Waste and Reactor-Related Greater Than Class C Waste," Title 10, "Energy" (10 CFR Part 72) that are relevant to the review areas addressed by this chapter. The NRC staff reviewer should read the exact regulatory language. Table 1-1 matches the relevant regulatory requirements associated with this chapter to the areas of review.

Table 1-1 Relationship of Regulations and Areas of Review

Areas of Review	10 CFR Part 72 Regulations					
	72.2(a)(1), (b)	72.122 (a), (h)(1)	72.140 (c)(2)	72.230 (a)	72.230 (b)	72.236(a), (c), (h),(m)
DSS Description and Operational Features	•	•		•		
Drawings	•			•		
DSS Contents	•					•
Qualifications of the Applicant	•					
Quality Assurance	•		•			
Consideration of 10 CFR Part 71 Certified Transportation Cask System Requirements	•				•	•

1.4   Acceptance Criteria

This section identifies the acceptance criteria for the material provided in the introduction. This initial aspect of the DSS review should contain sufficient information to allow all reviewers, regardless of their specific review assignments, to understand the principal functions and design features of the DSS.

1.4.1   DSS Description and Operational Features

The application should contain a broad overview and a general, non-proprietary description (including engineering drawings, sketches, and illustrations) of the DSS. This information should clearly identify the functions of all principal components and principal auxiliary equipment, and provide a list of those components classified as being "important to safety." Important aspects from all of the disciplinary areas should be summarized. If there are several versions of the cask because of design limitations of nuclear power plants and ISFSIs, the differences between the versions should be delineated. Typical operational sequences for loading and unloading procedures should be described.

If the potential exists that the DSS will be used to store damaged fuel, the SAR should include a discussion of how the sub-criticality requirement of 10 CFR 72.236(c) and the wet or dry loading and unloading requirements of 10 CFR 72.236(h) will be maintained.
The reviewer should verify that any documents submitted to the NRC in other applications and incorporated in whole or in part have been indexed, and a summary has been included in the appropriate section of the SAR.

1.4.2 Drawings

Drawings should be included in the first chapter of the SAR. The drawings should contain sufficient detail to allow the reviewer to understand the operation of the DSS and any special equipment used for loading, unloading, transportation, or long-term storage of the DSS. Also, the drawings should provide enough detail to allow the reviewer the option of developing an analysis model for confirmatory calculations.

Ideally, the drawings should be non-proprietary. However, in some cases, the applicant may request to have certain specific portions of the drawings classified as proprietary. Reviewers should note that any drawings relied on as the technical basis for adding the DSS design to the list of approved DSSs contained in Subpart K of 10 CFR 72 become part of the public record. Such drawings will not be treated as proprietary and will be made available to the public [10 CFR 2.390].

Any request for withholding from public disclosure subject to the provisions of 10 CFR 2.390 should be accompanied by an affidavit and must include information to support the claim that the material is proprietary. The NRC Project Manager will develop and administer public disclosure determinations, and the Office of the General Counsel will review them for compliance with the requirements of 10 CFR 2.390.

1.4.3 DSS Contents

The reviewer should ensure specifications are provided for the contents expected to be stored in the DSS (normally spent nuclear fuel [SNF]). These specifications may include, but not be limited to, type of SNF (i.e., boiling-water reactor [BWR], pressurized-water reactor [PWR], or both); number of SNF assemblies the cask can accommodate; maximum allowable enrichment of the fuel before any irradiation; burnup (i.e., MWd/MTU); minimum acceptable cooling time of the SNF before storage in the DSS (e.g., aged at least 1 year); maximum heat designed to be dissipated; maximum SNF loading limit; condition of the SNF (i.e., intact, undamaged, damaged, etc.); weight and nature of non-SNF contents; and inert atmosphere requirements.

1.4.4 Quality Assurance

Reviewers should verify that the application describes the proposed quality assurance (QA) program and cites the applicable implementing procedures. This description should satisfy all requirements of 10 CFR Part 72, Subpart G. A detailed review of the QA program to be described in the SAR is presented in Chapter 14, "Quality Assurance Evaluation," of this SRP.

1.4.5 Consideration of 10 CFR Part 71 Requirements Regarding Transportation

If the DSS has previously been evaluated for use as a transportation cask, the submittal should include the Part 71 Certificate of Compliance (CoC) and associated documents in accordance with 10 CFR 72.230(b). If application for storage is submitted, the transportability, per 10 CFR 72.236(m) should be addressed. (See Section 1.5.5).

1.5 Review Procedures

Figure 1-1 presents an overview of the evaluation process and a complete bulleted listing of pertinent information for each chapter. Figure 1-1 and the corresponding figures in each chapter of this Standard Review Plan (SRP) provide a means to coordinate the review among the NRC staff disciplines.

Regulatory requirements of 10 CFR Part 72 applicable to the general description review are delineated in the following subsections. Since the review of the General Description of the SAR is interdisciplinary, the reviewer should coordinate with other reviewers (e.g., structural, thermal, shielding, criticality, materials), as necessary, for identification of related issues.

1.5.1 DSS Description and Operational Features (MEDIUM Priority)

Reviewers should verify that the application provides a broad overview of the DSS design that is non-proprietary and may be used as a tool to familiarize interested parties with the features of the proposed DSS. This description should present the principal characteristics of the DSS including its dimensions, weight, and construction materials. In addition, the description should clearly identify all components considered important to safety. Features such as the confinement vessel, fuel basket, valves, lids, seals, penetrations, trunnions, closure mechanisms, shielding safety features, criticality control features, impact limiters, and cask identification should be identified and described. A clear definition of the primary confinement system is particularly important. Special design features of the DSS such as a non-passive heat-removal system, neutron poisons or monitoring instrumentation should be discussed.

Sketches and diagrams found throughout the SAR should be compared with the detailed drawings presented in SAR Chapter 1, "General Information". If the application includes proprietary drawings and descriptions that will remain proprietary upon approval of the license or certificate, the sketches, drawings, and diagrams that provide the general description and operational features need not show the proprietary features. This may be achieved by depicting less detail or by illustrating generic components that fulfill the design function. However, these representations should show the operational concept and features important to safety in sufficient detail to form an acceptable basis for public review and comment.

In addition to information on a single DSS, the application should describe any limitations on the arrangement of DSS arrays. For a particular DSS, these limitations may include the minimum spacing between the casks, maximum density of casks in an array, and/or total number of casks or amount of SNF that may be stored at a single ISFSI. The acceptable limitations should be included among the technical specifications in the Safety Evaluation Report (SER) (see Chapter 13, "Technical Specifications and Operating Controls and Limits Evaluation," of this SRP). For a DSS such as those with metal confinement vessels stored in a concrete vault, information on the configuration of vault compartments and horizontal/vertical arrangement is necessary. The operational sequences for loading and unloading the cask should be described.

Damaged fuel may require canning for storage and transportation. The purpose of canning is to confine gross fuel particles to a known, subcritical volume during off-normal and accident conditions, and to facilitate handling and ready retrieval of contents. Therefore, the reviewer should verify that a description of how damaged fuel would be canned, the characteristics of the can, and the means in which the can would be placed in the cask and either readily retrieved (recovered) or retrieved is in the application.

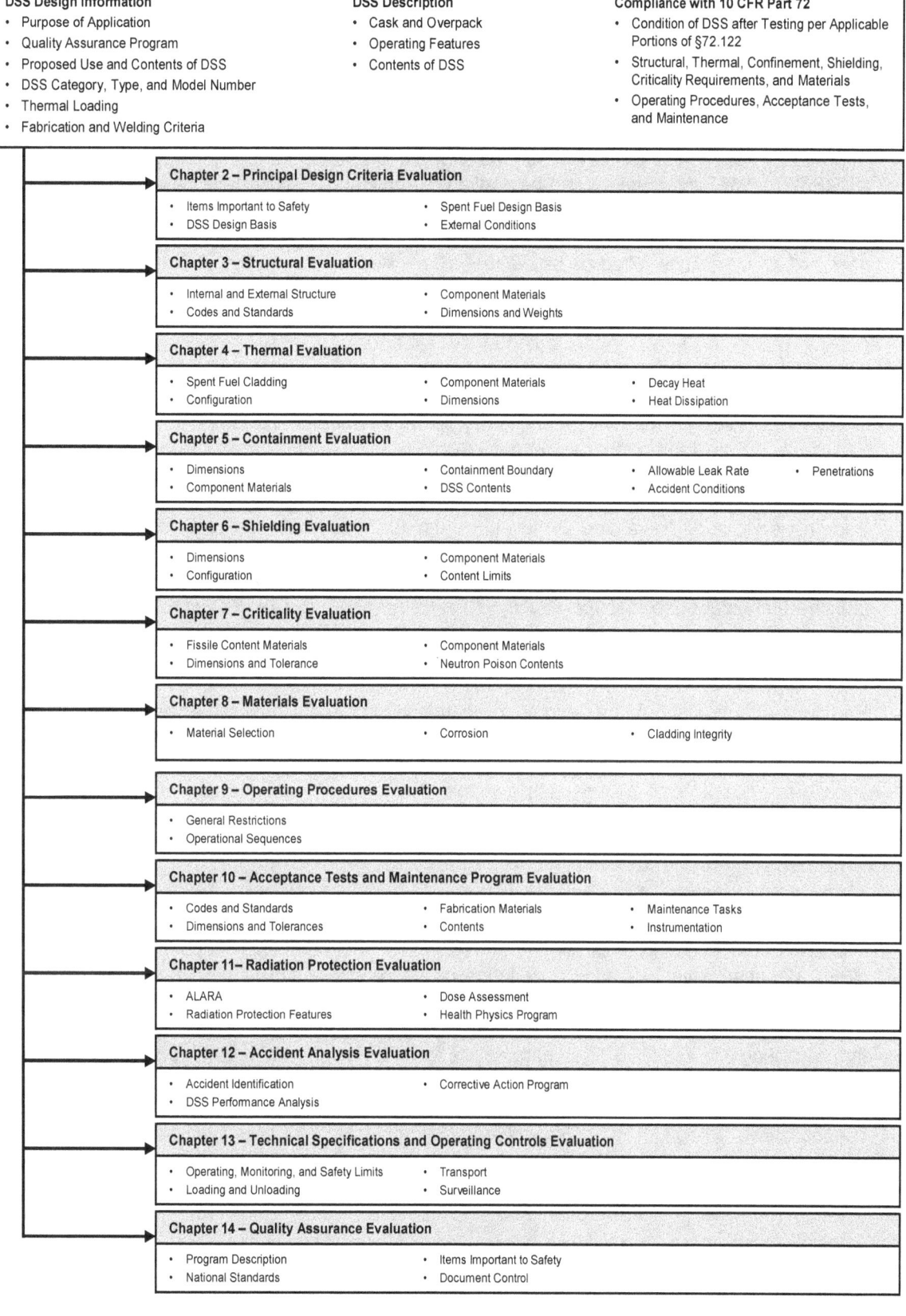

Chapter 1 – General Information Evaluation

DSS Design Information
- Purpose of Application
- Quality Assurance Program
- Proposed Use and Contents of DSS
- DSS Category, Type, and Model Number
- Thermal Loading
- Fabrication and Welding Criteria

DSS Description
- Cask and Overpack
- Operating Features
- Contents of DSS

Compliance with 10 CFR Part 72
- Condition of DSS after Testing per Applicable Portions of §72.122
- Structural, Thermal, Confinement, Shielding, Criticality Requirements, and Materials
- Operating Procedures, Acceptance Tests, and Maintenance

Chapter 2 – Principal Design Criteria Evaluation
- Items Important to Safety
- DSS Design Basis
- Spent Fuel Design Basis
- External Conditions

Chapter 3 – Structural Evaluation
- Internal and External Structure
- Codes and Standards
- Component Materials
- Dimensions and Weights

Chapter 4 – Thermal Evaluation
- Spent Fuel Cladding
- Configuration
- Component Materials
- Dimensions
- Decay Heat
- Heat Dissipation

Chapter 5 – Containment Evaluation
- Dimensions
- Component Materials
- Containment Boundary
- DSS Contents
- Allowable Leak Rate
- Accident Conditions
- Penetrations

Chapter 6 – Shielding Evaluation
- Dimensions
- Configuration
- Component Materials
- Content Limits

Chapter 7 – Criticality Evaluation
- Fissile Content Materials
- Dimensions and Tolerance
- Component Materials
- Neutron Poison Contents

Chapter 8 – Materials Evaluation
- Material Selection
- Corrosion
- Cladding Integrity

Chapter 9 – Operating Procedures Evaluation
- General Restrictions
- Operational Sequences

Chapter 10 – Acceptance Tests and Maintenance Program Evaluation
- Codes and Standards
- Dimensions and Tolerances
- Fabrication Materials
- Contents
- Maintenance Tasks
- Instrumentation

Chapter 11– Radiation Protection Evaluation
- ALARA
- Radiation Protection Features
- Dose Assessment
- Health Physics Program

Chapter 12 – Accident Analysis Evaluation
- Accident Identification
- DSS Performance Analysis
- Corrective Action Program

Chapter 13 – Technical Specifications and Operating Controls Evaluation
- Operating, Monitoring, and Safety Limits
- Loading and Unloading
- Transport
- Surveillance

Chapter 14 – Quality Assurance Evaluation
- Program Description
- National Standards
- Items Important to Safety
- Document Control

Figure 1-1 Overview of Safety Evaluation

1.5.2 Drawings (MEDIUM Priority)

Drawings are usually presented in Chapter 1, "General Information" of the SAR. Reviewers should be familiar with NUREG/CR-5502, "Engineering Drawings for 10 CFR Part 71 Package Approval." While NUREG/CR-5502 was written for transportation packages, the criteria in NUREG/CR-5502 for drawings can be applied to applications for storage casks.

Although some applications may contain drawings designated as "proprietary," reviewers should note that any drawings relied on as the technical basis for adding the DSS design to the "list of approved spent-fuel storage DSS" contained in Subpart K of 10 CFR 72 become part of the public record. Such drawings will not be treated as proprietary and will be made available to the public [10 CFR 2.390(a)]. Applicants may submit additional drawings showing greater detail to support their evaluations, and these may be exempted from the public record if they are not relied on by the staff as part of the technical basis for DSS design approval. The reviewer should verify that all structures, systems, and components (SSC) important to safety are sufficiently detailed to enable reviewers to evaluate their effectiveness. In addition, information on non-safety items may also be necessary to ensure they do not impede the safety systems.

Each reviewer should evaluate the level of detail furnished with the application. The drawings should specify those details of the cask design that affect its evaluation. Those design features that have a significant effect on safety if altered or modified, should be considered for inclusion into the technical specifications directly or by reference. If size reduction has rendered any information unclear or illegible, the Project Manager in the Division of Spent Fuel Storage and Transportation (SFST) should request that the applicant provide larger or full-size drawings.

Particular attention should be devoted to ensuring that dimensions, materials, and other details on the drawings are consistent with those described in both the text of the SAR and those used in supplementary analysis. The dimensions shown on the general arrangement drawing should specify the overall size of the cask and the location and configuration of the contents. All dimensions indicated on drawings should include tolerances that are consistent with the cask evaluation.

1.5.3 DSS Contents (MEDIUM Priority)

The application should present a general description of the contents proposed for storage in the DSS. Because a very detailed description of the proposed DSS contents or SNF is typically provided in Chapter 2, "Principal Design Criteria," of the SAR, the information presented in Chapter 1, "General Information" of the SAR is important only to the extent that it permits overall familiarization with the DSS. Key parameters for SNF include the type of fuel (i.e., PWR, BWR, or both), number of fuel assemblies, the radiation source terms associated with these fuel assemblies, preferential loading, and condition of the fuel assemblies (i.e., intact or consolidated). Chapter 1 may also include additional characteristics such as maximum burnup, initial enrichment, heat load, and cooling time as well as the assembly vendor and configuration (e.g., Westinghouse 17x17). These characteristics may also be repeated in Chapter 2. In addition, the cover gas, if any, should be identified.

If the applicant proposes the storage of damaged fuel or components that are associated with or integral to the fuel assembly that do not have an integral confinement boundary, the range of permissible conditions for the stored material should be defined. If the DSS system is intended to be used to store damaged fuel or components that are associated with or integral to the fuel assembly with an integral confinement boundary when placed in the confinement DSS, the possible range of conditions of the fuel or components should be stated. 10 CFR 72.122(h)(1) requires "canning" or use of other acceptable means for storing fuel with cladding that is not or may not remain intact and for unconsolidated assemblies (without intact cladding). 10 CFR 72.236(c) requires the damaged fuel be maintained in a subcritical condition, while 10 CFR 72.236(h) requires the damaged fuel to be compatible with wet or dry loading and unloading facilities. If damaged fuel is to be stored, the application should address how the following basic requirements will be met:

- Maintain subcriticality;
- Prevent unacceptable release of contained radioactive material;
- Avoid excessive radiation dose rates and doses;
- Maintain ready retrieval of the contents.

If the application requests approval to use the DSS system to store components that are associated with or integral to the fuel assembly (i.e., control spiders, burnable poison rod assemblies, control rod elements, thimble plugs, fission chambers, and primary and secondary neutron sources, or BWR channels that are an integral part of the fuel assembly that do not require special handling), the application should present summary descriptions of those components in Chapter 1, "General Information" of the SAR. The SFST staff has made a practice of carefully characterizing components as being "associated with or integral to" the fuel assembly because only those components listed above are acceptable at a geologic repository per 10 CFR 961.11, Appendix E, Section B.2. Components that are associated with or integral to the fuel assembly are reviewed in more detail as part of Chapter 2, "Principal Design Criteria Evaluation," of this SRP. Also, if the components are degraded (e.g., the component does not provide adequate confinement under design basis conditions to contain radioactive gas or other dispersible radioactive materials), the application should describe the possible conditions and alternative confinement methods, if any.

1.5.4 Quality Assurance Program (See Chapter 14 for Priority)

The application should describe the proposed QA program, citing all implementing procedures in a manner that satisfies the 18 criteria defined in 10 CFR Part 72, Subpart G, "Quality Assurance" (10 CFR §§ 72.142-72.176). The description need only refer to procedures that implement the QA program, and these procedures need not be explicitly included in the application. The QA program should address design, fabrication, construction, testing, operation, and modification activities regarding the SSCs that are important to safety. The application should also discuss the activities to be performed under the QA program and how these activities will be controlled to ensure compliance with all of the requirements of Subpart G. These controls may be applied to the various activities using a graded approach as presented in NUREG/CR-6407, "Classification of Transportation Packaging and Dry Spent Fuel Storage System Components According to Importance to Safety" (i.e., QA efforts expended for a given activity should be consistent with that activity's system classification and function).

Per 10 CFR 72.140(d), a QA program previously approved by the NRC and established, maintained, and executed for another DSS will be accepted as satisfying the requirements for a QA program for the purpose of this application. Additionally, previously approved QA programs that meet the requirements of Appendix B to 10CFR 50 or Subpart H to 10 CFR 71, will be acceptable provided they also meet the recordkeeping requirements of §72.174. Any reference to a previously approved QA program should identify the program by date of submittal to the NRC, docket number, and date of NRC approval. The reviewer should coordinate with the Chapter 14, "Quality Assurance Evaluation," review of this SRP.

1.5.5   Consideration of 10 CFR Part 71 Requirements (MEDIUM Priority)

Casks that have been certified for transportation of SNF under 10 CFR Part 71 may be approved for the storage of SNF under 10 CFR Part 72 provided the application contains:

- A copy of the CoC issued under 10 CFR Part 71,

- Copies of all drawings and other documents referenced in the 10 CFR Part 71 CoC, and

- Sufficient information in the SAR to demonstrate that the cask is suitable for the storage period of SNF as defined by 10 CFR 72.230(b).

Because applications for dual-purpose certification under 10 CFR Parts 71 and 72 are sometimes submitted jointly, the final (approved) version of such documents may not be available at the time of initial DSS SAR submission. Nonetheless, applicable documentation of the Part 71 certification (or application), including questions and responses from the related review, should be provided to the Part 72 review team, as appropriate.

Substantial coordination of the Part 71 and Part 72 reviews is necessary to ensure consistency and avoid duplication of effort. The reviewer should verify that a process for promptly informing each of the review teams about DSS system design changes precipitated by any concurrent safety reviews has been identified by the applicant. Provisions for communicating these changes should be addressed by, and discussed with, the applicant. In addition, transportability of storage-only or dual purpose casks, per 10 CFR 72.236(m) should be addressed. The applicant should address how it is planning to address the transportation requirements. The reviewer should verify that such considerations have been made and described in the SAR, when the SAR and/or accompanying documentation indicate plans to use the cask system for transportation purposes.

1.6   Evaluation Findings

The evaluation findings are prepared by the reviewer on satisfaction of the regulatory requirements in Section 1.3. These statements should be similar to the following examples, if the documentation submitted with the application supports positive findings for each of the regulatory requirements (the finding number is for convenience in reference within the SRP and SER):

F1.1   A general description and discussion of the DSS is presented in Section(s) of the SAR, with special attention to design and operating characteristics, unusual or novel design features, and principal considerations important to safety.

F1.2   Drawings for SSCs important to safety are presented in Section _____ of the SAR. A listing of those drawings (including dates and revision numbers) that were relied upon as a basis for approval appears in Section _____ of the SER.

F1.3   Specifications for the SNF to be stored in the DSS are provided in SAR Section _____. Additional details concerning these specifications are presented in Chapter ____ of both the SAR and SER.

F1.4   The quality assurance program and implementing procedures are described in Section _____ of the SAR.

F1.5   The [DSS system designation] [has been/is/is not being] certified under 10 CFR Part 71 for use in transportation. A copy of the SAR and CoC issued under 10 CFR Part 71 is on file with the NRC under Docket No. _____ [if applicable].

A summary statement similar to the following should be made:

"The staff concludes that the information presented in Chapter 1, "General Information" of the SAR satisfies the requirements for the general description under 10 CFR Part 72. This finding is reached on the basis of a review that considered the regulation itself, Regulatory Guide 3.61, and accepted practices."

2 PRINCIPAL DESIGN CRITERIA EVALUATION

2.1 Review Objective

The objective of evaluating the principal design criteria related to structures, systems, and components (SSCs) important to safety is to ensure that, in the view of the U.S. Nuclear Regulatory Commission (NRC) staff, the principal design criteria comply with the relevant general criteria established in U.S. Code of Federal Regulations (CFR) Part 72, "Licensing Requirements for the Independent Storage of Spent Nuclear Fuel, High-Level Radioactive Waste and Reactor-Related Greater Than Class C Waste," Title 10, "Energy" (10 CFR Part 72). Further guidance can be found in NUREG/CR-6407, "Classification of Transportation Packaging and Dry Spent Fuel Storage System Components According to Importance to Safety." Material provided in this chapter will form the basis for accepting the safety analysis report (SAR) for NRC staff review.

With regard to reviewing the principal design criteria, the applicant may take one of two approaches: (1) SAR Chapter 2, "Principal Design Criteria" may discuss these criteria in general terms with details provided in later sections or (2) SAR Chapter 2 may present detailed discussions of selected (or all) criteria. Past applicants have generally selected the latter approach. Subsequent chapters of this Standard Review Plan (SRP) provide detailed discussions of the design criteria applicable to each functional area (e.g., structural, thermal) without regard to those that may have been presented in SAR Chapter 2.

2.2 Areas of Review

The review of the principal design criteria should provide reasonable assurance that all design criteria are addressed in the SAR. The following areas of review have been adopted by the NRC staff:

Structures, Systems, and Components Important to Safety

Design Basis for Structures, Systems, and Components Important to Safety
Spent Nuclear Fuel (SNF) Specifications
External Conditions

Design Criteria for Safety Protection Systems
General
Structural
Thermal
Shielding/Confinement/Radiation Protection
Criticality
Material Selection
Operating Procedures
Acceptance Tests and Maintenance
Decommissioning

2.3 Regulatory Requirements

This section presents a summary matrix of the portions of U.S. Code of Federal Regulations (CFR) Part 72, "Licensing Requirements for the Independent Storage of Spent Nuclear Fuel, High-Level Radioactive Waste, and Reactor-Related Greater Than Class C Waste" Title 10,

"Energy" (10 CFR Part 72) that are relevant to the review areas addressed by this chapter. The NRC staff reviewer should read the exact regulatory language. Table 2-1 matches the relevant regulatory requirements associated with this chapter to the areas of review.

Table 2-1 Relationship of 10 CFR Part 72 Regulations and Areas of Review										
Areas of Review	10 CFR Part 72 Regulations									
	72.2 (a)(1)	72.104 (a), (b), (c)	72.106 (a), (b), (c)	72.122 (a), (b) (1)(2) (3), (c), (f)	72.122 (h)(1) (4)	72.122 (i), (l)	72.124 (a), (b)	72.126 (a)(1) (2)(3) (4)(5) (6)	72.236 (a), (b), (c), (d)	72.236 (e), (f), (g), (h), (i), (l), (m)
SSCs Important to Safety									●	
Design Bases for SSCs Important to Safety	●			●					●	
Design Criteria for Safety Protection Systems		●	●	●	●	●	●	●	●	●

2.4 Acceptance Criteria

The reviewer should verify that the applicant has provided either sufficient general or summary discussions of the SSC design features and of both operational and accident conditions. This demonstrates a clear and defensible case that they have met the design criteria. In evaluating the principal design criteria related to DSS SSCs that are important to safety, reviewers should seek to ensure that the given design fulfills the following acceptance criteria.

2.4.1 SSCs Important to Safety

The reviewer should verify that the applicant presents the general configuration of the DSS and provides an overview of specific components and their intended functions. In addition, the reviewer should ensure the applicant identifies those components deemed to be important to safety and addresses the safety functions of these components in terms of how they meet the general design criteria and regulatory requirements discussed above. Additional information concerning specific functional requirements for individual DSS components is addressed in subsequent chapters of this SRP.

2.4.2   Design Bases for SSCs Important to Safety

Detailed descriptions of each of the items listed below are generally found in specific sections of the SAR. However, a brief description of these areas, including a summary of the analytical techniques used in the design process, should also be captured in Chapter 2, "Principal Design Criteria" of the SAR. This description gives reviewers a perspective of how specific DSS components interact to meet the regulatory requirements of 10 CFR Part 72. This discussion should be non-proprietary since it may be used to familiarize interested persons with the design features and bounding conditions of operation of a given DSS.

2.4.2.1   SNF Specifications

The range and types of SNF or other radioactive materials that the DSS is designed to store should be specified. In addition, these specifications should include, but are not limited to:

- The type of SNF (i.e., boiling-water reactor (BWR), pressurized-water reactor (PWR), or both),

- Cladding material,

- Maximum assembly uranium mass loading,

- Weights of the stored materials,

- Dimensions and configurations of the fuel,

- The identification and limits on amount and position of damaged fuel, if damaged fuel is to be stored, and the dimensions of the "damaged-fuel can,"

- Maximum allowable enrichment of the fuel before any irradiation for criticality safety and minimum enrichment for the shielding evaluation,

- Assigned Burnup Loading Value (i.e., MWd/MTU),

- Loading Curves for each set of licensing conditions if Burnup Credit is used (required minimum burnup versus enrichment curve),

- Operational history parameters (e.g., average in-core soluble boron concentration, average moderator temperature, etc.) if burnup credit is used,

- Minimum acceptable cooling time of the SNF before storage in the DSS,

- Maximum heat to be dissipated,

- Maximum number of SNF elements,

- Condition of the SNF (i.e., intact assembly, damaged fuel or consolidated fuel rods),

    Inerting atmosphere requirements and the maximum amount of fuel permitted for storage in the DSS.

For DSSs that will be used to store components that are associated with or integral to fuel assemblies (e.g., control rods and BWR fuel channels), the reviewer should ensure the applicant specifies the types and amounts of radionuclides, heat generation, and the relevant source strengths and radiation energy spectra permitted for storage in the DSS. For other radioactive materials to be stored with the SNF assemblies, the SAR should specify the following:

- The design basis source term;

- The effects of gas generation on the cask internal pressure;

- The effects of the additional weight and length of the proposed material on structural and stability analyses;

- The impact of the added heat from these components, including the impact on heat transfer characteristics; and

- Credit for any negative reactivity from residual neutron absorbing material remaining in the control components.

2.4.2.2   External Conditions

The SAR should define the bounding conditions under which the DSS is expected to operate. Such conditions include both normal and off-normal environmental conditions as well as accident conditions. In addition, the reviewer should verify that the applicant has considered the effects of natural events such as tornadoes, earthquakes, floods, and lightning strikes.

2.4.3   Design Criteria for Safety Protection Systems

2.4.3.1   General

The SAR should define an expected lifetime for the cask design. The minimum licensing period is defined in 10 CFR 72.230(b). The reviewer should verify that the applicant has provided a brief description of the proposed quality assurance (QA) program, and applicable industry codes and standards, which will be applied to the design, fabrication, construction, and operation of the DSS. The applicant should also describe how the cask design reflects consideration of compatibility with removal from a reactor site, transportation, and ultimate disposition of the stored spent fuel.

In establishing normal and off-normal conditions applicable to the design criteria for DSS designs, applicants should account for actual facility operating conditions. Therefore, design considerations should reflect normal operational ranges, including any seasonal variations or effects.

2.4.3.2   Structural

The SAR should define how the DSS structural components are designed to accommodate combined normal, off-normal, and accident loads while preserving recover and protecting the DSS contents from significant structural degradation, criticality, and loss of confinement. This discussion is generally a summary of the analytical techniques and calculation results from the detailed analysis discussed in SAR Chapter 3, "Structural Evaluation," and should be presented in a non-proprietary form.

2.4.3.3   Thermal

The SAR should contain a general discussion of the proposed heat-removal systems, including the reliability and verifiability of such systems, and any associated limitations. All heat-removal systems should be passive and independent of intervening actions under normal and off-normal conditions.

2.4.3.4   Shielding/Confinement/Radiation Protection

The reviewer should ensure that the applicant describes those features of the cask that protect occupational workers and members of the public against direct radiation dosages and releases of radioactive material, and minimize the dose after any off-normal or accident-level conditions.

2.4.3.5   Criticality

The SAR should address the mechanisms and design features that enable the DSS to maintain SNF in a subcritical condition under normal, off-normal, and accident-level conditions.

2.4.3.6   Material Selection

The materials selected for the DSS must demonstrate adequate corrosion performance during normal operation, off-normal, and accident-level conditions in the environmental conditions of the ISFSI for the duration of the license.

The spent fuel cladding must be protected during storage against degradation that leads to gross ruptures, or the fuel must be otherwise confined such that degradation of the fuel during storage will not pose operational problems with respect to its removal from storage.

2.4.3.7   Operating Procedures

The reviewer should ensure that the applicant provides potential licensees with guidance regarding the content of normal, off-normal, and accident response procedures. Cautions regarding both loading, unloading, and other important procedures should be mentioned here. Retrievability should be provided for normal and off-normal conditions. Applicants may choose to provide model procedures to be used as aids in preparing detailed site-specific procedures.

2.4.3.8   Acceptance Tests and Maintenance

The reviewer should verify that the applicant identifies the general commitments and industry codes and standards used to derive acceptance, maintenance, and periodic surveillance tests used to verify the capability of DSS components to perform their designated functions. In addition, the reviewer should ensure the applicant discusses the methods used to assess the need for such tests with regard to specific components.

2.4.3.9   Decommissioning

Casks should be designed for ease of decontamination and eventual decommissioning. The reviewer should examine the SAR to ensure the applicant describes the features of the design that support these two activities.

2.5   Review Procedures

Chapter 2, "Principal Design Criteria" applies to all review disciplines. Figure 2-1 presents an overview of the evaluation process and may be used as a guide for the coordination of the review among review disciplines.

Reviewers for each section of the SAR should consider SAR Chapter 2 in combination with additional details presented later in the SAR. In this SRP, evaluations of design criteria applicable to each of the relevant chapters of the SAR are discussed in detail. Reviewers should coordinate the review of each chapter with the applicable disciplines to ensure that multi-disciplinary issues, which impact more than one chapter, have been addressed.

2.5.1   SSCs Important to Safety (MEDIUM Priority)

Reviewers should verify that the applicant has clearly identified all SSCs important to safety (see Glossary for the definition of "important to safety") and documented the rationale for this designation. Such information may be provided in tabular form. Reviewers should review the general DSS description presented in SAR Chapter 1, "General Description." Reviewers should ensure that the applicant has provided adequate justification for excluded SSCs.

Reviewers should pay particular attention to instrumentation and other equipment (e.g., lifting devices and transport vehicles). In general, the NRC staff accepts that monitoring systems need not be classified as being important to safety. For example, a failure in the functioning of the pressure monitoring system does not directly result in a release of radionuclides. Additional justification for not considering such systems as being important to safety may be presented in later sections of the SAR and summarized in SAR Chapter 2, "Principal Design Criteria".

Reviewers should consider adding to SAR Chapter 13 "Technical Specifications and Operating Controls and Limits" any design features that would have a significant effect on safety if altered or modified. Any such additions to Chapter 13 should be thoroughly discussed in their respective sections of the SER.

2.5.2   Design Bases for SSCs Important to Safety

The reviewer should verify that the applicant's design basis for DSS approval accurately identifies the range of SNF configurations and characteristics, the enveloping conditions of use, the bounding site characteristics, and is consistent with or bounds the DSS's Technical Specifications. These factors determine the bounds within which an ISFSI owner may use the SAR rather than provide additional proof regarding suitability of the covered topics.

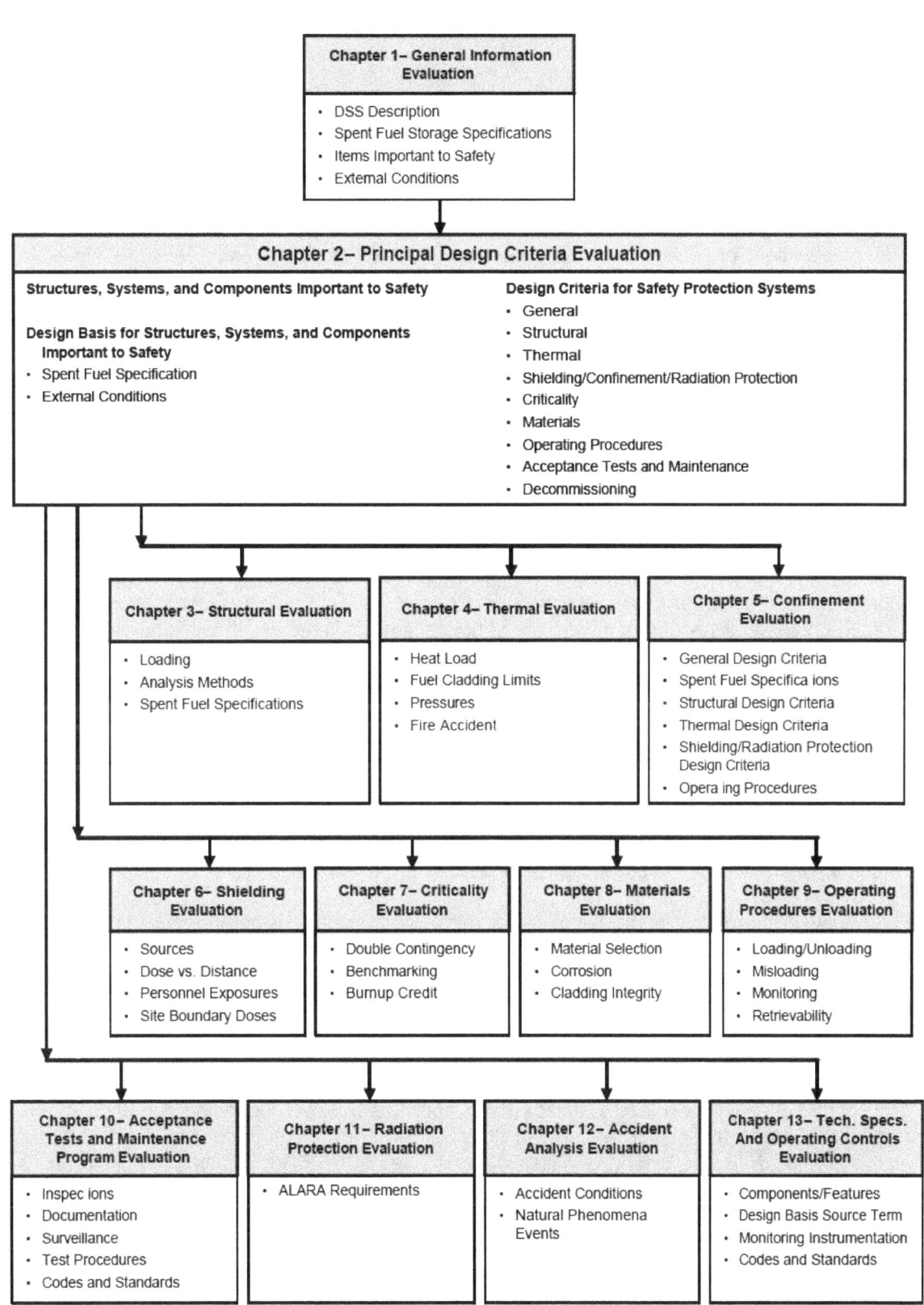

Figure 2-1 Overview of Principal Design Criteria Evaluation

2.5.2.1   SNF Specifications (MEDIUM Priority)

The reviewer should review the detailed specifications for the SNF to be stored in the DSS as presented in SAR Chapter 2, "Principal Design Criteria" and ensure that they are consistent with those specifications discussed in SAR Chapter 1, "General Information" and later in the SAR. The description of the range of SNF to be stored should include the type (PWR, BWR, or both); configuration (e.g., 17x17, 15x15, or 8x8); fuel vendor; number of assemblies per cask; enrichment; burnup and burnup profiles; minimum cooling time; decay heat generation rate; type of cladding; physical dimensions; total weight per assembly; and uranium weight per assembly. In addition, if components associated with fuel assemblies (e.g., control assemblies) will be stored with the fuel, the reviewer should ensure that combined weight, dimensions, heat load, and other appropriate information (e.g., number per cask) are specified.

The reviewer should examine any limitations regarding the condition of the SNF. If damaged fuel is allowed, the effects of such damage should be assessed in later sections of the SAR. Specific conditions that define damaged fuel are provided in Section 8.6, "Supplemental Information for Methods for Classifying Fuel," of this SRP with methods for classifying fuel identified in Section 8.4.17.2 of this SRP. If damaged rods have been removed from a fuel assembly, and they have/have not been replaced with solid dummy rods, the criticality reviewer should determine whether the intended loading configuration has been adequately analyzed to show sub-criticality. Note, the presence of additional moderating material will need to be addressed in the criticality analysis in SAR Chapter 7, "Criticality". Coordination with the structural reviewer is necessary if there are structural defects in the assembly hardware.

The release of fill and fission product gases from failed fuel rods increases the pressure in the cask cavity and the potential source term in the event of confinement failure. Consequently, the reviewer should verify that the applicant provides information regarding the fill/fission product gas present in the fuel as well as the free volume in the cask cavity to enable reviewers to evaluate the pressure in the cask cavity resulting from cladding failure during storage. For the purpose of calculating internal cask pressures, the NRC staff has accepted the bounding assumptions given in SRP Section 4.5.4.6, "Pressure Analysis" regarding the minimum percentages of fuel rods have failed (and released their gases).

The reviewer should pay particular attention to the specification of burnup, cooling time, and decay heat generation rate. These parameters are generally not independent, and the manner in which they are specified and combined can significantly affect the maximum allowed cladding temperature as discussed in SRP Chapter 4, "Thermal Evaluation."

The SAR will typically list various fuel assemblies that can be stored in the DSS. It is not expected that one type of fuel assembly will be bounding for all analyses. The reviewer should ensure that the applicant has justified which specifications are bounding for each of the evaluations presented in subsequent sections of the SAR. Specifications used in these analyses should also be clearly identified or referenced in SAR 13, "Technical Specifications and Operational Controls and Limits".

If the applicant requests permission for the storage of components that are associated with or integral to the fuel assembly in the cask, the reviewer should examine the relevant detailed specifications, conditions, and constraints presented in the SAR. These specifications should be as detailed as the applicable information presented for the fuel designs to provide the reviewer with a basis for determining that the relevant safety functions of the DSS will be maintained. The reviewer should ensure that the applicant also considers the storage of these components in the analyses.

If the applicant requests burnup credit, the reviewer should examine the relevant detailed specifications of the contents to which burnup credit is being applied. These specifications include those that are already considered in criticality analyses for fresh fuel (e.g., maximum initial enrichment). Additional specifications that must be reviewed include the cooling time, the burnup, the requested amount of credit (i.e., the specific actinides), and operational history parameters (e.g., core average boron concentration and assembly average moderator temperature).

2.5.2.2   External Conditions (MEDIUM Priority)

The SAR should identify those external conditions that significantly affect, or could potentially affect, the performance of the DSS. These design-basis conditions will generally restrict either the sites at which the DSS can be used for SNF storage or the manner in which the DSS can be handled. For example, by selecting the design earthquake, the SAR limits the use of the cask being reviewed to sites that are bounded by this seismic limit. By establishing a design-basis drop, the SAR defines the maximum height to which a cask can be lifted without additional safety analysis or design changes (e.g., addition of impact limiters) by the applicant.

Reviewers should note that movement of cask system components within a reactor building may not meet the NRC's criteria described in the NRC Bulletin 96-02, "Movement of Heavy Loads over Spent Fuel, over Fuel in the Reactor Core, or over Safety Related Equipment," for movement of heavy loads within the reactor building. As such, if a potential user (licensee) has been identified, coordination with the appropriate project manager or technical lead from the NRC's Office of Nuclear Reactor Regulation (NRR) should occur during the early stages of DSS design review.

At a minimum, the NRC staff has generally addressed the conditions discussed below; however, other conditions may be relevant depending on specific details of the DSS design. Reviewers should pay particular attention to special design features and how these might be affected both by other external conditions and other DSS components. Reviewers should ensure all required information is provided in the SAR for the design earthquake accident analysis.

"Normal" conditions (including conditions involving handling and transfer) and the extreme ranges of normal conditions are presumed to exist during design-basis accidents or design-basis natural phenomena with the exception of irrational or readily avoidable combinations. For example, an earthquake or tornado may occur at any time and in combination with any "normal" condition. By contrast, it can be presumed that transfer, loading, and unloading operations would not be conducted during a flood.

"Off-normal" conditions and events are presumed to occur in combination with normal conditions that are not mutually exclusive. Nonetheless, it is not required that the SAR analyze or the system be designed for the simultaneous occurrence of independent off-normal conditions or events, design-basis accidents, or design-basis natural phenomena.

Conditions involving a "latent" equipment or instrument failure or malfunction (that is, one that occurs and remains undetected) should be presumed to exist concurrently with other off-normal or design-basis conditions and events. Typical latent malfunctions include a misreading instrument that is not detected as part of routine procedures, an undetected ventilation blockage, or undetected damage from an earlier design-basis event or condition if no provisions exist for detection, recovery, or remediation of such conditions.

For normal, off-normal, and accident-level conditions, reviewers should verify that the applicant has defined appropriate operating and accident scenarios. For these scenarios, the reviewer should verify the applicant includes in the SAR a comprehensive evaluation of the effects of such scenarios on the SSCs important to safety. The analyses of such events are addressed in individual chapters of the SRP. For example, the analyses of an earthquake on the DSS structural components are addressed in SRP Chapter 3, "Structural Evaluation." The applicant's evaluations should demonstrate that the requirements of 10 CFR §§ 72.104 and 72.106 as well as 10 CFR Part 20 have been met.

If appropriate, the following design bases should be included as operating controls and limits in SAR Chapter 13, "Technical Specifications and Operational Controls and Limits Evaluation":

(1)   Normal Conditions

For a given SNF specification, the primary external conditions that affect DSS performance are the ambient temperatures, insolation, and the operational environment experienced by the DSS.

The NRC accepts as the maximum and minimum "normal" temperatures the highest and lowest ambient temperatures recorded in each year, averaged over the years of record. For the SAR, the applicant may select any design-basis temperatures as long as the restrictions they impose are acceptable to both the applicant and the NRC. If the cask is also designed for transportation, the temperature requirements of 10 CFR Part 71 could determine the design-basis temperatures for storage.

For storage casks, the NRC staff accepts a treatment of insolation similar to that prescribed in 10 CFR Part 71.71 for transportation casks. If the applicant selects another design approach, the alternative approach should be justified in the SAR.

The operational environment experienced by the DSS under normal conditions includes the manner in which the cask is loaded, unloaded, and lifted. Occupational dose rates will, in part, depend on whether the cask is sealed in a wet or a dry environment. Fuel cladding temperatures may also be affected. The manner in which the cask is lifted will determine the load on the trunnions and/or lifting yoke. The orientation of the cask (vertical or horizontal) and its height above ground during transport to the ISFSI will establish initial conditions for the drop accidents discussed below.

(2)   Off-Normal Conditions

An applicant's SAR generally addresses several off-normal conditions. These should include variations in temperatures beyond normal, failure of 10 percent of the fuel rods combined with off-normal temperatures, failure of one of the confinement boundaries, partial blockage of air vents, human error, out-of-tolerance equipment performance, equipment failure, and instrumentation failure or faulty calibration.

(3)   Accident Conditions

The staff has generally considered that the following accidents should be evaluated in the SAR. These do not constitute the only accidents that should be addressed if the SAR is to serve as a reference for accidents for a specific application. Other credible accidents that may be derived from a hazard analysis could include accidents resulting from operational error, instrument failure, lightning, and other occurrences. Post-accident recovery of damaged fuel may require such systems as overpacks or dry- transfer systems since ready retrieval of the fuel is required only for normal and off-normal conditions. Accident situations that are not credible because of design features or other reasons should be identified and justified in the SAR. Chapter 3, "Structural Evaluation" of this SRP provides more detail regarding accidents.

(a)   Cask Drop

The SAR should identify the operating environment experienced by the cask as well as the drop events (i.e., end, side, corner) that could result. Generally, the design basis is established either in terms of the maximum height to which the cask may be lifted when handled outside the reactor site SNF building or in terms of the maximum acceleration that the cask could experience in a drop.

(b)   Cask Tipover

Although cask system supporting structures may be identified and constructed as being important to safety (i.e., designed to preclude cask tipovers), the NRC considers that cask tipover events should be analyzed. In some cases, cask tipover may be determined to be a credible hazard, and the associated analysis should reflect the conditions (e.g., heights and accelerations) associated with that hazard.

The NRC staff has accepted an unyielding surface for determining the bounding cask deceleration loads. Prototype or scale model testing and analytical modeling can be used. In the analytical approach, the hard receiving surface need not be unyielding.

(c)   Fire

The fire conditions postulated in the SAR should provide an "envelope" for subsequent comparison with site-specific conditions. The NRC accepts the methods discussed in 10 CFR 71.73(c)(4). In addition, the NRC staff accepts that the applicant may consider a fire based upon the limited availability of flammable material at an ISFSI (e.g., only that associated with vehicles transporting or lifting the cask, or sources of nearby combustible materials). Regardless of which approach the applicant takes, the SAR should specify and justify the bounding conditions for a "design-basis" fire.

(d)   Fuel Rod Rupture

The regulations require that the cask be designed to withstand the effects of accident conditions and natural phenomena events without impairing its capability to perform safety functions. Consequently, during the cask analysis for conditions resulting from design-basis accidents and natural phenomena, the NRC has asserted and the applicant should assume a release of 100 percent of the initial rod fill gases and a release of 30 percent of the fission product gases from the fuel rods into the cask interior. The remaining 70 percent of the fission product gases is presumed to be retained within the fuel pellet.

(e)   Leakage of the Confinement Boundary

Casks are designed to provide the confinement safety function under all credible conditions.

(f)   Explosive Overpressure

The conditions under which a DSS may be exposed to the effects of an explosion vary greatly among individual sites. Generally, explosive overpressure is postulated to originate from an industrial accident. The NRC separately evaluated the effects of various sabotage methods on cask systems in developing appropriate regulations in 10 CFR Part 73. Consequently, this SRP does not consider explosive overpressures from sabotage events.

The extent to which explosive overpressure is addressed in the SAR directly affects the degree of site-specific review required. The principal concern in the SAR should be the effects of explosive overpressure on the storage system rather than descriptions of hypothesized causes. Design parameters for blast or explosive overpressures should identify pressure levels as reflected ("side-on") overpressure and provide an assumed pulse length and shape. This discussion should provide sufficient information for licensees to determine if the effects of their site-specific hazards are bounded by the cask system design bases.

(g)   Air Flow Blockage

For storage systems with internal air flow passages, the reviewer should verify the applicant considers blockage of air inlets and outlets in an accident condition. The NRC staff considers that the effects of such an assumption should be utilized in determining the appropriate inspection intervals, and/or monitoring systems, for the DSS.

(4)   <u>Natural Phenomena Events</u> (LOW Priority)

The NRC staff has generally considered that the following events should be evaluated as design-basis accidents in the SAR:

(a)   Flood

The SAR should establish a design-basis flood condition. This condition may be determined on the basis of the presumption that the cask cannot tip over and the yield strength of the cask will not be exceeded. Alternatively, the SAR can show that credible flooding conditions have negligible impact on the cask design.

If the SAR establishes parameters for a design-basis flood, all of the potential effects of flood water and ravine flood byproducts should be recognized. Serious flood consequences can involve effects such as blockage of ventilation ports by water and silting of air passages. Other potential effects include scouring below foundations and severe temperature gradients resulting from rapid cooling from immersion.

(b)   Tornado

The NRC staff accepts design-basis tornado wind loading as defined by RG 1.76, "Design Basis Tornado and Tornado Missiles for Nuclear Power Plants" (Region 1) and RG 1.117, "Tornado Design Classification." Design criteria should be established for the cask on the basis of these wind-loading and missile-impact definitions. The cask should not tip over, and the capability to perform the confinement safety function should not be impaired. The NRC staff considers that tornados and tornado missiles may occur without warning. The review should note that, in general, the effects of a tornado missile bound those of a light general aviation aircraft directly impacting a DSS.

(c)   Earthquake

The SAR should state the parameters of the design earthquake. For use of a DSS at reactor sites, this is equivalent to the SSE used for analysis of nuclear facilities under 10 CFR Part 50. An analysis for an Operating-Basis Earthquake (OBE) is not required for a DSS SAR prepared in accordance with 10 CFR Part 72. Cask tipover accidents are analyzed, but tipover caused by an earthquake may not be a credible event. The reviewer should verify that the SSCs meet appropriate guidance in RG 1.29, "Seismic Design Classification," RG 1.61, "Damping Values for Seismic Design of Nuclear Power Plants," and RG 1.92, "Combining Modal Responses and Spatial Components in Seismic Response Analysis."

(d)   Burial Under Debris

Debris resulting from natural phenomena or accidents that may affect cask system performance may be addressed in the SAR or left to the site-specific application. Such debris can result from floods, wind storms, or land slides. The principal effect is typically on thermal performance.

(e)   Lightning

Lightning typically has a negligible effect on cask systems; however, the requirements of the Lightning Protection Code and National Electric Code should be applied to the design of the cask system structures. The applicant should cite these codes as part of the general design criteria for the cask system (see Section 2.4.3.1). In addition, the SAR should address lightning as a natural phenomenon if cask-system performance may be impacted by the effect of lightning on a component that is important to safety.

(f)   Other

10 CFR Part 72 identifies several other natural phenomena events (including seiche, tsunami, and hurricane) that should be addressed for SNF storage. The SAR may include these natural phenomena as design-basis events or show that their effects are bounded by other events. If these events are not addressed in the SAR and they prove to be applicable to a specific site, a safety analysis is required prior to approval for use of the DSS under either a site-specific or general license.

2.5.3   Design Criteria for Safety Protection Systems (MEDIUM Priority)

Cask system components that are to be used in facility areas subject to review under 10 CFR Part 50 should satisfy both the requirements in 10 CFR Part 72 (with review guided by this SRP) and 10 CFR Part 50 (with review guided by NUREG-0800). Acceptance of the cask system in areas covered by 10 CFR Part 50 license requirements is not addressed in this SRP for approval under 10 CFR Part 72. If the applicant states that the storage system will be used at a specific reactor site, then the Division of Spent Fuel Storage and Transportation (SFST) project manager should inform the appropriate NRR project manager. The reviewer is reminded that heavy loads are a likely matter of interest to NRR.

Table 2-2 presents a summary of design criteria (and design bases) that should generally be identified during the initial stages of the review. The applicability of Table 2-2 may vary depending on the details of the storage system design.

Regardless of where the descriptions and associated criteria are located in the SAR, reviewers should include a description and evaluation of the safety protection systems in SER Chapter 2, "Principal Design Criteria." The system descriptions should address the functions of the various system components in providing confinement, cooling, subcriticality, radiation protection of the public and workers, and SNF retrievability. Summary criteria for the performance of the system as a whole in providing for these capabilities or functions should also be described and evaluated. Reviewers should verify that the design-basis assumptions presented are consistent with and reasonable for actual site or facility conditions. Reviewers should also include a description and evaluation of the cask system design's compatibility with removal from a reactor site, transportation, and ultimate disposition of the stored spent fuel.

Table 2-2 Outline of Design Criteria and Bases for DSS

Design Life	• Limited to the requested term in the application

Design Bases	• SNF Specifications
	   (1) Type
	   (2) Configuration/Vendor
	   (3) Enrichment (Maximum and Minimum)
	   (4) Weight or Range of Weights
	   (5) Burnup
	   (6) Type of Cladding
	   (7) Assemblies/Cask
	   (8) Dimensions
	• Decay Heat/Assembly
	   (1) Minimum Decay/Cooling Time (e.g., 5 years, 10 years, etc.)
	   (2) Maximum Kilowatts per assembly
	• Gas Volume (at Temperature)
	• Fuel Condition/Damage Allowed
	• Burnup Credit
	   (1) Credit Amount (specific actinides)
	   (2) Operational History Parameters
	• Non-Fuel Hardware

Normal Design Event Conditions	• Ambient Temperature
	   (1) Maximum
	   (2) Minimum
	• Loading
	   (1) (Wet/Dry)
	• Storage Handling Orientation
	   (1) (Vertical/Horizontal)
	• Maximum Lift Height
	• Maximum Cladding Temperature
	• Other Conditions Considered in 2.5.2.2 (1)

Off-Normal Design Event Conditions	• Summarize Events Considered in 2.5.2.2 (2)

Design-Basis Accident Design Events and Conditions	• End Drop
	   (1) Lift Height (or Maximum Acceleration)
	• Side Drop
	   (1) Lift Height (or Maximum Acceleration)
	• Tip-Over
	   (1) Acceleration (if applicable)
	• Fire
	   (1) Duration
	   (2) Temperature
	• Other Events Considered in 2.5.2.2 (3)
	(As Applicable)

Table 2-2 Outline of Design Criteria and Bases for DSS	
Design-Basis Natural Phenomena Design Events and Conditions	• Flood • Earthquake • Tornado • Other Events Considered in 2.5.2.2 (4) (As Applicable)

Structural	• Design Code (e.g., ASME, AISC)
	  (1) Containment
	  (2) Noncontainment
	  (3) Basket
	  (4) Trunnions
	  (5) Storage Radiation and Protective Shielding and Enclosure
	  (6) Transfer Radiation and Protective Shielding and Enclosure
	  (7) Cooling Structure or System
	• Design Weight
	• Design Cavity Pressure
	  (1) Normal/Off-Normal/Accident
	• Response and Degradation Limits
	  (1) Normal/Off-Normal/Accident

Thermal	• Maximum Design Temperatures
	  (1) Cladding
	  (2) Other Components
	• Insolation (Side/Top/Bottom)
	• Fill Gas
	  (1) Type (e.g., helium, etc.)
	  (2) Initial Fill Pressure (at temperature)
	• Modes of Heat Transfer Utilized in the Design

Confinement	• Description of Confinement Boundary
	• Redundant Seals for Closure
	• Maximum Leak Rate for Confinement Boundary
	  (1) Normal/Off-Normal/Accident
	  (2) Justification of Leakage Rate if not Leaktight
	• Monitoring System Specifications

Table 2-2 Outline of Design Criteria and Bases for DSS	

Radiation Protection/Shielding	• Confinement Cask
	  (1) Surface Position
	     Normal/Off-Normal/Accident
	• Exterior of Shielding
	  (1) Transfer Mode Position
	  (2) Storage Mode Position
	     Normal/Off-Normal/Accident
	• ISFSI Controlled Area Boundary
	  (1) Dose Rate
	  (2) Annual Dose
	     Normal/Off-Normal/Accident

Criticality	• Method of Control Geometry, Fixed Poison, Soluble Poison • Minimum Boron Concentration (Fixed and/or Soluble Poison) • Maximum k_{eff} • Burnable Neutron Absorber Credit • Burnup Credit Analysis
Materials	• Cladding Hoop Stress • Corrosion
Operating Procedures	• Normal and Off-Normal • After Accident-level Conditions
Acceptance Tests and Maintenance	• Industry codes and standards
Tech Specs	• Operational Controls and Limits

Criteria relating to redundancy and allowable levels of response by the DSS under normal, off-normal, and accident-level conditions and events should be described and evaluated. In general, no unacceptable degradation in physical condition or functional performance should result from normal or off-normal conditions. The design criteria regarding limits of permissible system response and degradation resulting from an accident condition should be evaluated against the SSC capabilities to perform the principal safety functions. Considerations of permissible responses should include detect-ability and corrective actions that may be proposed as conditions of system use.

The staff accepts that both routine surveillance programs and active instrumentation meet the intent of "continuous monitoring" as required in 10 CFR 72.122(h)(4).

Reviewers should note that some DSS designs may contain a component or feature whose continued performance over the licensing period has not been demonstrated to staff with a sufficient level of confidence (e.g., rubber "O" rings). Therefore, staff may require the use of active instrumentation if the failure of that system or component causes an immediate threat to the public health and safety, and if that failure would not be detected by any other means. In some cases, to demonstrate compliance with 10 CFR 72.122(h)(4), the vendor or NRC staff may propose a technical specification requiring such instrumentation as part of the first use of a cask system. After first use, and if warranted and approved by staff, such instrumentation may be discontinued or modified.

The staff should verify that the applicant has met the intent of continuous monitoring so that the applicant can determine when corrective action needs to be taken to maintain safe storage conditions.

2.6 Evaluation Findings

The reviewer will prepare evaluation findings on satisfaction of the regulatory requirements in Section 2.3. If the documentation submitted with the application supports positive findings for each of the regulatory requirements (the finding number is for convenience in reference within the SRP and SER), these statements should be similar to the following examples:

F2.1 The SAR and docketed materials adequately identify and characterize the SNF to be stored in the DSS in conformance with the requirements given in 10 CFR 72.236.

F2.2 The SAR and the docketed materials relating to the design bases and criteria meet the general requirements as given in 10 CFR 72.122(a), (b), (c), (f), (h)(1), (h)(4), (i), and (l).

F2.3 The SAR and docketed materials relating to the design bases and criteria for structures categorized as important to safety meet the requirements given in 10 CFR 72.122(a), (b)(1), (b)(2) and (b)(3), (c), (f), (h)(1), (h)(4), and (i); and 10 CFR 72.236.

F2.4 The SAR and docketed materials meet the regulatory requirements for design bases and criteria for thermal consideration as given in 10 CFR 72.122 (a), (b)(1), (b)(2) and (b)(3), (c), (f), (h)(1), (h)(4), and (i).

F2.5 The SAR and docketed materials relating to the design bases and criteria for shielding, confinement, radiation protection, and ALARA considerations meet the regulatory requirements as given in 10 CFR 72.104(a) and (b); 10 CFR 72.106(b); 10 CFR 72.122(a), (b), (c), (f), (h)(1), (h)(4), and (i); 10 CFR 72.126(a).

F2.6 The SAR and docketed materials relating to the design bases and criteria for criticality safety meet the regulatory requirements as given in 10 CFR 72.124(a) and (b).

F2.7 The SAR and docketed materials relating to the design bases and criteria for retrievability meet the regulatory requirements as given in 10 CFR 72.122(a), (b)(1), (b)(2), and (b)(3), (c), (f), (h)(1), (h)(4), and (l).

F2.8 The SAR and docketed materials relating to the design bases and criteria for other SSCs not important to safety but subject to NRC approval meet the general regulatory requirements as given in the following subparts of

10 CFR Part 72: Subpart E, "Siting Evaluation Factors" 72.104 and 72.106; Subpart F, "General Design Criteria" 72.122, 72.124, and 72.126; and Subpart L, "Approval of Spent Fuel Storage Casks."

The reviewer should provide a summary statement similar to the following:

"The staff concludes that the principal design criteria for the [cask designation] are acceptable with regard to meeting the regulatory requirements of 10 CFR Part 72. This finding is reached on the basis of a review that considered the regulation itself, appropriate regulatory guides, applicable codes and standards, and accepted engineering practices. A more detailed evaluation of the design criteria and an assessment of compliance with those criteria are presented in Chapters 3 through 14 of the SER."

3 STRUCTURAL EVALUATION

3.1   Review Objective

In this portion of the dry storage system (DSS) review, the U.S. Nuclear Regulatory Commission (NRC) evaluates aspects of the DSS design and analysis related to structural performance under normal and off-normal operations, accident conditions, and natural phenomena events. In conducting this evaluation, the NRC staff seeks a high degree of assurance that the cask system will maintain confinement, subcriticality, radiation shielding, and retrievability or recovery of the fuel, as applicable, under all credible loads for normal and off-normal conditions accidents, and natural phenomenon events.

3.2   Areas of Review

This chapter of the DSS Standard Review Plan (SRP) provides guidance for use in evaluating the design and analysis of the proposed cask system with regard to its structural performance. All DSSs include a confinement cask that may have both internal components and integral external components. In addition, some DSSs have a variety of other components that are subject to NRC evaluation and approval under the cask certification provisions of Subpart L of 10 CFR Part 72.

Recognizing the diversity of the various cask system components, the NRC has broadly categorized the applicable review procedures and acceptance criteria as follows:

- Structural Capability of the Confinement boundary and Internals,
- Other structural system components important to safety,
- Other structural components subject to NRC approval.

Within these broad categories, the NRC focuses the DSS structural evaluation, as described in Section 3.5, "Review Procedures," using the following areas of review as appropriate:

Scope

Structural Design Criteria and Design Features
    Design Criteria
        General Structural Requirements
        Applicable Codes and Standards
    Structural Design Features

Materials Related to Structural Evaluation

Structural Analysis
    Load Conditions
        Normal Conditions
        Off-normal Conditions
        Natural Phenomena and Accident Conditions
    Structural Analysis Methods
        Finite-element Analysis
        Closed-form Calculations
        Structural Analysis for Specific Cask Components
    Structural Evaluation

Structural Capability
Fabrication and Construction

3.3   Regulatory Requirements

Table 3-1 presents a matrix that shows the primary relationship of the regulations provided in this section to the specific areas of review associated with this SRP chapter. The NRC staff reviewer should verify the association of regulatory requirements with the areas of review presented in the matrix to ensure that no requirements are overlooked as a result of unique applicant design features.

Table 3-1 Relationship of Regulations and Areas of Review				
Areas of Review	**10 CFR Part 72 Regulations**			
	72.124(a)	72.234(a), (b)	72.236(b),(c), (d), (l)	72.236(g), (h)
Scope	•	•	•	
Structural Design Criteria and Design Features	•	•	•	•
Materials Related to Structural Evaluation			•	
Structural Analysis		•	•	
Structural Evaluation		•	•	•

3.4   Acceptance Criteria

The most important function of the structural analysis is to ensure sufficient structural capability for every applicable section of the cask system to withstand the worst-case loads under accident conditions and natural phenomena events. Withstanding such loads enables the cask system to successfully preclude the following negative consequences:

- Unacceptable risk of criticality,
- Unacceptable release of radioactive materials,
- Unacceptable radiation levels,
- Impairment of retrievability or recovery, as applicable.

Because of the diversity of cask system components and various materials that are subject to NRC evaluation and approval, it is not possible to define objective structural review criteria that address all possible component configurations. No single structural code currently accepted by the NRC (such as the American Society of Mechanical Engineers [ASME] Boiler and Pressure Vessel [B&PV] Code, Section III, Division 1 [ASME B&PV]) or Section III, Division 2 may cover the design of all spent nuclear fuel (SNF) storage systems. Consequently, the acceptability of any given structure will be contingent upon a combination of adherence to applicable portions of multiple codes and a review of the functional performance of the structure taken as a whole.

This combined approach allows the designer to request relief, or provide alternatives, and the reviewer to impose additional restrictions when warranted by specific design features.

In general, the DSS structural evaluation seeks to ensure that the proposed design and analysis fulfill the following acceptance criteria that reflect the industry codes and standards the NRC staff has accepted in past DSS structural evaluations. The American National Standards Institute's (ANSI) "Design Criteria for an Independent Spent Fuel Storage Installation (Dry Storage Type)" (ANSI/ANS-57.9) generally applies to the design and construction of an ISFSI but contains some criteria/design requirements relative to dry storage systems.

3.4.1   Confinement Cask and Metallic Internals

3.4.1.1   Steel Confinement Cask

The structural design, fabrication, and testing of the confinement system and its redundant sealing system should comply with an acceptable code or standard such as ASME B&PV Code. (The NRC has accepted use of either Subsection NB or Subsection NC of Section III, Division 1 of this code.) Division 3 of Section III of the ASME B&PV Code, addressing storage of spent nuclear fuel, has been published, but currently no NRC position has been established on that standard. Other design codes or standards may be acceptable depending on their application. An applicant must justify the use of new criteria where no NRC staff position has been established.

    i.   The NRC staff evaluates the proposed limitations on allowable stresses and strains in the confinement cask, steel parts important to safety and subject to review by comparison with those specified in applicable codes and standards. Where certain proposed load combinations will produce values that exceed the accepted limits for localized points on the structure, the applicant should provide adequate justification to show that the deviation will not affect the functional integrity of the structure. Under certain conditions limiting strains and limiting deformations may form part of the acceptance criteria.

    ii.   The NRC has accepted the use of applicable subsections of the ASME B&PV Code, Section III, Division 1, such as Subsections NF and NG, for components used in the cask system. This includes the "basket" structure used in casks to restrain and position multiple fuel elements in the storage system in which Subsection NG has been used.

3.4.1.2   Steel-Lined Concrete Confinement Cask

    i.   The American Concrete Institute (ACI) and ASME's "Code for Concrete Reactor Vessels and Containments" (ACI 359), also known as Section III, Division 2 of the ASME B&PV Code, constitutes an acceptable standard for prestressed and reinforced concrete structures that are an integral component of a steel-lined concrete confinement cask that must withstand internal pressure in operation or testing and constitutes a confinement cask. The minimum functional requirements of ANSI/ANS-57.9 for subject areas not specifically addressed in ACI 359 shall be met.

    ii.   The NRC will review the use of applicable subsections of the ASME B&PV Code, Section III, Division 1, such as Subsections NF and NG, for components used within the confinement cask but not integrated with it. This includes Subsection NG for the "basket" structure used in casks to restrain and position multiple fuel elements in the storage system.

3.4.2 Other Structural System Components and Structures Important to Safety

The NRC accepts the use of ANSI/ANS-57.9 (together with the codes and standards cited therein) as the basic reference for the ISFSI dry storage systems important to safety that are not designed in accordance with accepted provisions or alternatives to applicable portions of Section III, Division 1 or 2 (ACI-359) of the ASME B&PV Code. Structures and components that are important to safety which are related to lifting and handling cask systems should comply with American National Standards Institute (ANSI) Standard, "American National Standards for Radioactive Material Lifting Devices for Shipping Containers Weighing 10,000 lbs (4500 kg) or More" (N14.6). The loadings defined in American Society of Civil Engineers, "Minimum Design Loads for Buildings and Other Structures," (ASCE 7) can be used when load combinations are considered on the basis of ANSI/ANS-57.9.

3.4.2.1 Steel Structures

The principal codes and standards include the following references that may be applied to steel structures and components:

a. American Institute of Steel Construction (AISC), "Specification for Structural Steel Buildings — Allowable Stress Design and Plastic Design."

b. AISC, "Load and Resistance Factor Design Specification for Structural Steel Buildings."

c. American Welding Society, "Structural Welding Code Steel," (AWS D1.1).

3.4.2.2 Reinforced Concrete Structures

ACI's "Code of Requirements for Nuclear Safety Related Concrete Structures," ACI 349 can be applied to reinforced concrete structures and components.

3.4.3 Other Structural Components Subject to NRC Approval

For structural design and construction of other components subject to NRC approval, the principal codes and standards include the following:

a. American Society of Civil Engineers (ASCE), "Minimum Design Loads for Buildings and Other Structures" (ASCE 7).

b. International Building Code (IBC) 2006 from International Code Council.

c. AISC, "Specification for Structural Steel Buildings—Allowable Stress Design and Plastic Design."

d. AISC, "Code of Standard Practice for Steel Buildings and Bridges."

e. ASME B&PV Code, Section VIII.

f.   ACI 318, "Building Code Requirements for Structural Concrete."

3.5   Review Procedures (HIGH Priority)

The SAR documentation should be reviewed to confirm that the design of the cask structure provides for satisfactory functional performance. This includes operating suitability within specified limiting conditions and satisfaction of the basic safety criteria under all credible events and environmental conditions.

The SAR should clearly identify the confinement system and other structures important to safety, and each component should have sufficient structural capability for every applicable section to withstand the worst-case loads under accident-level events and conditions to successfully preclude the following:

- Unacceptable risk of criticality.

- Unacceptable release of radioactive materials to the environment.

- Unacceptable radiation dose to the public or workers.

- Significant impairment of retrievability or recovery, as applicable, of stored nuclear materials (the NRC has accepted some degradation of retrievability under accident conditions and severe natural phenomena events that are treated as design bases events).

This position does not necessarily require that all confinement system and other structures important to safety survive all design-basis accidents and extreme natural phenomena without any permanent deformation or other damage. Some load combination expressions for the design basis event (DBE) and conditions for structures important to safety permit stress levels that exceed yield. The SAR should include computations of the maximum extent of potentially significant accident deformations and any permanent deformations, degradation, or other damage that may occur. The reviewer should verify that the applicant has performed computations, analyses, and/or tests and that both the tests and results are acceptable to the NRC to clearly demonstrate that any permanent deformations, degradation, or other damage that may occur does not render the system performance unacceptable.

Structures important to safety are not required to survive accidents to the extent that they remain suited for use for the life of the cask system without inspection, repair, or replacement. If the service life of structures important to safety may be degraded by accident-level conditions, there must be SAR commitments and procedures for determining and correcting the degradation and performing other acceptable remedial action.

The proposed technical specifications should be reviewed to ensure that they include adequate restrictions on cask handling and operations to preclude the possibility of damage to the structure or the confined nuclear material. Operating controls and limits of the technical specifications (reviewed under Chapter 13 of this SRP) should be included in both the SAR and the SER, and should describe actions to be taken and inspections to be conducted upon occurrence of events that may cause such damage.

Figure 3-1 presents an overview of the evaluation process and can be used as a guide to assist in coordinating with other review disciplines.

In evaluating the structural design and performance of a proposed DSS, the reviewer should select and emphasize aspects of the following review procedures, as appropriate for the particular DSS, in relation to the acceptance criteria summarized in Section 3.4.

Description of Structures, Systems, and Components Important to Safety

The reviewer should verify that the applicant's safety analysis report (SAR) clearly identifies the proposed structural design and construction of structures, systems, and components (SSCs) that are important to safety and necessary for effective functional performance and safety of the DSS. The SAR and supplemental material submitted by the applicant should be reviewed to assess compliance with the applicable scope and content requirements defined in 10 CFR 72.230. The reviewer should focus in particular on requirements and conditions of use related to design, construction, implementation, operation, and maintenance of structural SSCs.

Applicable Codes, Standards, and Specifications

NRC guidelines recommend that the safety evaluation report (SER) prepared by the NRC staff include a table (in the design criteria evaluation section) summarizing the applicable reference sources. This table should identify all source documents cited in the SAR, their usage (e.g., description of model, prior NRC approval of cask system elements, design code, construction code), and acceptability for that usage. The sources of interest include documents directly referenced in the SAR; sources of material incorporated by reference; and codes, standards, specifications, and other sources of criteria that further define the design and construction of the proposed structures. If not tabulated, the consolidated review and assessment of reference sources should otherwise be included in the SER.

Loads and Load Combinations

The reviewer should verify that the loads and load combinations are as specified in Chapter 2, "Principal Design Criteria Evaluation," of this SRP. If the applicant has not adequately justified any deviations from the acceptance criteria for loads and load combinations, the reviewer should identify the deviations as unacceptable and transmit them to the applicant for further justification. If components associated with or integral to the fuel assembly are to be stored in the cask, then the reviewer should ensure these components are considered by the applicant in the structural analyses.

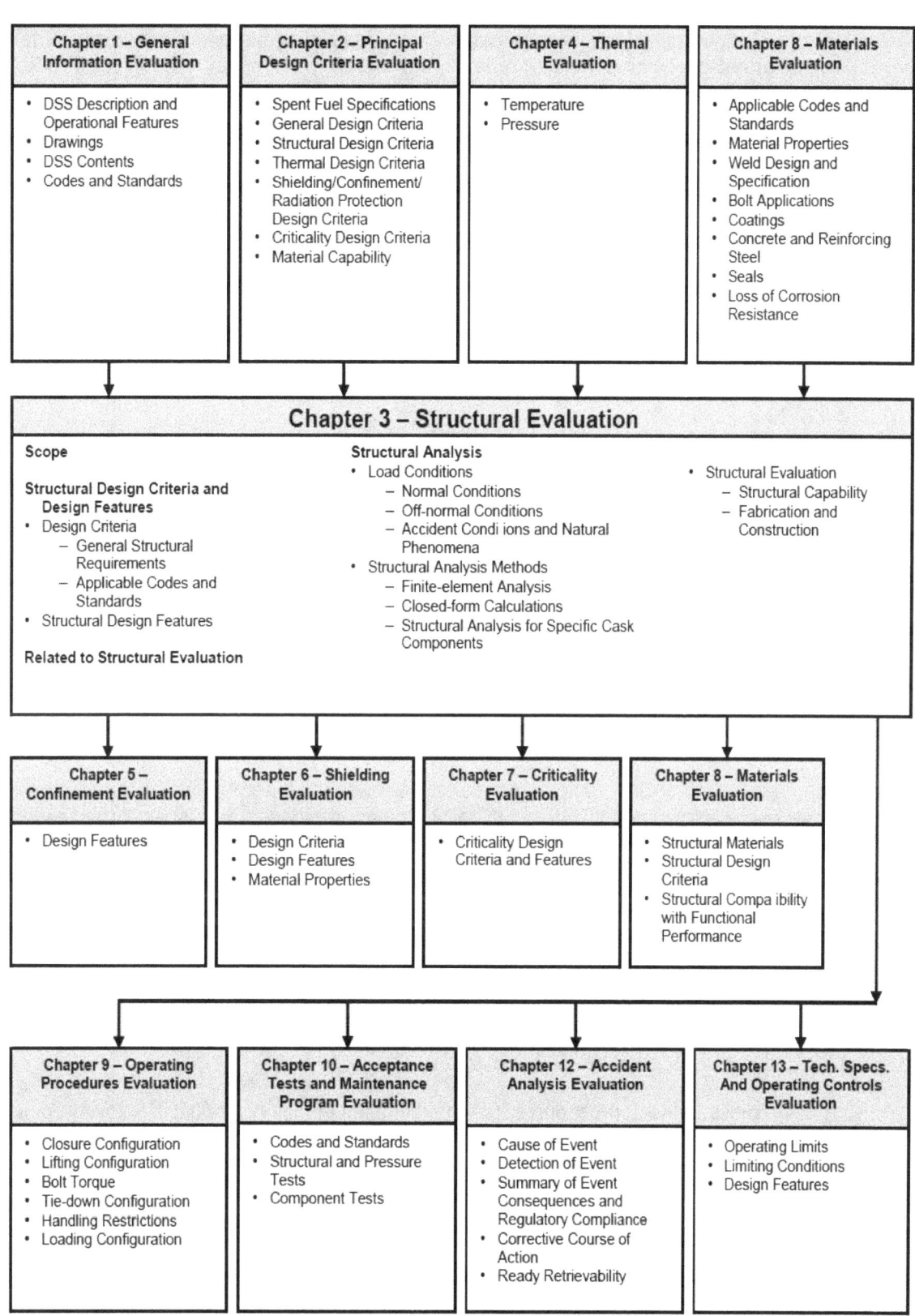

Figure 3-1 Overview of the Structural Evaluation

The SAR should include a comprehensive table of load combinations and safety margins for selected structural sections of components important to safety (or otherwise subject to NRC evaluation). The summary table should include sufficient structural sections and forms of loadings (e.g., shear, flexure, axial, and combined stress situations) to verify that the lowest margins of safety are represented for the various components. In addition, this table can be used to summarize the structural capacity evaluation.

Design and Analysis Procedures

The reviewer should determine whether the applicant's design and analysis procedures and assumptions are conservatively defined on the basis of accepted engineering practice. The behavior of the structure under various loads, and the manner in which these loads are treated in conjunction with other coexistent loads should be reviewed, while compliance with the acceptance criteria, defined in Section 3.4 of this SRP should be assessed.

Structural Acceptance Criteria

The proposed limitations on allowable stresses, strains, or deformations in the confinement cask, its internals, system components important to safety, and other components subject to review should be analyzed. The reviewer should compare the proposed limitations with those specified in the applicable codes and standards. Where the applicant proposes to exceed the accepted limits for certain load combinations at localized points on the structure, the reviewer should evaluate the justification provided to ensure that the deviation will not affect the functional integrity of the structure. If the justification is not acceptable, the reviewer should request additional justification and bases.

Materials, Quality Control, and Special Fabrication Techniques

Information provided in the SAR regarding materials is reviewed under the guidance of Chapter 8, "Materials Evaluation" of this SRP. Quality control methods, and special fabrication techniques, if any, related to the structural evaluation should be reviewed in coordination with the materials discipline and QA. The QA program is reviewed under Chapter 14 "Quality Assurance Evaluation" of this SRP. If the applicant proposes to use a new material not addressed in prior approvals, the applicant must provide sufficient data regarding the material's structural properties to establish the acceptability of the material. Similarly, the reviewer should evaluate any new quality control programs or construction techniques to ensure that they will not degrade the structural quality, integrity, or function of the DSS.

Testing and In-Service Surveillance Requirements

The proposed pressure test procedures for the confinement cask should be reviewed in comparison with the procedures described in ASME Code, Section III, Division 1, Subsection NB-6000, and in conjunction with Chapter 10, "Acceptance Tests and Maintenance Program Evaluation" of this SRP. Also, the proposed acceptance test and maintenance requirements for trunnions should be reviewed in comparison with those described in the ASME Code and ANSI N14.6, as applicable for load bearing components. Any other proposed testing and in-service surveillance programs should be reviewed on a case-by-case basis. Also, the reviewer should read SAR Section 10 to verify that the applicant has included all appropriate acceptance tests and addressed all required evaluations in Section 10 of the SER.

Conditions for Use of Structures

The structural evaluation should be reviewed to determine if conditions of use or technical specifications should be associated with the structural design or proposed fabrication and construction methods. The reviewer should determine the appropriateness of and need for any proposed technical specifications related to structural design and construction. Also, the reviewer should determine whether any additional technical conditions related to structural performance are needed and, if so, provide input to the conditions of use discussed in the SER. In addition, the reviewer should describe the basis for the suggested conditions in the structural evaluation section of the SER. Structure-related conditions of use may be linked to evaluations performed under other sections (such as a field verification that maximum concrete temperatures predicted from thermal analysis will not be exceeded).

The remainder of this section provides specific review procedures for each of the three categories of cask system components including the confinement cask and steel internals, other structures important to safety, and other components subject to NRC approval. Within each of these broad categories, the specific review procedures focus the DSS structural evaluation using the areas of review identified in Section 3.2 of this SRP.

3.5.1   Confinement Cask and Metallic Internals

The structural review of the confinement cask addresses drawings, plans, sections, supporting computations, and specifications for those structural components comprising confinement barriers. The review also addresses structural and sealing interfaces, and connections that are necessary to complete the confinement system (as defined in 10 CFR Part 72). In addition, this review includes evaluation of components that serve no structural function to confirm that they do not impair the functioning of the confinement cask. The review also encompasses the evaluation of the metallic internals that constitute the "basket" structure.

3.5.1.1   Scope

The SAR must describe all components of the confinement cask and internals important to safety in sufficient detail to allow evaluation of their structural behavior and effectiveness under the imposed design conditions. In addition, the SAR must identify all codes and standards applicable to the components.

The discussion in the SAR must demonstrate that all components of the confinement cask and internals important to safety will be designed and fabricated to quality standards commensurate with the importance to safety of the function to be performed. In addition, components of the confinement cask and internals important to safety must be designed to accommodate the combined loads anticipated during normal, off-normal, accident, and natural phenomenon events with an adequate margin of safety.

3.5.1.2   Structural Design Criteria and Design Features

i.   Design Criteria (MEDIUM Priority)

The cask-related design criteria presented in SAR Chapter 2, "Principal Design Criteria Evaluation" should be reviewed as well as the criteria provided herein. The NRC generally considers the following design criteria to be acceptable to meet the structural requirements of 10 CFR Part 72:

(1)   General Structural Requirements

The proposed cask must maintain confinement of radioactive material under normal and off-normal operations, accident conditions, and natural phenomenon events. In addition, neither the cask nor any basket within the cask may deform under credible loading conditions in a manner that would jeopardize the subcritical condition and recovery or retrievability of the fuel, as applicable.

The design must adequately protect the fuel cladding against gross rupture caused by degradation resulting from design or accident conditions. In addition, the design must ensure that the SNF will not experience accelerations/decelerations that would damage its structural integrity or jeopardize its subcritical condition or retrievability under normal and off-normal design conditions.

The applicant must analyze the cask to show that it will not tip over or drop in its storage condition as a result of a credible natural phenomenon event. A tipover or drop is always assessed as a bounding condition during handling operations.

Radiation shielding in the cask system is required to protect the public and workers involved with spent fuel handling and storage, and such shielding must not degrade under normal or off-normal conditions or events. The shielding function may degrade as a result of an accident (e.g., displacement of source or shielding, reduction in shielding). However, the loss of function must be readily visible, apparent, or detectable. (Any permissible degradation in shielding must be shown to result in dose rates sufficiently low to permit recovery of the damaged cask including unloading, if necessary). The necessary structural criteria to assure adequate shielding remains in-place should be clearly identified.

(2)   Applicable Codes and Standards

The structural design, fabrication, and testing of the confinement system and any necessary redundant sealing system should comply with acceptable codes or standards. Use of codes and standards previously accepted by the NRC expedites the evaluation process. Use of other codes and standards, definition of criteria composed of extracts from multiple codes and standards with overlapping scopes, or substitution of other criteria, in whole or in part, in place of acceptable published codes or standards requires a custom NRC review and may delay the evaluation process.

Section III, Division 1, of the ASME B&PV Code is an accepted code for design, fabrication, and test of steel confinement casks. Specifically, the NRC has accepted use of either Subsection NB or NC. Other design codes or standards may be acceptable depending on their application. The NRC has accepted use of the applicable subsections of the ASME

Code, Section III, Division 1, for cask system components used within the confinement cask but not integrated with it. This includes the "basket," which is a structure used in casks to restrain and position multiple fuel elements. Section III, Division 3 of the ASME B&PV Code is also available and addresses storage cask systems, but NRC has not endorsed its use at the current time.

Also, the NRC has accepted applicable subsections of Division 1, of the ASME Code, for structural external integral elements of the confinement (e.g., Subsection NF for integral supports) cask.

Commitments for structures important to safety to ASME Code Section III, with proposed alternatives to the Code, should be documented in the application. Likewise, NRC staff-approved alternatives to the Code should be incorporated, either directly or referenced, in the certificate of compliance (or in the technical specifications attached to the certificate) issued by the NRC. In the event that alternatives to codes are required during fabrication and the alternatives do not impact the quality or safety of the component, an alternative to the requirements of the certificate of compliance or technical specification may be granted with approval of the NRC.

Applicants should propose a condition to the certificate of compliance or technical specification, either directly or referenced, describing the alternatives to the referenced codes. The condition or technical specification should also describe a process to address one-time alternatives from the ASME Code that may occur during fabrication. The information provided should include the identification of the component, the reference to the ASME Code (code edition, addenda, section or article), description of the Code requirement, and a description of the alternative. In addition, the applicant should justify the alternative, including a description of how the alternative would provide an acceptable level of quality and safety. Additionally, the applicant should describe how compliance with the code provisions would result in hardship or difficulty without a compensating increase in the level of quality or safety.

For a steel-lined concrete confinement cask system, NRC accepts ACI 359, also designated Section III, Division 2, of the ASME Boiler and Pressure Vessel Code. This Code is acceptable for prestressed and reinforced concrete that is an integral component of a radioactive material containment vessel that must withstand internal pressure in operation or testing. ACI 359, as endorsed by RG 1.136, Rev. 3, "Design Limits, Loading Combinations, Materials, Construction, and Testing of Concrete Containments," and Section 3.8.1, "Concrete Containments" of NUREG-0800, "Standard Review Plan for Review of Safety Analysis Reports for Nuclear Power Plants," should be applied on the basis of containment function regardless of whether the concrete structure is fixed or portable and regardless of where the concrete structure is fabricated. ACI 359 also applies to structural concrete supports constructed as an integral part of the containment. If ACI 359 and RG 1.136 apply to the structure, the Code applies to the entire design, material selection, fabrication, and construction of that structure.

As an alternative to the requirements of Section CC-3440 of ACI 359, the NRC also accepts the following. These criteria are an alternative to the temperature requirements of ACI 349, A.4, but only for the specified uses and temperature ranges:

a. If concrete temperatures of general or local areas are 93°C (200°F) in normal or off-normal conditions/ occurrences, no tests to prove capability for elevated temperatures or reduction of concrete strength are required.

b. If concrete temperatures of general or local areas exceed 93°C (200°F) but would not exceed 149°C (300°F), no tests to prove capability for elevated temperatures or reduction of concrete strength are required if Type II cement is used and aggregates are selected which are acceptable for concrete in this temperature range. The following criteria for fine and coarse aggregates are acceptable:

   1) Satisfy ASTM C33, ("Standard Specification for Concrete Aggregates") requirements and other requirements referenced in ACI 349 for aggregates.

   2) Satisfy ASTM C150, ("Standard Specification for Portland Cement") requirements and other requirements referenced in ACI 349 for cement.

   3) Have demonstrated a coefficient of thermal expansion (tangent in temperature range of 20°C to 38°C [70°F to 100°F]) no greater than 11×10^{-6} mm/mm/°C (6×10^{-6} in./in./°F), or be one of the following minerals: limestone, dolomite, marble, basalt, granite, gabbro, or rhyolite.

c. If concrete temperatures of general or local areas in normal or off-normal conditions or occurrences do not exceed 107°C (225°F), the requirements of 1 and 2 apply to the coarse aggregate, but fine aggregate that meets 1 and is composed of quartz sands or sandstone sands may be used in place of compliance with 2.

ii. Structural Design Features (LOW Priority)

The cask-related descriptive information presented in SAR Chapter 1, "General Information Evaluation" should be reviewed as well as any related information provided in SAR Chapter 3 "Structural Evaluation". The drawings, figures, tables, and specifications included in the SAR should fully define the structural features of the cask. These may include the cask system that could include an inner shell, an outer shell, and a gamma shield, inner and outer lids and bolts, port covers and bolts, vent port covers to be welded in place, neutron shields and shell, trunnions, fuel basket, and impact limiters (if used).

The reviewer should coordinate with the confinement review (Chapter 5, "Confinement Evaluation," of this SRP) to verify that the SAR clearly identifies the confinement boundaries. These boundaries include the primary confinement vessel; its penetrations, seals, welds, and closure devices; and the redundant sealing system as provided by the system.

The list of weights and calculation of the cask center of gravity should be reviewed. The reviewer should verify that the applicant used the appropriate limiting cases in the structural evaluations.

3.5.1.3   Materials Related to Structural Evaluation (HIGH Priority)

The structural reviewer should coordinate with the materials reviewer to determine the impact of corrosion, reviewed in Chapter 8, "Materials Evaluation" of this SRP, on structural integrity. The reviewer should ensure that the applicant used appropriate corrosion allowances for the structural analyses. The reviewer should consider the static and dynamic (where appropriate) stresses, strains, deformations, and response, and the limits used for the structural design and evaluations.

A DSS serves to confine and maintain safe storage conditions throughout its service life. Design and construction codes (e.g., ASME B&PV Code Section III) give reasonable assurance that the as-fabricated material will provide the necessary integrity. It is noted that the ASME Code Section III, Division 1, applies specifically to maintaining pressure boundaries and supporting structures in nuclear power plants, and may not necessarily be totally applicable to all DSS. However, designers may choose to cite it as the code to which selected components are to be fabricated. Codes such as the ASME B&PV are not likely to address all the potential performance conditions (e.g., cracking, creep, corrosion, etc.) that may arise from environmental, electrochemical, or dynamic-loading. These and other effects are specific to the individual application and should be addressed to meet the guidance of Chapter, 8, "Materials Evaluation" of this SRP.

The reviewer should verify that the properties used are appropriate for the load conditions of interest (e.g., static or dynamic, impact loading, hot or cold temperature, wet or dry conditions). SAR Chapter 12, "Accident Analyses Evaluation" should be reviewed to ensure that the applicant considered any appropriate restrictions regarding temperature or environmental conditions for the materials under accident conditions.

The reviewer should coordinate with the thermal and material disciplines to determine the appropriate temperatures at which allowable stress limits should be defined. For most cask materials, the stress limits should be defined at the maximum temperature for each material as established by the SAR thermal analysis. Further discussion of the background for the temperature limits can be found in Chapters 4, "Thermal Evaluation" and 8, "Materials Evaluation" of this SRP.

The reviewer should coordinate with the materials, criticality, and shielding reviews to ensure that, during storage and accident conditions, any structural materials considered as neutron absorbers and/or gamma shields will perform safety functions as intended.

If the cask has impact limiters, used in the transfer and storage operations, the applicant should thoroughly evaluate and verify their nonlinear impact characteristics. In addition, the applicant should tabulate and describe the crush characteristics and properties of the limiters in the directions that are to be used.

3.5.1.4   Structural Analysis

i.   Load Conditions

(MEDIUM Priority) To meet the structural requirements of 10 CFR Part 72, the DSS design must accommodate the full spectrum of load conditions including all anticipated normal, off-normal, and accident-level conditions (including natural phenomenon events).   The system should not experience any permanent deformation or loss of safety function capability during normal or off-normal operating conditions.   However, the system may experience some permanent deformation, but no loss of safety function capability, in response to an accident.

(1)   Normal Conditions (LOW Priority)

Normal conditions and events are those associated with cask system operations, including storage of nuclear material, under the normal range of environments. The SAR should state the assumed limits of normal use environments to support evaluation by a user of the certified cask system suitability for use at a specific site under a general license.

Loads normally applicable to a confinement cask include weight, internal and external pressures, and thermal loads associated with operating temperature.   The loads experienced may vary during loading, preparation for storage, transfer, storage, and retrieval operations.   The weight is the maximum or design weight (including tolerances) of the cask as it is stored and loaded with SNF.   However, depending on the operation and procedures, the weight should also include water fill.   The applicant should evaluate all orientations of the cask body and closure lids during normal operations and storage conditions including loads associated with loading, transfer, positioning, and retrieval of the confinement cask.

Internal pressures result from hydrostatic pressure, cask drying and purging operations, filling with non-reactive cover gas, out-gassing of fuel, refilling with water, radiolysis, and temperature increases. Temperature variations and thermal gradients in the structural material may cause additional stresses in the cask and closure lids.   The reviewer should coordinate with the thermal review (Chapter 4, "Thermal Evaluation," of this SRP) to determine the conservative (or enveloping) values and combinations of the cask internal pressures and temperatures for both hot and cold conditions.   The reviewer should use the temperature gradients calculated in SAR Chapter 4 to determine thermal stresses.   Note that if the confinement system has several enclosed areas; all areas may not have the same internal pressures.   In some casks, enclosed areas consist of the cask cavity and the region between the inner and outer lids.

Required evaluations include weight plus internal pressures and thermal stresses from both hot and cold conditions.   The reviewer should verify that the applicant included the maximum thermal gradient as determined in the thermal analysis, when evaluating thermal stresses.

(2)   Off-Normal Conditions (LOW Priority)

The review should identify and evaluate all off-normal events and conditions described in Chapter 12, "Accident Analyses Evaluation," of this SRP. The off-normal conditions and events should be reviewed for those that affect the confinement cask structure. The confinement cask components should satisfy the same structural criteria required for normal conditions, as discussed above.

The SAR should clearly identify anticipated off-normal conditions and events that may reasonably be expected to occur during the life of the cask system at the proposed site. In addition, the SAR should state the environmental limits to support comparison of the cask system design bases with specific site environmental data. Off-normal conditions and events can involve potential mishandling, simple negligence of operators, equipment malfunction, loss of power, and severe weather (short of extreme natural phenomena).

(3)   Accident-Level Events and Conditions

The reviewer should follow the guidance below in reviewing the structural response to accident-level conditions. Note that the SAR must address, at a minimum, each of the following accidents. However, this discussion may not address all of the potential events or accidents that apply to a cask (Chapter 12 of this SRP addresses the identification and evaluation of accidents).

(a)   Cask Drop and Tipover (HIGH Priority)

The reviewer should ensure the applicant performs a cask drop and tipover analysis or demonstrates that this scenario is not credible. The SAR should identify the operating environment experienced by the cask and the drop events (end/side/tipover) that could result. Generally, applicants establish the design basis in terms of the maximum height to which the cask is lifted outside the building or the maximum deceleration that the cask could experience in a drop. The design-basis drops should be determined on the basis of the actual potential handling and transfer accidents.

If the analytical approach described in the LLNL report UCID-21246 (Chun, R., et al., 1986) for axial buckling is used to assess fuel integrity for the cask drop accident, the analysis should use the irradiated material properties and should include the weight of fuel pellets.

Alternatively, an analysis of fuel integrity which considers the dynamic nature of the drop accident and any restraints on fuel movement resulting from cask design is acceptable if it demonstrates that the cladding stress remains below yield. If a finite element analysis is performed, the analysis model may consider the entire fuel rod length with intermediate supports at each grid support (spacer). Irradiated material properties and weight of fuel pellets should be included in the analysis.

The NRC will accept cask tipover about a lower corner onto a hard receiving surface from a position of balance with no initial velocity. The NRC has also accepted analysis of cask drops with the longitudinal axis horizontal which, together with analysis of a vertical drop, could bound a non-mechanistic tipover case.

NRC staff has accepted an unyielding surface for determining the bounding cask deceleration loads that can far exceed the decelerations experienced by a cask dropping onto or tipping over the concrete storage pad that will bend and deform. Prototype or scale model testing can be used to obtain more realistic cask deceleration or equivalent load for quasi-static analyses. Alternatively, applicants can develop an analytical model to calculate cask deceleration loads. In the analytical approach, the hard receiving surface for a drop or tipover accident need not be an unyielding surface, and its flexibility may be included in the modeling.

The structural discipline should review validation of the analytical model. The staff has completed a series of low-velocity impact tests of a steel billet from which a model validation approach and corresponding acceptance criteria have been developed. These tests and analytical evaluations are summarized in NUREG/CR-6608, *Summary and Evaluation of Low-Velocity Impact Tests of Solid Steel Billet Onto Concrete Pads* (Witte, 1998). On the basis of the report, the following model validation acceptance criteria apply to a cask-pad-soil analytical model for predicting impact responses of the cask:

- When solid steel billet is used to replace the cask in the cask-pad-soil analytical model, it should predict a pulse amplitude slightly higher than the recorded pulse amplitude from the billet test.

- The calculated pulse duration and shape should be similar, but not necessarily identical, to those recorded from the billet test.

The validated billet-pad-soil model is considered adaptable to a cask-pad-soil analysis model if relevant attributes of the cask are used to replace those of the billet.

(b)   Explosive Overpressure (LOW Priority)

Explosion-induced overpressure and reflected pressure may result from explosion hazards associated with explosives and chemicals transported by rail or on public highways, natural gas pipelines, and vehicular fires of equipment used in the transfer of casks. Explosions may result from detonation of an air-gaseous fuel mixture. With the exception of transfer vehicle accidents, the explosion hazards are typically similar to those for facilities subject to reviews under 10 CFR Part 50, "Domestic Licensing of Production and Utilization Facilities."

The SAR should state the level of overpressure that the cask system can withstand for this accident condition. This overpressure level would then serve as the quantitative envelope for future comparison with hazards for specific site installations. The pressure criteria for the assumed design-basis wind or tornado may also serve as an envelope for the explosive pressures for comparison with actual site hazards of a general licensee's facility.

If the SAR includes bounding explosion effects for which the cask system is to be approved, the reviewer should verify that the applicant also provided structural analyses of those effects for cask system structures that may be affected. The SAR should identify the maximum response determined. The maximum response includes pressure-induced maximum stresses at critical cask locations and governing structural performance modes for the cask components important to safety. That response should be sufficiently low such that while damage may occur, it would not impair the capability of the component to perform its safety functions. In addition, the SAR should identify any post-event inspection and remedial actions that may be necessary.

(c)   Fire (LOW Priority)

Chapter 4, "Thermal Evaluation" of this SRP addresses potential fire conditions. Fire-related structural evaluation considerations include increased pressures in the confinement cask, changes in material properties, stresses caused by different coefficients of thermal expansion and/or temperatures in interacting materials, and physical destruction.

The reviewer should evaluate the discussion in the SAR concerning the treatment of structural effects associated with the presumed fire. The reviewer should evaluate the appropriateness of the applicant's analysis of those structural effects for the assumed parameters of the design-basis fire. The reviewer should confirm that the applicant defined the confinement cask pressure capacity on the basis of the cask material properties at the temperature resulting from the fire. Spalling of concrete that may result from a fire is generally considered acceptable and need not be estimated or evaluated. Such damage is readily detectable, and appropriate recovery or corrective measures may be presumed. The NRC accepts concrete temperatures that exceed the temperature limits of ACI349 for accidents, providing that the temperatures result from a fire. However, corrective actions may need to be taken for continued safe storage.

(d) Flood (LOW Priority)

The applicant's evaluation of the DSS design should be reviewed with regard to the structural consequences of a flood event. The SAR may stipulate an assumption that the DSS not be used at any site where there is potential for flooding. In this case, the DSS would have to be placed at an ISFSI site above the maximum probable flood level (SAR Chapter 12, "Accident Analyses Evaluation" should state this condition). Alternatively, an application for a certificate of compliance to use a DSS at a site with flooding potential would require a full analysis for a defined flood event.

If a design flood event is defined for the certificate of compliance the reviewer should verify that the SSCs meet appropriate guidance in RG 1.59, Rev. 2 and 1.102, Rev. 1 for that level of flood protection.

One possible structural consequence of a flood is that a vertically stored cask may tip over or translate horizontally (slide) because of the water velocity. Another possible consequence is that external hydrostatic pressure will exceed the capacity of the cask. The applicant may state the critical water velocity and hydrostatic pressure as bounds for the SAR flood analysis.

The NRC accepts the evaluation for flooding events when the flood conditions for overturning and sliding of stored confinement casks and other cask system structures (with a safety factor of 1.1 for accidents cases) have been applied. The applicant should state the basis for estimation of lateral pressure on a structure as a result of water velocity.

The NRC accepts the use of Hoerner's *Fluid-Dynamics Drag* (Hoerner, 1965) for estimating drag coefficients and net lateral water pressure. An approach for calculating the velocity corresponding to the cask stability limit is to assume that the cask is pinned at the outer edge of the cask bottom and rotates about that outer edge, and the pinned edge does not permit sliding. The overturning moment from the velocity of the flood water can be compared to the stability moment of the cask (with buoyancy considered). The structural consequences of the flood event are typically bounded by analyses for the drop or tipover accident cases.

The analysis of the confinement cask should be reviewed for flood-related hydrostatic pressure. The analysis should include the combined effects of weight, external hydrostatic pressure, internal pressure(s), and thermal stress. Resistance of the confinement cask to flood-related hydrostatic pressure should be analyzed in accordance with Section III, Subsection NB or NC, of the ASME B&PV Code (depending on the subsection used for design).

Additional flood consequences include potential scouring under a foundation, damage to access routes, temporary blockage of ventilation passages with water, blockage of ventilation passages and interstitial spaces between the confinement cask and shielding structure with mud, and steep temperature gradients in the shielding structure and confinement cask. The consequences of these conditions may be analyzed in the SAR and identified in the certificate of compliance so a general licensee will be able to consider these factors when siting an ISFSI.

(e)   Tornado Winds (LOW Priority)

The reviewer should verify that the SAR addresses the potential structural consequences of design-basis tornado or extreme wind effects. The load combination analyses should be reviewed for acceptable inclusion of tornadoes and tornado missiles. Current NRC guidance provided in RG 1.76, Rev. 1, recognizes three regions in the contiguous United States each with distinct design-basis tornado parameters. The applicant for a certificate of compliance must clearly define the boundary conditions of the proposed cask system with respect to these regions or utilized Region 1.

Confinement casks may be vulnerable to overturning and/or translation caused by the direct force of the drag pressure while in storage or during transfer operations. Criteria for resistance to overturning or sliding should be provided in the SAR.

Confinement casks are generally not vulnerable to damage from overpressure or negative pressure associated with tornadoes or extreme winds. However, they may be vulnerable to secondary effects, such as wind-borne missiles (see (f), below) or collapse of a weather enclosure, if used. The capability and behavior of the cask system under the collapse of any such external structure, if allowed by the Certificate of Compliance should be identified in the SAR.

Tornadoes typically produce the greatest "design-level" wind effects for American sites. However, there are some potential American sites at which high winds may be more severe than the credible tornado. The SARs for a limited set of potential sites could reflect high wind effects as a basis for structural analysis. If the certificate is to include proven design resistance to tornadoes or extreme winds, the SAR documentation must identify the wind levels (e.g., in miles or kilometers per hour), source (tornado or high wind), and specific wind-driven missiles (shape, weight, and velocity) for which the design is to be evaluated.

RG 1.76, Rev. 1, "Design-Basis Tornado for Nuclear Power Plants," provides applicable tornado-related parameters. The NRC accepts the use of ASCE 7 for conversion of wind speed to pressure and for typical building shape factors. Conversion of tornado or other wind speeds to pressure in the SAR documentation should assume that the cask system is at sea level.

The reviewer should verify that the cask system design meets appropriate guidance in the RG 1.76, Rev. 1, and 1.117, Rev. 1, and NUREG-0800 "Standard Review Plan for Power Reactors," Section 3.3.2, Rev. 3 for tornado protection.

Tornadoes and high winds can produce a significant negative pressure differential between interior spaces and the outside in a storage cask system that should be considered. This is a function of wind speed and factors relating to the structure. The magnitude of negative pressure depends on other parameters of the tornado or wind, and on wall pressure coefficients (as expressed in ASCE 7). There is no need for the SAR to separately state negative pressure to establish an envelope for approval since negative pressure is insignificant with regard to confinement cask accident pressure analysis.

The NRC does not accept the presumption that there will be sufficient warning of tornadoes that operations such as transfer between the fuel pool facility and storage site may never be exposed to tornado effects. Overturning during onsite transfer is considered by the staff to be a design-basis event. The tornado analysis should determine if tornado-induced overturning is bounded by drop and tipover cases. In addition, the SAR should show that the cask system will continue to perform its intended safety functions (i.e., criticality, radioactive material release, heat removal, radiation exposure, and retrievability).

(f)   Tornado Missiles (LOW Priority)

The applicant's evaluation of the cask system design should be reviewed with regard to the structural consequences of wind-driven missile impact (RG 1.76, Rev. 1 and NUREG-0800, "Standard Review Plan for Power Reactors," Section 3.5.1.4 (Rev. 3) and Section 3.5.3 (Rev. 3) describe the effects of tornado missiles). The SAR should define the missile parameters for which the cask system is to be evaluated based on the three tornado regions currently identified in the RG 1.76, Rev. 1.

Among the possible missile effects, the SAR should address those that may result in a tipover and those that may cause physical damage as a result of impact. The damage should not result in unacceptable radiation dose or significantly impair either criticality control, heat removal, or the retrievability of the fuel.

The NRC has accepted use of the analytical approaches given in U.S. Reactor Containment Technology, ORNL-NSIC-5, Volume 1, Chapter 6 (Cottrell and Savolainen), for estimating the potential effects of missile impact on steel sheets, plates, and other structures. Further guidance on analytical acceptable approaches for use in ISFSI design is provided in NUREG-0800, Section 3.5.3, "Barrier Design Procedures." In addition, for analysis and design regarding the ability of reinforced concrete structures to resist missiles, the NRC has accepted use of "Review of Procedures for the Analysis and Design of Concrete Structures to Resist Missile Impact Effects" (Kennedy, 1975).

Cask systems are not required to survive missile impacts without permanent deformation. However, the maximum extent of damage from a design-basis event must be predicted and should be sufficiently limited. Moreover, the capability of the SSC to perform their safety functions should not be impaired.

(g)   Earthquake (MEDIUM Priority)

The applicant's evaluation of the cask design should be reviewed with regard to the structural consequences of the earthquake event. Cask designs must satisfy the load combinations that encompass earthquake, including those for sliding and overturning. The applicant should demonstrate that no tipover or drop will result from an earthquake. In addition, impacts between casks should either be precluded, or should be considered an accident event for which the cask must be shown to be structurally adequate.

Appendix H of ANSI/ANS-57.9-1992 provides guidance for seismic analysis. Implicit in this guidance is the assumption that the ISFSI concrete pad, supported by soil, behaves as a rigid mat and therefore possesses no out-of-plane flexibility. This is valid for the majority of nuclear power plant structures where relatively thick mats support integral reinforced concrete walls. However, ISFSI pads are usually relatively thin structures (i.e., small thickness to length ratio) and generally do not incorporate integral walls to stiffen the pad. While the cask itself is relatively rigid, the rigid cask resting on a flexible pad has a lateral mode frequency that is generally low enough to fall within the amplified range of most design earthquake spectra. Thus, in determining the inertia forces that act at the center of gravity of the cask for the purpose of evaluating the onset of sliding or tipping, the reviewer should ensure that the applicant has either accounted for the out-of-plane flexibility of the pad in the seismic analysis or demonstrated that it is not an important parameter in determining the response of the cask, ("Influence of ISFSI Design Parameters on the Seismic Response of Dry Storage Casks," Bjorkman & Moore, 2001).

The reviewer should verify that the cask system design meets appropriate guidance in RGs 1.29, Rev. 4, 1.60, Rev. 1, 1.61, Rev. 1, and 1.92, Rev. 2, for protection against seismic events.

The SAR documentation should include analysis of the potential for impacts between components of the cask system. These could include contact between the confinement shell and its inner components or outer shield and the rocking and fall back of a vertically or horizontally oriented confinement cask on its supports.

Cask systems are not required to survive a design earthquake without permanent deformation. However, the maximum extent of damage from a design earthquake must be predicted, and the capability to provide principal safety functions should not degrade.

ii.   Structural Analysis Methods

(LOW Priority) The applicant's structural analysis of various loading combinations and the resulting stresses, strains, and deformations from different loads should be reviewed. The reviewer should verify that the applicant properly used acceptable analytical approaches and tools. In addition, the applicant should have performed and reviewed the associated computations internally under an acceptable independent design review (equivalent to ASME NQA-1) and quality assurance procedures. The scope of the staff's review may include performing detailed parallel computations (such as finite element analyses) to validate submitted computations or their results. The reviewer may perform separate, less extensive calculations when these could most readily evaluate any suspected problems.

The applicant's analysis of loads and load combinations resulting from different structural conditions should be consistent with the code or criteria requirements used in designing the component.

Subsection NB or NC of the ASME B&PV Code defines the requirements for categorizing stresses and determining allowable stress limits for the confinement boundary of the cask. For the fuel basket, Subsection NG of the Code applies. These references also provide definitions of stress categories and stress intensity limits for normal and off-normal operating conditions. For Level D or accident conditions, Appendix F to the ASME B&PV Code provides definitions of the stress intensity limits.

In accordance with these references, stress intensity is defined on the basis of the maximum shear stress theory for ductile materials. Since the maximum shear stress is not identical to the maximum octahedral shear stress, octahedral shear stresses should not be compared with the stress intensity limits. Values for the stress intensity limits are defined in Appendices I and III of the ASME

Code. Stresses resulting from inertial and pressure loads should be considered primary stresses. Thermal stresses resulting from temperature gradients may be considered secondary stresses if they are self-limiting and do not cause structural failure. Stresses due to thermal gradients in fuel baskets may not be self-limiting and should be considered by the applicant because of the possibility of uneven heat loadings of adjacent assemblies as well as the effects of asymmetry in the basket structure.

(1)　Finite-Element Analyses (HIGH Priority)

Because of the complexity of many structural design considerations and load conditions, structural design computations are often performed using finite-element analysis.

The applicant should perform the finite-element analyses using a general-purpose program that is well benchmarked and widely used for many types of structural analyses.

Consistent with the provisions of ASME Code, Section III, Appendix F, inelastic material properties may be used for the storage cask design analysis evaluation for accident loads. The SAR should identify the sources used for the inelastic material properties.

Lead shielding can be modeled either with elastic or inelastic properties. The elastic modulus and limit used for lead in the elastic analysis should be determined on the basis of the potential temperature of the material. An appropriate plasticity model of lead can be used to account for its inelastic behavior.

Nonstructural components of the confinement cask are generally not included in finite element models. However, the models should include any influence these nonstructural components may have on the structural performance of the cask. Possible influences include the nonstructural components' inertial weight, restraint to motion of the structural components, and localized influence on load applications because of geometrical effects.

Bolted connections can be modeled either discretely or with contact conditions. To discretely model the bolted connections, the applicant should use appropriate element types and material properties. With contact conditions, the interfaces joined by the bolts can be modeled as tied.

Verify that the applicant has provided information on any computer-based modeling as described in Appendix 3A to this chapter, and review the structural analyses submitted by the applicant in accordance with the Appendix.

(2)　Closed-Form Calculations (MEDIUM Priority)

The applicant should perform closed-form calculations for relatively simple structural load conditions or conditions for which a formula has been developed. Closed-form calculations are also typically used to check the results of finite-element analyses. In addition, this type of calculation can be used for analyses involving principles of conservation of energy and comparisons of overturning moments.

One source of closed-form equations accepted by the NRC is *Formulas for Stress and Strain* (Roark, 1965). Use of a particular equation or formulation for the load conditions should be justified. The most important aspect of the calculations to evaluate is the basis for the assumptions used in the calculations. In many cases, the calculations are faulty in that they fail to include portions of the cask, or the load conditions are idealized inappropriately.

To be consistent with the provisions in Section III of the ASME Code, the analyses should use linear material properties. Linear analysis should be the basis for all closed-form calculations.

(3)   Structural Analysis for Specific Cask Components

The following paragraphs present a few specific examples of structural analysis for some of the confinement cask components of a cask storage system.

(a)   Fuel Basket (HIGH Priority)

The fuel basket design should be reviewed to assess the applicant's analysis of the combined effects of weight, thermal stresses, and cask-drop impact forces that could arise during spent fuel transfer and storage operations. The weight supported by the basket should be the maximum or design weight of the SNF to be stored. In addition, the applicant should evaluate all credible potential orientations of the cask and basket during cask transfer and handling drops while transferring the spent fuel into storage. End or side drops typically produce the greatest structural demand on various basket components. In an end drop, the basket is supported by the bottom of the confinement cask cavity upon impact. In the side drop, the basket structure and points of contact with the confinement cask must support the mass of the basket and loaded fuel.

In previous DSS evaluations, the NRC has accepted two approaches for analyses regarding the structural capability of the basket to acceptably survive a cask drop during transfer and storage. The first approach uses dynamic analyses in a two-step process. In Step 1, the applicant performs a dynamic analysis of the cask body impacting a target surface and assesses the performance of the cask body, including determining the time-history response from the cask drop impact. In Step 2, this time-history response can be translated into a forcing function that can be applied to the supporting contact points of an appropriate model of the fuel basket.

The second approach uses a quasi-static analysis of the basket subjected to the equivalent acceleration inertial load derived from the cask-drop impact analysis. In this analysis, the applicant should apply the equivalent acceleration inertial load using an appropriate model of the basket with the location(s) most vulnerable to the impact. Support provided by the inside surface of the cask cavity should be represented by the appropriate boundary conditions on the outside edge of the basket. In addition, the applicant should conservatively select the equivalent acceleration inertial load such that it bounds the possible inertial loads resulting from a cask-drop accident onto the bounding target surfaces. If applicable, the inertial load should also account for dynamic amplification effects by using a dynamic amplification factor.

The applicant should also evaluate the buckling capacity of the cask basket materials. Acceptable guidance for this evaluation is provided in Section III of the ASME B&PV Code and NUREG/CR-6322, "Buckling Analysis of Spent Fuel Basket," (Lee and Bumpas, 1995). For this evaluation, the applicant should select the appropriate end conditions used in the buckling capacity equations on the basis of sensitivity studies. These studies can bound the range of conditions that are typically either fixed for a welded connection or free if there is no rigid connection.

(b)   Closure Lid Bolts of Confinement Boundary (MEDIUM Priority)

The design analysis for the closure-lid bolts should be reviewed to ensure that it properly includes the combined effects of weight, internal pressure(s), thermal stress, O-ring compression force, cask impact forces, and bolt pre-load. Typically, applicants specify the pre-load and bolt torque for the closure bolts on the basis of bolt diameter, and the coefficient of friction between the bolt and the lid. Externally applied loads (such as the internal pressure and impact force) produce direct tensile force on the bolts as well as an additional prying force caused by lid rotation at the bolted joint. The tensile bolt force obtained by adding together the pressure loads, impact forces, thermal load, and O-ring compression force should then be compared with the tensile bolt force computed from the pre-load and operating temperature load alone. The larger of the two calculated tensile forces should control the design. The maximum design bolt force should then be obtained by combining the larger direct tensile bolt force with the additional prying force. The weight is derived from the maximum or design weight of the closure lids and any cask components supported by the lids. Acceptable analytical methods for closure bolts are given in NUREG/CR-6007, "Stress Analysis of Closure Bolts for Shipping Casks" (Mok and Fischer, 1993).

The bolt engagement lengths should be reviewed. If the lids are fabricated from relatively non-hardened materials, threaded inserts may be used in the closure lids to accommodate the hardened material of the bolts.

(c)   Trunnions (LOW Priority)

The design of the trunnions, their connections to the cask body, and the cask body in the local area around the trunnions should be reviewed. The design basis for the trunnions can be either non-redundant or redundant. In either case, the design should meet the requirements of ANSI N14.6 for critical loads and the requirements of NUREG-0612, "Control of Heavy Loads at Power Plants."

Non-redundant lifting systems should be designed for not less than 6 times the material yield strength and 10 times the material ultimate strength given the design lift weight of the loaded cask. Redundant lifting systems should be designed for not less than 3 times the material yield strength and 5 times the material ultimate strength given the design loaded lift weight of the cask. Acceptance testing requirements for trunnions are discussed in Chapter 10, "Acceptance Tests and Maintenance Program Evaluation," of this SRP.

For a typical trunnion design, the maximum stress occurs at the base of the trunnion as a combination of bending and shear stresses. A conservative technique for computing the bending stress is to assume that the lifting force is applied at the cantilevered end of the trunnion, and that the stress is fully developed at the base of the trunnion. If other assumptions, including ASME Section III stress limits by the finite element design analysis and slight material yielding at localized regions, are considered, the applicant should provide adequate justifications.

iii.   Structural Evaluation

(1)   Structural Capability (LOW Priority)

The applicant's structural analyses should be reviewed to assess the information regarding margins of safety or compliance with ASME Code stress limits, overturning margins, and other criteria appropriate for the division of the ASME Code being used. The comparisons of capability versus demand for the various applicable loading conditions should be presented in the same terms used in the design code (e.g., type of stress). In addition, margins of safety should be included on the basis of comparisons between capacity and demand for each of structural component analyzed. The minimum margin of safety for any structural section of a component should be included for the different load conditions.

(2) Fabrication and Construction (MEDIUM Priority)

The NRC has accepted fabrication of metallic confinement casks in accordance with Section III, Division 1 of the ASME B&PV Code. If the fabrication, construction, or assembly deviate in any way from the subsection of this standard used for design, the SAR must explicitly state the applicant's justification for the deviation, and the justification must be acceptable to the NRC.

If the design of the confinement cask is proposed to be governed by ASME, Section III, Division 2, similar to a metallic-lined concrete pressure vessel NRC would expect the fabrication/construction of such a cask to also be governed by the Division 2 requirements. Any deviations from the Code requirements should be addressed as noted for Division I above for metallic containment.

If the design of the confinement cask is proposed to be governed by ASME, Section III, Division 3, the applicant will have to provide supplemental details to the Code provisions since Subsection WC does not provide guidance to address all construction details for classic containments.

3.5.2 Other System Components and Structures Important to Safety

3.5.2.1 Scope

This portion of the DSS structural review provides guidance by addressing procedures for evaluating all structures that are important to safety (as defined in 10 CFR Part 72.3), whether steel, concrete or other material not addressed as the confinement cask and internals (Subsection 3.5.1). Structures may include items such as gamma and neutron shielding, overpack material, any respective encasement foundations, structural supports, ventilation passages, weather enclosures, earth retention structures, and protective structures. This evaluation should include drawings, plans, sections, and technical specifications for these SSCs.

3.5.2.2 Structural Design Criteria and Design Features

i. Design Criteria (MEDIUM Priority)

(1) General Structural Requirements

Structural requirements are driven by the functional roles of the system components and the need to maintain safety. Safety requirements are expressed in the referenced rules, standards, and codes and as criteria specific to the component. The basic safety requirements are that the structural and functional design must preclude the following:

• Unacceptable risk of criticality.

- Unacceptable release of radioactive materials to the environment.

- Unacceptable radiation dose to the public or workers.

- Significant impairment of retrievability of stored nuclear materials during normal and off-normal conditions.

The applicant should consider the potential for liquefaction and other soil instabilities attributable to vibrating ground motion, for any structure or system component such as a cask system support pad.

Reinforced concrete pads that support confinement casks in storage do not constitute "pavements." As such, they should be designed and constructed as foundations under an applicable code such as, ACI 349, ACI 318, or IBC. Such pads typically are not classified as important to safety; however, in some cases they may be.

Steel embedments in reinforced concrete structures must satisfy the requirements of the design code applicable to the reinforced concrete structure. Similarly, structural steel must satisfy the requirements of the applicable steel design code (e.g., ASME B&PV Code, AISC, or other identified code).

(2) Applicable Codes and Standards

The codes and standards identified in the SAR should be reviewed as well as their proposed applications. This subsection addresses the codes and standards that the NRC has accepted for structures important to safety categorized by application that are not confinement casks or the steel internals.

The NRC accepts the use of ANSI/ANS-57.9 (together with the codes and standards cited therein) as the basic reference for the structures important to safety that are not designed in accordance with the Section III, Division 1 or Division 2 of the ASME B&PV Code. However, both the lifting equipment design and the devices for lifting system components that are important to safety must comply with ANSI Standard N14.6. The NRC accepts the load combinations shown in Table 3-3 for structures not designed under either Section III of the ASME B&PV Code Section III, Division 1 or 2 (ACI 359). See Table 3-2 for loads and their descriptions.

The reviewer should review the suitability of the applicant's identification of codes and standards that are to be met by the structural design and construction of other components subject to NRC approval. The principal codes and standards include the following references that may apply to steel structures and components as well as concrete portions of the cask system:

- AISC, "Specification for Structural Steel Buildings – Allowable Stress Design and Plastic Design." The NRC has not yet received any applications that propose a steel design on the basis of the AISC's "Load and Resistance Factor Design (LRFD) Specification for Structural Steel Buildings." If such a design was received, the NRC would evaluate the proposal for compliance with the load combinations summarized in Table 3-3 and for consistent application of the LRFD design methodology.

- To date, the NRC has not required applicants to design or build structural steel components of a cask system important to safety in compliance with ANSI/ANS N690, "Nuclear Facilities — Steel Safety-Related Structures for Design Fabrication and Erection."

- AWS D1.1, "Structural Welding Code Steel."

- ASCE 7, "Minimum Design Loads for Buildings and Other Structures."

- ACI 349, Appendix D, for anchoring to concrete or Section 10.14 for composite compression sections, as applicable, when constructed of structural steel embedded in reinforced concrete. Where requirements do not conflict, the steel must also comply with the requirements of the codes stated above. In addition, ACI 349 defines constraints for obtaining ductile response to extreme loads by ensuring that the strength of steel embedments controls the design; these constraints must not be subverted by over-design of the steel.

- For reinforced concrete the NRC has not accepted the use of a set of criteria selected from multiple standards and codes, except when the selected criteria meet the most limiting requirements of each code. However, in recognizing a graded approach to quality assurance, the NRC has approved the use of ACI 349 for design and material selection for reinforced concrete structures important to safety (not confinement). The NRC has allowed the optional use of ACI 318 as an alternative standard for construction as described below.

- In both cases, however, the design, material selection and specification, and construction must also meet any additional or more stringent requirements given in ANSI/ANS-57.9.

The following paragraphs identify the portions of ACI 349 that apply to design (including material selection) and must be met by applicants who choose to use ACI 318 for construction. (The paragraph references are as in ACI 349-06.). Unlisted and excepted sections address construction requirements for which the NRC accepts substitution of ACI 318.

Chapter 1   "General Requirements," Sections 1.1 and 1.5 (except references to construction), and Sections 1.2 and 1.4.

3-29

Chapter 2	"Definitions."
Chapter 3	"Materials" (except Sections 3.1, 3.2.3, 3.3.3, 3.5.3.1.1, 3.6.1.0, and 3.7).
Chapter 4	"Durability Requirements"
Chapter 6	"Form Work, Embedded Pipes, and Construction Joints," Sections 6.3.13, 6.3.14, and 6.3.15.
Chapter 7	"Details of Reinforcement."
Chapter 8	"Analysis and Design General Considerations."
Chapter 9	"Strength and Serviceability Requirements."
Chapter 10	"Flexure and Axial Load."
Chapter 11	"Shear and Torsion."
Chapter 12	"Development and Splices of Reinforcement."
Chapter 13	"Two-way Slab Systems."
Chapter 14	"Walls."
Chapter 15	"Footings."
Chapter 16	"Precast Concrete."
Chapter 17	"Composite Concrete Flexural Members."
Chapter 18	"Prestressed Concrete."
Chapter 19	"Shells."
Appendix A	"Strut-and-Tie Models."
Appendix D	"Anchoring to Concrete."
Appendix E	"Thermal Considerations."
Appendix F	"Special Provisions for Impulsive and Impactive Effects" (except that the load combinations included herein, must be used.

For fluid systems used with a cask system that may be connected to a penetration of the confinement barrier outside an enclosing structure licensed under 10 CFR Part 50 (e.g., the fuel pool building), the NRC accepts construction consistent with requirements comparable to those used for Quality Group C, as shown in RG 1.26, "Quality Group Classifications and Standards for Water-, Steam-, and Radioactive Waste-Containing Components of Nuclear Power Plants," Rev. 4 and NUREG-0800," Section 3.2.2, "Standard Review Plan for Nuclear Power Plants." In this context, "construction" includes materials, design, fabrication, examination, testing, inspection, and certification required in the manufacture and installation of components. Quality Group D may, under some circumstances be justified.

Quality Group C requires construction of piping, pumps, valves, atmospheric storage tanks, and 0-15 psig storage tanks in conformance with Section III of ASME B&PV Code 1, Class 3 (Subsection ND). In addition, Quality Group C requires that supports for these components meet the requirements of Subsection NF.

By contrast, Quality Group D requires compliance with the following codes, as a minimum:

Piping: ANSI/ASME B31.1, "Power Piping."

Pumps: Manufacturer's Standards.

Valves: ANSI/ASME B31.1 and ANSI B16.34, "Valves."

Atmospheric Storage Tanks:
American Water Works Association (AWWA), "Standard for Steel Tanks — Standpipes, Reservoirs, and Elevated Tanks for Water Storage" (AWWA D100) or ANSI/ASME B96.1, "Specification for Welded Aluminum-Alloy Field-Erected Storage Tanks."

0–15 psig Storage Tanks:
American Petroleum Institute's (API) "Recommended Rules for Design and Construction of Large, Welded, Low-Pressure Storage Tanks" (API 620).

The NRC accepts the "Boundaries of Jurisdiction" applicable to Section III, Subsections NB-1130 and NC-1130, of ASME B&PV Code. These boundaries apply to attachments to penetrations of the confinement barrier outside an enclosure licensed under 10 CFR Part 50. Specifically, these boundaries define whether the attachments must be designed, fabricated, and installed in accordance with Section III, Subsection NB or NC, of ASME B&PV Code.

Note that codes, other than those discussed herein (e.g., the "Electric, Life Safety, and Lightning Protection Codes" promulgated by the National Fire Protection Association [NFPA]), may apply to the design and construction of the cask system. It is acceptable to include such codes in the design by inclusion in the SAR. Where designs of structures subject to approval are also covered by such other codes, the review should include evaluation of compliance with those codes.

The NRC has not yet received any applications for licensing or approval of a cask system that included masonry important to safety. Masonry is not considered suitable for confinement, but it may be acceptable for enclosures and physical or radiation-shielding applications.

ii.   Structural Design Features (MEDIUM Priority)

The design description in the SAR documentation should be reviewed to ensure that it defines the functional performance required of the structures. The design description of the non-confinement safety-related structures of the cask system should provide a clear understanding to be reached by the reviewer of the significance of the safety-related features to the required performance.

The SAR documentation should also be reviewed regarding the physical design of the structures important to safety. This should include the following as a minimum. As appropriate to the specific structure the following information should be provided.

- Dimensioning of all structural elements.

- Locations, sizes, configuration, spacing, welding, fasteners etc. of the safety-related non-confinement structures should be provided.

- Locations and specifications for controls, that will be necessary in fabrication and construction.

- Structural materials with defining standards or specifications summarized or references to Chapter 8, "Materials Evaluation" of this SRP herein should be reviewed.

- Information on the physical design of attachments, embedments, and other structural elements should be provided.

Auxiliary cask system equipment important to safety has often been specially designed. In particular, the structural design features that provide for safety should be supported by design or operational analysis. This analysis should demonstrate that the equipment will meet the basic safety criteria, regardless of problems that may occur in mechanical, electrical, human operator, or other operations.

The NRC has accepted and approved cask system designs that depend on the operation of new mechanical systems for system use. NRC approval does not certify that the mechanical systems will operate as projected but rather that proper functioning is necessary to successfully complete a specified operation. Such approval reflects a finding by the NRC staff that, regardless of the system's success (or lack thereof) in mechanical operation, the basic safety criteria will be met, as stated above.

The proposed system design should be reviewed against planned normal and off-normal, operations and accidents. The reviewer should determine whether the structural design of the equipment provides for continuing satisfaction of the basic safety criteria. The reviewer should consider that the equipment could fail to operate at any time (i.e., during operations at the physical limits of speed or range, or during a credible, off-normal, or accident-level event).

3.5.2.3   Structural Analysis

Subsections 3.5.1.4 (i) and (ii) provide guidance regarding structural analysis for the confinement cask and metallic internals of cask systems. These subsections provide supplemental guidance primarily related to steel and concrete structures, other than the confinement cask and its contents and integral components that are important to safety. The appropriateness, completeness, and correctness of the applicant's proposed implementation of these load conditions and combinations for the metallic and reinforced concrete structures should be reviewed.

    i.   Load Conditions (MEDIUM Priority)

The load definitions and combinations shown in Tables 3-2 and 3-3 have been accepted by the NRC for analysis of steel and reinforced concrete ISFSI structures that are important to safety. These load combinations are included in or derived from ANSI/ANS 57.9 and ACI 349.

Structures that are important to safety should have sufficient capability for every section to withstand the worst-case loads under normal and off-normal conditions. Such capability ensures that these structures will not experience permanent deformation or degradation of the capability to withstand any future loadings.

The NRC accepts the load combinations in Table 3-3 that implement and supplement those of ANSI/ANS-57.9.

    (1)   Normal Conditions

The SAR documentation should be reviewed to ensure adequate inclusion of the following conditions that may be of particular concern for concrete structures important to safety if the loading condition is appropriate:

- Live and dynamic loads associated with transfer of the confinement cask to and from its storage position and in its storage location for its service lifetime.

- Live and dynamic loads associated with installing closures.

- Load or support conditions associated with potential differential settlement of foundations over the life of the cask system.

- Thermal gradients associated with the normal range of operations and ranges of ambient temperature.

- Thermal gradients that may result from impingement of precipitation on highly heated concrete.

    (2)   Off-Normal Conditions

The SAR should be reviewed to ensure adequate inclusion of the following off-normal operations and events:

- Live and dynamic loads associated with equipment or instrument malfunctions, or accidental misuse during transfer of the confinement cask to and from its storage position.

- Situations in which a confinement cask is jammed or moved at an excessive speed into contact with a reinforced concrete structure.

- The impact of reinforced concrete structures by a suspended transfer, confinement, or storage cask.

- Off-normal ambient temperature conditions (although they may be less severe than accident conditions, these may be of concern because of different sets of factors in the off-normal and accident load combinations, and because concrete temperature limits for off-normal conditions are the same as for normal conditions. Note that greatly elevated concrete temperatures are allowed for accident conditions in accordance with ACI 349, Section A.4).

(3)   Accident Conditions and Natural Phenomena Events

The SAR should be reviewed for adequate inclusion of the following conditions associated with accident and conditions that may be of special concern for reinforced concrete structures:

- Loads associated with accidental drops or other impacts during transfer of the confinement cask to and from its storage position.

- Events that produce extreme thermal gradients in the concrete.

- Contact caused by earthquake between the confinement cask and the reinforced concrete structures.

- Drop of a closure into position or onto the structure.

The ACI codes are intended to ensure ductile response beyond initial yield of structural components. ACI 349 also imposes conditions on design (beyond those of ACI 318) that effectively increase ductility. In particular, the reviewer should review the proposed reinforced concrete design to ensure that it provides code levels of ductility by satisfying the pertinent ACI 349 provisions. Seismic loads are considered to be "impulsive" and, therefore, are subject to the additional design constraints of Appendix F to ACI 349. Other accident conditions or natural phenomenon events may also produce impulsive or impactive loadings requiring the additional requirements of Appendix F to ACI 349.

Reviewers should check the steel reinforcement schedules and drawings to ensure that any reinforcing steel quantities, sizes, and locations are consistent with the design analysis.

In particular, consider the following aspects of the design:

- Upper limit (60 ksi, 4219 kgf/cm2) on the specified yield strength of reinforcement, lower limit (3 ksi, 211 kgf/cm2) on concrete specified compressive strength (f'c), and upper limit on concrete strength, as analyzed and specified for the ISFSI cask storage pads.

- Limit on the amount (cross-section area) of compressive reinforcement in flexural members.

- Requirements on continuation and development lengths of tensile reinforcement.

- Specifications for confinement and lateral reinforcement in compression members, in other compressive steel, and at connections of framing members.

- Aspects of the design that ensure flexure controls (and limits) the response.

- Requirements for shear reinforcement.

- Limitations on the amount of tensile steel in the flexural members relative to that which would produce a balanced strain condition.

- Projected maximum responses to design-basis loads within the permissible ductility ratios for the controlling structural action.

- Embedments designed for ductile failure and to fail in the steel before pullout from the concrete.

In addition, the construction specifications or descriptions (to the extent included in the SAR documentation) should be reviewed to ensure that substitution of materials, use of larger sizes, or placement of larger quantities of steel will be precluded, and that provisions for splicing or development of reinforcing steel will not reduce ductility of the members.

ii.   Structural Analysis Methods (HIGH Priority)

The applicant should select and use analytical methods that are appropriate for the proposed type of materials and construction. In certain instances, however, the applicant may have to adapt existing analytical methods, codes, and models for highly specialized cask system equipment designs. Such instances require special review attention. In particular, the reviewer should ensure that the adapted approach is fully documented, supported, and acceptable. In addition, the reviewer should consider the potential for safety-related risk associated with a possible error in the design of special cask system equipment. The degree of risk indicates the suitability and acceptability of the adapted approach. Subsection 3.5.1.4.ii provides acceptable analytical methods of analysis that can be utilized. Appendix 3A addresses the application of computational modeling software.

iii.   Structural Evaluation (LOW Priority)

In evaluating the variety of cask system equipment and structures that may be important to safety, the reviewer should ensure compliance with the basic safety criteria in Subsection 3.5.2.2 (i)(1) and that the specified parameters for acceptability such as stress, strain or deflection are within the permitted values identified in Subsection 3.5.2.2.i.(2).

The NRC accepts strength design as presented in the current revision of ACI 349 for reinforced concrete structures important to safety that are not within the scope of ACI 359. If the applicant uses another design approach, the review conducted within the scope of the DSS SAR evaluation should include in-depth comparison of that approach with the provisions of ACI 349.

The NRC accepts the use of guidance in NUREG-0800 for analysis of natural phenomena, as related to the conditions that apply to the design of cask systems. However, the load combinations shown in Table 3-3 and the design and construction requirements of the codes cited above take precedence. The NRC accepts the American Society of Civil Engineers' "Seismic Analysis of Safety Related Nuclear Structures" (ASCE 4) and ASCE 7 as the standards for seismic analysis. In addition, the NRC accepts tornado missile impact analysis in accordance with Kennedy's *Review of Procedures for the Analysis and Design of Concrete Structures to Resist Missile Impact Effects.*

(1)   Structural Capability (LOW Priority)

Section 3.5.1.4.iii (1) addresses the assessment of the structures capability with respect to the ASME Code stress limits which are appropriate for metallic structures under Division 1 and for concrete structures under Division 2.

For other safety related structural concrete, strength (or "ultimate strength") design is the approach usually used in reinforced concrete design. Strength design is the only design approach that has been accepted for reinforced concrete structures that are part of cask systems not within the scope of ACI 359, and it is the approach used in the current revisions of ACI 349. This design code was tested and developed on the basis of extensive empirical experience with concrete construction. The current strength design approach, as presented in this code, includes empirically derived requirements and constraints. Determination that a reinforced concrete structure designed by another approach satisfies ACI 349 typically requires clause-by-clause review of the code for compliance. Allowable stress design was formerly used as the basis for ACI codes related to reinforced concrete design. However, those codes do not reflect additional experience gained through observations of structural performance and experimental testing that has since been included in the current approach to strength design.

With respect to structural steel or other metallic structures important to safety, but not to the confinement structure or internals, the structural capability of the design may be based on the ASME Code with the use of the appropriate subsections as identified in Section 3.5.2.2 (i)(2) herein, or the AISC specifications also identified. Allowable stress, plastic design, and load and resistance factor methods of design are acceptable for use when there is justification for the method used provided in the application.

(2) Fabrication and Construction (MEDIUM Priority)

For structures and structural components analyzed and designed based on ASME B&PV Code requirements of Section III, Division 1 or Division 2, the fabrication and construction provisions of these documents should form the basis for the production and installation of the structures and components of the cask storage system.

NRC accepts construction in accordance with ACI 349 or ACI 318. Selection and validation of the proper concrete mix to meet design requirements are considered a construction function. By contrast, specification of cement type, aggregates, and special requirements for durability and elevated temperatures is considered a design or material selection function and is, therefore, governed by ACI 349 (and/or ACI 359, if applicable).

The following sections of ACI 318 (chapters, appendix, and paragraphing per ACI-318-02) have been accepted by the NRC for construction of ISFSI reinforced concrete structures that are not within the scope of ACI 359:

Chapter 1   "General Requirements," Sections 1.1.1, 1.1.2, 1.1.3, and 1.1.5 (except references to design and material properties), and Section 1.3.
Chapter 2   "Definitions" (use ACI 349, Chapter 2).
Chapter 3   "Materials," Sections 3.1 and 3.8 (except A-616, A-617, A-767, A-775, A-884, and A-934).
Chapter 4   "Durability Requirements."
Chapter 5   "Concrete Quality, Mixing, and Placing."
Chapter 6   "Form Work, Embedded Pipes, and Construction Joints" (except references to design and material properties, which are governed by ACI 349).

3.5.3   Other Structural Components Subject to NRC Approval (MEDIUM Priority)

3.5.3.1   Scope

The cask system description provided in the SAR may include a variety of components that are not important to safety such as transporters, ram systems, vacuum drying systems, drain and fill quick disconnects, support pads and other concrete structures not important to safety. These components should be reviewed to ensure proper functioning to the extent that the structures represent required elements of the total cask system. In particular, the reviewer should evaluate all structures that are proposed for approval in a cask system design acceptable to the NRC. This evaluation should ensure that the SAR provides sufficient information to confirm the proper functioning of the components and the overall system. For each system element that is not important to safety, the reviewer should address the potential response to accidents and natural phenomenon events to ensure that the given element will not jeopardize the safety provided by other system elements.

3.5.3.2   Structural Design Criteria and Design Features

i.   Design Criteria

(1)   General Structural Requirements

Structures subject to approval but not important to safety should be reviewed on the basis of determining whether the structures can properly perform their intended function(s). In addition, the NRC review should ensure that the response of the structures to credible off-normal and accident conditions will not create secondary hazards for cask system components or the stored nuclear materials.

(2)   Applicable Codes and Standards

The reviewer should review the suitability of the applicant's identification of codes and standards to be met by the structural design and construction of other components subject to NRC approval. The principal codes and standards include the following references although any of the previously identified codes in Sections 3.5.1.2.ii(2) and 3.5.2.2.i(2) may be used.

- ASCE 7.

- International Building Code (IBC).

- AISC, "Specification for Structural Steel Buildings—Allowable Stress Design and Plastic Design."

- AISC, "Code of Standard Practice for Steel Buildings and Bridges."

- ASME B&PV Code, Section VIII.

- ACI 318.

ii.   Structural Design Features

The reviewer should examine the adequacy of the applicant's descriptions of cask system components that are not important to safety but are subject to NRC approval. These descriptions should adequately identify the intended function(s) of each component.

Although the components evaluated in this portion of the DSS review are not directly important to safety, a credible possibility may exist that the structural response or failure of these components may cause a secondary risk to other components that *are* important to safety or to the subject nuclear material. For example, under tornado or seismic event conditions, the components may impact other components that are important to safety. When such a possibility exists, the applicant must provide more extensive structural information and greater assurance of acceptable fabrication and construction.

3.5.3.3 Materials Related to Structural Evaluation

The identification of structural materials should be reviewed in coordination with the materials discipline in Chapter 8 to the extent appropriate to determine if they are adequate for their intended function(s). The reviewer should determine the required level of review and extent of information in relation to the possibility and consequences of secondary effects on components that are important to safety. Materials should be as permitted or specified in the applicable code(s).

3.5.3.4 Structural Analysis

i. Load Conditions

The load definitions and combinations shown in Tables 3-2 and 3-3 have been accepted by the NRC for analysis of steel and reinforced concrete ISFSI structures that are important to safety. These load combinations may also be used for structures not important to safety.

In addition, for structures not important to safety, the NRC accepts the use of load combinations given in the IBC as well as ACI 349, ANSI/ANS 57.9, and ASCE 7.

The NRC also accepts the load descriptions, combinations, and analytical approaches given in the ASME B&PV Code, Section VIII, for pressure systems, vessels, and casks that do not form elements of the confinement cask.

ii. Structural Analysis Methods

The reviewer should evaluate the applicant's selection and use of structural analysis methods, codes, and models and ensure that these are consistent with and appropriate for the design code applicable to the component (as discussed above).

iii. Structural Evaluation

The reviewer may determine that an NRC structural evaluation of certain other components is not necessary for approval of the cask system. Similarly, the NRC may determine that approval of the cask system does not need to include specific components that are not important to safety, even though the applicant seeks approval of those components as part of the application.

The SER should identify the system components that are excluded from the approval, stating the rationale for exclusion of each. As a corollary, the SER should also identify the components that are included, stating any limitations on the scope of the NRC review (e.g., "reviewed for functionality only").

3.6   Evaluation Findings

The structural evaluation must provide reasonable assurance that the cask system will allow safe storage of SNF. This finding should be reached on the basis of a review that considered the regulation, appropriate RG, applicable codes and standards, and accepted engineering practices. Acceptance of the structural design of a storage cask system therefore implies that the design meets the relevant requirements of the following regulations:

F3.1   The SAR adequately describes all SSCs that are important to safety, providing drawings and text in sufficient detail to allow evaluation of their structural effectiveness.

F3.2   The applicant has met the requirements of 10 CFR Part 72.236(b). The SSCs important to safety are designed to accommodate the combined loads of normal or off-normal operating conditions and accidents or natural phenomena events with an adequate margin of safety. Stresses at various locations of the cask for various design loads are determined by analysis. Total stresses for the combined loads of normal, off-normal, accident, and natural phenomena events are acceptable and are found to be within limits of applicable codes, standards, and specifications.

F3.3   The applicant has met the requirements of 10 CFR Part 72.236(c), for maintaining subcritical conditions. The structural design and fabrication of the DSS includes structural margins of safety for those SSCs important to nuclear criticality safety. The applicant has demonstrated adequate structural safety for the handling, packaging, transfer, and storage under normal, off-normal, and accident conditions.

F3.4   The applicant has met the requirements of 10 CFR 72.236(l), "Specific Requirements for Spent Fuel Storage Cask Approval." The design analysis and submitted bases for evaluation acceptably demonstrate that the cask and other systems important to safety will reasonably maintain confinement of radioactive material under normal, off-normal, and credible accident conditions.

F3.5   The applicant has met the requirements of 10 CFR 72.236 with regard to inclusion of the following provisions in the structural design:

-    Design, Fabrication, Erection, and Testing to Acceptable Quality Standards.

-    Adequate Structural Protection Against Environmental Conditions and Natural Phenomena, Fires, and Explosions.

-    Appropriate Inspection, Maintenance, and Testing.

-    Adequate Accessibility in Emergencies.

-    A Confinement Barrier that Acceptably Protects the Cladding During Storage.

- Structures that are Compatible with Appropriate Monitoring Systems.

- Structural Designs that are Compatible with Retrievability of SNF.

F3.6 The Applicant has met the specific requirements of 10 CFR 72.236(g) and (h) as they apply to the structural design for spent fuel storage cask approval. The cask system structural design acceptably provides for the following required provisions:

- Storage of the Spent Fuel for a Minimum Required Years.

- Compatibility with Wet or Dry Loading and Unloading Facilities.

The reviewer should provide a summary statement similar to the following:

"The staff concludes that the structural properties of the structures, systems, and components of the [cask designation] are in compliance with 10 CFR Part 72, and that the applicable design and acceptance criteria have been satisfied. The evaluation of the structural properties provides reasonable assurance that the [cask designation] will allow safe storage of SNF for a licensed (certified) life of ____ years. This finding is reached on the basis of a review that considered the regulation itself, appropriate regulatory guides, applicable codes and standards, and accepted engineering practices."

3.7   Designations and Descriptions of Loads

Definitions of terms used in the following table are as accepted by the NRC. Many definitions are expanded with their intended applications more fully described and implemented than in the referenced sources.

Tables 3-2 and 3-3 do not apply to the analysis of confinement casks and other components designed in accordance with Section III of the ASME B&PV Code.

Capacities ("S" and "U" terms) and demands (factored or unfactored loads may be loads, forces, moments, or stresses caused by such loads. Usage must be consistent among the terms used in the load combination. Units of force, rather than mass, are to be used for loads.

Definitions of terms used in the load combination expressions for reinforced concrete and steel are derived from ANSI 57.9, ACI 349, AISC specifications, or another source. Where used in an expression related to steel analysis, definitions derived from ACI 349 are not limited in application to reinforced concrete analyses.

The load combinations defined on the basis of allowable stress apply to total stresses (that is, combined primary and secondary stresses). The load and stress factors do not change if secondary stresses are included.

Table 3-2 Loads and Their Descriptions

Symbol	Capacity or Load Term	Capacity or Load (or Demand) Description
S	Steel ASD strength	Strength of a steel section, member, or connection computed in accordance with the "allowable stress method" of the AISC "Specification for Structural Steel Buildings."
S_v	Steel ASD shear strength	Shear strength of a section, member, or connection computed in accordance with the "allowable stress method" of the AISC "Specification for Structural Steel Buildings."
U_s	Steel plastic strength	Strength (capacity) of a steel section, member, or connection computed in accordance with the "plastic strength method" of the AISC "Specification for Structural Steel Buildings."
U_c	reinforced concrete available strength	Minimum available strength (capacity) of reinforced concrete section, member, or embedment to meet the load combination, calculated in accordance with the requirements and assumptions of ACI 349 and, after application of the strength reduction factor, $\emptyset$, as defined and prescribed at §9.2, "Design Strength," of ACI 349. If strength may be reduced during the design life by differential settlement, creep, or shrinkage, those effects shall be incorporated in the dead load, D (instead of by subtraction from minimum available strength) reinforced concrete footing and foundation sections whose demand loads are dominated by the maximum soil reaction may be designed and evaluated using U_f.
U_f	Strength of foundation sections	Minimum available strength of reinforced concrete footing and foundation sections whose demand loads are dominated by the maximum soil reaction, and after the strength reduction factor, $\emptyset$, as defined and prescribed at §9.3, "Design Strength," of ACI 349 is applied. Structural elements interface with columns, walls, grade beams, or footings and foundations should be evaluated by using load factors and load combinations for U_c. These interface elements include anchor bolts and other embedments, dowels, lugs, keys, and reinforcing extended into the footing or foundation.
U_g	Soil reaction or pile capacity	Minimum available soil reaction or pile capacity is determined by foundation analysis (expressed in a SAR for approval of a cask system as a required minimum for the cask system design). U_g is derived using the same load factors and load combinations as shown for determination of U_c.
O/S	Overturning/ sliding resistance	Required minimum available resistance capacity of structural unit against both overturning or sliding. Capacities for resistance of overturning and sliding are checked against the factored load combination separately, although the minimum margins of safety may occur concurrently. O/S is not determined by strength capacities of structural elements. Stress or strength demands resulting from an overturning or sliding situation are evaluated in load combinations involving S, S_v, U_s, U_c, and U_f.

Table 3-2 Loads and Their Descriptions

Symbol	Capacity or Load Term	Capacity or Load (or Demand) Description
	All loads used in combination	If any load reduces the effects of the combination of the other loads and that load would always be present in the condition of the specific load combination, the net coefficient (factor) for that load shall be taken as 0.90. If the load may not always be present, the coefficient for that load shall be taken as zero. Each load that may not always be present in the load combinations is to be varied from 0 to 100 percent to simulate the most adverse loading conditions (to the extent of proving that the lowest margins of safety have been determined).
D	Dead load	Dead load of the structure and attachments including permanently installed equipment and piping. The weight and static pressure of stored fluids may be included as dead loads when these are accurately known or enveloped by conservative estimates. Loads resulting from differential settlement, creep, and/or shrinkage, if they produce the most adverse loading conditions, are included in dead load. If differential settlement, creep, or shrinkage would reduce the combined loads, it shall be neglected. D includes the weight of soil vertically over a footing or foundation for the purposes of determining U_g, U_f, and O/S. Regardless of the load combination factor applied, D is to be varied by +5 percent if that produces the most adverse loading condition.
L	Live loads	Live loads, including equipment (such as a loaded storage cask) and piping not permanently installed, and all loads other than dead loads that might be experienced that are not separately identified and used in the load combination, and that are applicable to the situation addressed by the load combination. Typically includes the gravity and operational loads associated with handling equipment and routine snow, rain, ice, and wind loads, and normal and off-normal impacts of equipment. Loads attributable to piping and equipment reactions are included. Depending on the case being analyzed, may include normal or off-normal events not separately identified, as may be caused by handling (not including drop), equipment or instrument malfunction, negligence, and other man-made or natural causes. Live loads attributable to casks with stored fuel need only be varied by credible increments of loading of an individual cask. Live loads attributable to multiple casks should be varied for the presence and positioning of one or more cask(s), as necessary and varied to determine the lowest margins of safety.

Table 3-2 Loads and Their Descriptions

Symbol	Capacity or Load Term	Capacity or Load (or Demand) Description
L	Live load for precast structures before final integration in-place	Live loads for precast structures shall consider all loading and restraint conditions from initial fabrication to completion of the structure including form removal, storage, transportation, and erection. The NRC is concerned with analysis of loading of reinforced concrete structures before use to the extent that the structures should not have suffered hidden damage from construction live loads, thereby jeopardizing the capacity of the structures when in use. If the damage would be visibly obvious before installation, analysis of capacity versus pre-completion demands is not required.
DB	"Design-basis" (accident-level) loads	Design-basis loads are controlling bounds for the following external event estimates: (1) Extreme credible natural events to be used for deriving design bases that consider historical data or rated parameters, physical data, or analysis of upper limits of the physical processes involved. (2) Extreme credible external man-induced events used for deriving design bases on the basis of analysis of human activity in the region taking into account the site characteristics and associated risks. Design-basis loads include credible accidents and extreme natural phenomena. Presumption of concurrent independent accidents or severe natural phenomena producing compounding design-basis loads is not required. Capacity to resist design basis loads can be assumed to be that of a structure that has not been degraded by previous design basis loads unless prior significant degradation in structural capacity may credibly occur and remain undetected.
T	Thermal loads	Thermal loads, including loads associated with "normal" condition temperatures, temperature distributions, and thermal gradients within the structure; expansions and contractions of components; and restraints to expansions and contractions with the exception of thermal loads that are separately identified and used in the load combination. Thermal loads shall presume that all loaded fuel has the maximum thermal output allowed at time of initial loading in the cask system. Thermal loads shall be determined for the most severe of both steady-state and accident conditions. For multiple cask storage facilities, thermal loads shall be determined for the worst-case loadings on potentially critical sections (e.g., all in place, only one cask in place, alternating casks in place).

Table 3-2 Loads and Their Descriptions

Symbol	Capacity or Load Term	Capacity or Load (or Demand) Description
T_a	Accident- level thermal loads	Thermal loads produced directly or as a result of *off-normal or design-basis* accidents, fires, or natural phenomena. [Note: Although off-normal and design-basis thermal loads are treated the same in the load combinations, there is a distinction between off-normal and design-basis temperature limits for concrete. Off-normal temperature limits are the same as for "normal" conditions.] For multiple cask storage facilities, thermal loads shall be determined for the worst-case loadings on potentially critical sections.
A	Accident loads	Loads attributable to the direct and secondary effects of an off-normal or design-basis accident as could result from an explosion, crash, drop, impact, collapse, gross negligence, or other man-induced occurrences; or from severe natural phenomena not separately defined. Loads attributable to direct and secondary effects may be assumed to be nonconcurrent unless they might be additive. The capacity for resistance to the demand resulting from secondary effects would be that residual capacity following any degradation caused by the direct effect.
H	Lateral soil pressure	Loads caused by lateral soil pressure as would exist in normal, off-normal, or design-basis conditions corresponding to the load combination in which used. H includes lateral pressure resulting from ground water, the weight of the earth, and loads external to the structure transmitted to the structure by lateral earth pressure (not including earthquake loads, which are included in E, see below). H does not include soil reaction associated with attempted lateral movement of the structure or structural element in contact with the earth.
G	Loads attributable to soil reaction	Used only in load combinations for footing and foundation structural sections for which demand is limited by the soil reactions. G represents loads attributable to the maximum soil reaction (horizontal (passive pressure limit) and vertical (soil or pile bearing limit) that would exist in normal, off-normal, or design-basis conditions corresponding to the load combination used. G is a function of U_g (i.e., $G = f(U_g)$).
W	Wind loads	Wind loads produced by normal and off-normal maximum winds. Pressure resulting from wind and with consideration of wind velocity, structure configuration, location, height above ground, gusting, importance to safety, and elevation may be calculated as provided by ASCE 7.

Table 3-2 Loads and Their Descriptions

Symbol	Capacity or Load Term	Capacity or Load (or Demand) Description
W_t	Tornado loads	Loads attributable to wind pressure and wind-generated missiles caused by the design-basis tornado or design-basis wind (for sites where design-basis wind rather than tornado produces the most severe pressure and missile loads). Pressure resulting from wind velocity and elevation may be calculated as provided for these factors in ASCE 7. Tornado wind velocity or pressure does not have to be increased for structure importance, gusting, location, height above ground, or importance to safety (these do apply for design-basis wind).
E	Earthquake loads	Loads attributable to the direct and secondary effects of the design earthquake or off-normal flood, including flooding caused by severe and extreme natural phenomena (e.g., seiches, tsunamis, storm surges), dam failure, fire suppression, and other accidents.

3.7.1 Load Combinations for Steel and Reinforced Concrete Non-Confinement Structures

The reinforced concrete structure load combinations apply to reinforced concrete structures important to safety that are not within the scope of ACI 359 (ASME B&PV Code, Section III, Division 2). The load combinations apply to steel structures important to safety that are not within the scope of the ASME B&PV Code, Section III, Division 1. The NRC accepts, but does not require use of these load combinations for steel and reinforced concrete structures that are not important to safety. The NRC accepts steel analyses that reflect allowable stress design or plastic strength design. Steel load combinations may be determined on the basis of the set of load combination expressions involving either "S" or "U_s."

Table 3-3 Load Combinations for Steel and Reinforced Concrete Non-Confinement Structures

Load Combination	Acceptance Criteria
Reinforced Concrete Structures — Normal Events and Conditions	
$U_c > 1.4\,D + 1.7\,L$	Capacity/demand >1.00 for all sections.
$U_c > 1.4\,D + 1.7\,(L + H)$	Capacity/demand >1.00 for all sections.
Reinforced Concrete Structures — Off-Normal Events and Conditions	
$U_c > 1.05\,D + 1.275\,(L + H + T)$	Capacity/demand >1.00 for all sections.
$U_c > 1.05\,D + 1.275\,(L + H + T + W)$	Capacity/demand >1.00 for all sections.
Reinforced Concrete Structures — Accidents and Conditions	
$U_c > D + L + H + T + (\,E \text{ or } F)$	Capacity/demand >1.00 for all sections.

3-46

Table 3-3 Load Combinations for Steel and Reinforced Concrete Non-Confinement Structures

Load Combination	Acceptance Criteria
$U_c > D + L + H + T + A$	Capacity/demand >1.00 for all sections. An overturning accident for a cask in transfer or in separate storage on a pad is to be assumed unless more severe overturning also occurs as a result of a natural phenomenon.
$U_c > D + L + H + T_a$	Capacity/demand >1.00 for all sections.
$U_c > D + L + H + T + W_t$	The load combination (capacity/demand >1.00 for all sections) shall be satisfied without missile loadings. Missile loadings are additive (concurrent) to the loads caused by the wind pressure and other loads; however, local damage may be permitted at the area of impact if there will be no loss of intended function of any structure important to safety.
Reinforced Concrete Footings/Foundations — Normal Events and Conditions	
$U_f > D + (L + G)$	Capacity/demand >1.00 for all sections. For footing and foundation sections with load limited by soil reaction.
$U_f > D + (L + H + G)$	Capacity/demand >1.00 for all sections. For footing and foundation sections with load limited by soil reaction.
Reinforced Concrete Footings/Foundations — Off-Normal Events and Conditions	
$U_f > D + (L + H + T + G)$	Capacity/demand >1.00 for all sections. For footing and foundation sections with load limited by soil reaction.
$U_f > D + (L + H + T + W + G)$	Capacity/demand >1.00 for all sections. For footing and foundation sections with load limited by soil reaction.
Reinforced Concrete Footings/Foundations — Accident-Level Events and Conditions	
$U_f > D + L + H + T + E + G$	Capacity/demand >1.00 for all sections. For footing and foundation sections with load limited by soil reaction.
$U_f > D + L + H + T + A + G$	Capacity/demand >1.00 for all sections. For footing and foundation sections with load limited by soil reaction.
$U_f > D + L + H + T_a + G$	Capacity/demand >1.00 for all sections. For footing and foundation sections with load limited by soil reaction.
$U_f > D + L + H + T + W_t + G$	Capacity/demand >1.00 for all sections. For footing and foundation sections with load limited by soil reaction.
$U_f > D + L + H + T + F + G$	Capacity/demand >1.00 for all sections. For footing and foundation sections with load limited by soil reaction.
Steel Structures Allowable Stress Design — Normal Events and Conditions	
$(S \text{ and } S_v) > D + L$	Factored strength/demand >1.00 for all sections.
$(S \text{ and } S_v) > D + L + H$	Factored strength /demand >1.00 for all sections.

Table 3-3 Load Combinations for Steel and Reinforced Concrete Non-Confinement Structures

Load Combination	Acceptance Criteria
Steel Structures Allowable Stress Design — Off-Normal Events and Conditions	
1.3 (S and S_v) > D + L + H + W	Factored strength /demand >1.00 for all sections.
1.5 S > D + L + H + T + W	Factored strength/demand >1.00 for all sections. Thermal loads may be neglected when analysis shows that they are secondary and self-limiting in nature, and when the material is ductile.
1.4 S_v > D + L + H + T + W	Factored strength/demand >1.00 for all sections. Thermal loads may be neglected when analysis shows that they are secondary and self-limiting in nature, and when the material is ductile.
Steel Structures Allowable Stress Design — Accidents and Conditions	
1.6 S > D + L + H + T + (E or W_t or F)	Factored strength/demand >1.00 for all sections. Thermal loads may be neglected when analysis shows that they are secondary and self-limiting in nature, and when the material is ductile.
1.4 S_v > D + L + H + T + (E or W_t or F)	Factored strength (allowable stress design)/demand >1.00 for all sections. Thermal loads may be neglected when analysis shows that they are secondary and self-limiting in nature, and when the material is ductile.
1.7 S > D + L + H + T + A	Factored strength/demand >1.00 for all sections. Thermal loads may be neglected when analysis shows that they are secondary and self-limiting in nature, and when the material is ductile.
1.4 S_v > D + L + H + T + A	Factored strength/demand >1.00 for all sections. Thermal loads may be neglected when analysis shows that they are secondary and self-limiting in nature, and when the material is ductile.
1.7 S > D + L + H + T_a	Factored strength/demand >1.00 for all sections.
1.4 S_v > D + L + H + T_a	Factored strength/demand >1.00 for all sections.
Steel Structures Plastic Strength Design — Normal Events and Conditions	
U_s > 1.7 (D + L)	Plastic capacity/demand >1.00 for all sections.
U_s > 1.7 (D + L + H)	Plastic capacity/demand >1.00 for all sections.
Steel Structures Plastic Strength Design — Off-Normal Events and Conditions	
U_s > 1.3 (D + L + H + W)	Plastic capacity/demand >1.00 for all sections.
U_s > 1.3 (D + L + H + T + W)	Plastic capacity/demand >1.00 for all sections. Thermal loads may be neglected when analysis shows that they are secondary and self-limiting in nature, and when the material is ductile.

Table 3-3 Load Combinations for Steel and Reinforced Concrete Non-Confinement Structures

Load Combination	Acceptance Criteria
Steel Structures Plastic Strength Design — Accidents and Conditions	
$U_s > 1.1 (D + L + H + T +$ $(E$ or W_t or $F))$	Plastic capacity/demand >1.00 for all sections. Thermal loads may be neglected when analysis shows that they are secondary and self-limiting in nature, and when the material is ductile. The load combination (capacity/demand >1.00 for all sections) shall be satisfied without missile loadings. Missile loadings are additive (concurrent) to the loads caused by the wind pressure and other loads; however, local damage may be permitted at the area of impact if there will be no loss of intended function of any structure important to safety.
$U_s > 1.1 (D + L + H + T + A)$	Plastic capacity/demand >1.00 for all sections. An overturning accident for a cask in transfer or in separate storage on a pad is to be assumed unless more severe overturning also occurs as a result of a natural phenomenon. Thermal loads may be neglected when analysis shows that they are secondary and self-limiting in nature, and when the material is ductile.
$U_s > 1.1 (D + L + H + T_{a)}$	Plastic capacity/demand >1.00 for all sections.
Overturning and Sliding — Normal and Off-Normal Events and Conditions	
$O/S \geq 1.5 (D + H)$	Capacity/demand $\geq$1.00 for structure to be satisfied for both overturning and sliding.
Overturning and Sliding — Accidents and Conditions	
$O/S \geq 1.1 (D + H + E)$	Capacity/demand $\geq$1.00 for structure to be satisfied for both overturning and sliding.
$O/S \geq 1.1 (D + H + W_t)$	Capacity/demand $\geq$1.00 for structure to be satisfied for both overturning and sliding.

APPENDIX 3A - COMPUTATIONAL MODELING SOFTWARE

Technical Review Guidance:

Computational Modeling Software (CMS) Application

The staff does not endorse the use of any specific type or code vendor of CMS. Any appropriate CMS application could be used for analyses of cask or package components; however, for any CMS to demonstrate that a particular cask design satisfies regulatory requirements, adequate validation of that CMS must be demonstrated by the applicant. Descriptions of CMS validations can be contained within a given application or incorporated by reference.

The reviewer should verify that the following information is provided in the SAR or related documentation (such as proprietary calculation packages or benchmark reports):

(1)   details of the methodology used to assemble the computational models and the theoretical basis of the program used;

(2)   a description of benchmarking against other codes or validation of the CMS against applicable published data or other technically qualified and relevant data that is appropriately documented;

(3)   standardized verification problems analyzed using the CMS, including comparison of theoretically predicted results with the results of the CMS; and

(4)   release version and applicable platforms.

Once the information described above has been docketed, it need not be submitted with each subsequent application, but can be referred to in subsequent SARs or related documents. If an applicant changes their analysis methodology or changes the type or vendor of the CMS used, the applicant should submit either a revision of previously submitted information or include a clear explanation of the methodology changes, and their effects on the analysis in question, in subsequent SAR submittals.

Modeling Techniques and Practices

Modeling techniques and practices used by applicants may need to be verified to demonstrate adequacy of the model.

•   The reviewer should verify that the CMS and the options used by the applicant are appropriate for adequately capturing the behavior of a cask, package, or any components.

Relevant input and results files or an equivalent detailed model description and output should be submitted with the original application.

•   Analysis input files should be submitted in an electronic format that would most easily allow the solution to be executed by the staff, should the staff desire to do so. In-depth review of CMS models is most easily done with input files that contain individual commands used to develop the model and apply the various boundary conditions; therefore, a text input file format (versus database format) is preferred.

- Input files should be annotated in a way that clearly demonstrates the process behind building and solving models developed using CMS. A well annotated input file will expedite staff review and preclude the need for further clarification questions by the staff.

- Appropriate electronic media should be used for submitting case and support files.

Computer Model Development

The reviewer should verify that the computer model used for the analysis is adequately described, either in the SAR or in other documentation, is geometrically representative of the cask design being analyzed, has addressed how material and manufacturing uncertainties might affect the analysis, has appropriate boundary conditions, and has no significant analysis errors.

- The reviewer should verify that the model description includes an adequate basis for the selection of parameters and/or components used in the analysis model (e.g., why was a particular element type applied in the analysis model?)

- The reviewer should verify that models sufficiently represent cask or package geometry and that adequate justification is provided for simplifications used. Models created with CMS are often simplified to reduce computer processing time. Models can often omit geometric details or use homogenized or smeared material properties to represent complex geometry or material combinations and still retain analytic accuracy.

- The reviewer should verify that the applicant has discussed how manufacturing and/or assembly tolerances and contact resistances will affect the analyses that have been conducted, if at all, in both the structural and thermal disciplines. The reviewer should also verify that the applicant has described how tolerances and/or contact resistances are accounted for, if applicable, in the cask or package analysis models that are submitted for review.

- The reviewer should verify that the applicant has provided a general discussion of how error, warning, or advisory messages generated by the software affect the analysis result (if applicable). When processing a computer model developed using CMS, the software will frequently provide error, warning, or advisory messages indicating a possible problem with the model that may or may not be sufficient to terminate processing. If the error/warning function has been disabled during processing, an explanation of why this is appropriate should be provided.

- The reviewer should verify that, within the specific disciplines, the dimensions and physical units used in the models developed are clearly labeled and mutually consistent. The fundamental units of time, mass, and length should be clearly identified. All other physical units derived must be consistent with the basic units adopted. For example, if the unit of length is the millimeter (mm), time in milliseconds (ms), and mass in gram (g), then, the mechanical force will have units of Newton (N), energy in milliJoule (mJ), and stress in megapascal (MPa). Verify that the input parameters are expressed in the units as assigned. If an applicant chooses not to adopt this uniformity of units, the appropriate conversion must be applied prior to processing input into CMS. Similar assurances must be provided for the output for the analysis solution.

Computer Model Validation

- The reviewer should verify that model validation done with applicable experiments or testing is properly documented and appropriate references are provided.

- The reviewer should ensure that if the applicant takes credit for modeling conservatisms, those conservatisms have been demonstrated through validation of the model or analysis methodology. For example, accounting for certain conditions that occur during the hypothetical accident condition (HAC) fire, such as combustion of materials, the turbulent flow of hot gasses in the pool fire environment, and material anomalies that may manifest themselves in a fire can be done with specialized CMS codes (specifically, coupled CFD-FEA codes such as Sandia National Lab's CAFÉ code), high performance computer hardware and extended compute times. Each of these conditions can be treated in a conservative fashion using standard CMS; however, validation of the CMS against actual data (such as open pool fire test data or material combustion data), to demonstrate the applicability of the CMS under the HAC fire, for a configuration similar to that which is being modeled, would be necessary.

Justification of Bounding Conditions/Scenario for Model Analysis

The applicant must determine the most damaging orientation and worst-case conditions for a given design and document how the analytic model was configured for the scenario.

The reviewer should verify that the applicant provided sufficient justification for selecting the most damaging orientation and worst-case conditions.

Description of Boundary Conditions and Assumptions

- The reviewer should verify, as necessary, that boundary conditions and assumptions are addressed in the textual description included in the SAR or other documents (e.g., emissivity values, absorptivity values, convective coefficients, radiation view factors, symmetry planes, and rigid surfaces). This information should be presented in either tabular form or in a complete textual manner. Justifications and bases for such items should also be included in the textual description.

- Values or quantities indicating performance enhancements, i.e., increasing material conductivity values to mimic internal convection or substantially reduced design load factors (DLFs) reflecting an unusually high degree of impact damping, should be accompanied with justifications and should be closely reviewed and independently verified, if needed, by staff.

Documentation of Material Properties

As needed, the reviewer should assess that:

(1)   units for material properties are consistent throughout the individual SAR chapters.

(2)   material properties for all applicable temperature ranges are included.

(3)   references to materials used by the CMS application and specific material properties based on geometry (e.g., conductivity in the X, Y and Z directions), are listed in the SAR or related documents.

Description of Model Assembly

•   The reviewer should verify that the types of elements used in the model are listed in the SAR, preferably in tabular format, along with the corresponding materials or components in which they are used in the analysis model. (i.e., the reviewer should quickly be able to discern what elements and materials are associated with specific components of the analysis model.)

•   The reviewer should verify that a sufficient explanation of the logic behind the creation of each specific computer model is provided, for effective confirmatory calculations to be performed.

•   The reviewer should verify that the applicant has provided annotated input files (as appendices to the SAR or in related documents), that clearly outline the various steps in building the computer models submitted. If input files are not provided or do not adequately describe model assembly, the applicant should provide an adequate explanation of how computer models were assembled using the CMS in the appropriate SAR chapters or related documents.

Loads and Time Steps

•   The reviewer should verify that loads, load combinations, and, if used by the analytical code, the load steps utilized in the computer model are clearly explained by the applicant. The staff should evaluate all loads, how they are placed on the computer models, load combinations, and if used, the time steps applied in the analysis.

•   The reviewer should verify that the time steps specified for the solution of the analysis are sufficiently small to accurately capture the behavior of the structures, systems, or components being modeled.

•   The reviewer should verify that incremental time steps (or sub-steps) are adequately converged. Information of convergence may be obtained from the output generated by the execution of the analysis solution.

Sensitivity Studies

The discussion of sensitivity studies should be included in the general Computer Model Development discussion, as noted above, with relevant references to examples included in the SAR or related documents.

- The reviewer should verify that the applicant has completed sensitivity studies for relevant CMS modeling parameters. This includes element type and mesh density, load step size, interfacing gaps or contact friction, material models and model parameters selection, and property interpolation, if applicable. For example, a mesh sensitivity study should be conducted not only for mesh density but also for mesh density/refinement in areas of thermal or structural concern or where performance of the material is crucial, such as seal areas, lid bolts, etc. A mesh sensitivity is also needed to make sure the analysis results are mesh independent.

- The reviewer should verify that the results of applicable sensitivity studies are clearly described in the SAR or related documentation and can be independently verified, if necessary.

- The reviewer should verify that the applicant's documentation includes at least a brief discussion of the different models used in their mesh sensitivity studies.

Results of the Analysis

- The reviewer should verify that the SAR, or related document(s), include all relevant results (tabular and computer plots) for applicable load cases and load combinations evaluated for design code compliance, and that all governing results (stresses/deformation) are clearly identified in the tables and on plots.

- The reviewer should verify that results are consistent throughout the SAR, and that the correct results are used in calculations of other cask or package performance parameters (e.g., calculated temperatures used in the internal pressure calculation should be verified).

4 THERMAL EVALUATION

4.1 Review Objective

The thermal review ensures that the cask and fuel material temperatures of the dry storage system (DSS) will remain within the allowable values or criteria for normal, off-normal, and accident conditions. This objective includes confirmation that the temperatures of the fuel cladding (fission product barrier) will be maintained throughout the storage period to protect the cladding against degradation that could lead to gross rupture. Also confirmed is the use by the applicant of acceptable analytical and/or testing methods in the Safety Analysis Report (SAR) when evaluating the DSS thermal design.

4.2 Areas of Review

As defined in Section 4.5, "Review Procedures," a comprehensive thermal evaluation should encompass the following areas of review:

Decay Heat Removal System

Material and Design Limits

Thermal Loads and Environmental Conditions

Analytical Methods, Models, and Calculations

> Configuration
> Material Properties
> Boundary Conditions
> Computer Codes
> Temperature Calculations
> Pressure Analysis
> Confirmatory Analysis

4.3 Regulatory Requirements

This section presents a summary matrix of the portions of the U.S. Code of Federal Regulations (CFR) Part 72, "Licensing Requirements for the Independent Storage of Spent Nuclear Fuel, High-Level Radioactive Waste, and Greater Than Class C Waste," Title 10, "Energy" (10 CFR Part 72) that are relevant to the review areas addressed by this chapter. The NRC staff reviewer should be familiar with the regulatory language in these sections. Table 4-1 matches the relevant regulatory requirements associated with this chapter to the areas of review.

Table 4-1 Relationship of Regulations and Areas of Review

Area of Review	10 CFR Part 72 Regulations	
	72.122 (h)(1), (l)	72.236 (b), (f), (g), (h)
Decay Heat Removal Systems	•	•
Material and Design Limits		•
Thermal Loads and Environmental Conditions	•	•
Analytical Methods, Models, and Calculations	•	•

4.4 Acceptance Criteria

4.4.1 Decay Heat Removal System

The applicant must provide a detailed description of the proposed cask heat removal system and its passive cooling characteristics. All major components are to be clearly identified and their contribution to heat-removal from the fuel thoroughly explained. The mechanism of heat removal (i.e., conduction, convection, radiation) for each component should also be discussed.

Evidence must be provided by the applicant that the decay heat removal system will operate reliably under normal and loading conditions.

All instrumentation used to monitor cask thermal performance should also be described.

4.4.2 Material and Design Limits

Cask components and fuel materials should be maintained between their minimum and maximum temperature limits for normal, loading, off-normal, and accident-level conditions to enable all components to perform their intended safety function.

To guarantee cladding integrity of zirconium-based alloys, the maximum calculated fuel cladding temperature should not exceed 400°C (752°F) for normal conditions of storage and short-term loading operations, including cask drying and backfilling. A higher temperature limit may ONLY be used for low burnup spent nuclear fuel (SNF) (less than 45 GWd/MTU), as long as the applicant can demonstrate that the best estimate cladding hoop stress is equal to or less than 90 MPa (13.1 ksi) for the temperature limit that is proposed. During loading operations, repeated thermal cycling should be limited to less than 10 cycles, with cladding temperature variations more than 65°C (149°F). For off-normal and accident conditions, the maximum zirconium based cladding temperature should not exceed 570°C (1058°F).

To guarantee stainless steel cladding integrity, the maximum calculated fuel cladding temperature should not exceed 570°C (1058°F) for off-normal and accident conditions and the maximum calculated fuel cladding temperature should not exceed 400°C (752°F) for normal conditions of storage and short-term loading operations, including cask drying and backfilling.

The applicant must clearly identify the operational temperature limits for all important-to-safety component materials under normal, loading, unloading, off-normal and accident-level conditions. The applicant shall provide reliable basis for all the temperature limits.

The maximum internal pressure of the fuel container should remain within its design pressures for normal, off-normal, and accident-level conditions assuming rupture of 1 percent, 10 percent, and 100 percent of the fuel rods, respectively. Assumptions for pressure calculations include release of 100 percent of the initial fill gas and 30 percent of the fission product gases generated within the fuel rods during operation.

The applicant must clearly identify the design pressure limits for the fuel container under normal, off-normal and accident-level conditions.

4.4.3   Thermal Loads and Environmental Conditions

Identification and justification of the design basis thermal load must be made by the applicant as well as the insolation and ambient temperature assumptions used as boundary conditions for the normal, loading, off-normal, and accident scenarios.

4.4.4   Analytical Methods, Models, and Calculations

The applicant shall present a thermal analysis that clearly demonstrates the storage system's ability to manage design heat loads and have the various materials and components remain within temperature limits. The analysis shall be conducted for normal, loading, draindown/reflood, off-normal, and accident-level conditions. Resulting temperature profile and internal pressure information are necessary to support the structural analysis (Chapter 3) and the confinement analysis (Chapter 5) of the SAR.

The applicant shall specify the analytical methods used in the thermal evaluations including any computational modeling software, (i.e., heat transfer or computational fluid dynamics computer analysis codes) and shall discuss the basis for the parameters and options selected for the analysis. All models should be clearly described. Material thermal properties for all cask components shall be provided and justified. The applicant must discuss, quantify, and report in the SAR any conservatism associated with the proposed thermal models. The level of detail of the discussion should be comparable with sections of the SAR that describes the analytical thermal models. A table of results should be provided in the SAR showing how the associated conservatisms affect the safety parameters (e.g. calculated peak cladding temperature, confinement seal temperatures, etc.). The table of results must be supported with fully documented analytical models and calculations.

The computer codes used in the thermal evaluation should be well-verified and validated. The applicant must provide acceptable basis (e.g., benchmark efforts, published results) for the accuracy of the chosen computer code(s) and justification for its use in the proposed evaluation. A discussion of the resulting level of convergence and conservatism achieved as a function of the modeling options (e.g., meshing, time-differencing) must be provided by the applicant.

To facilitate confirmatory analyses, electronic copies of the most significant input and output files should be provided. Further guidance on the review of analytical methods, models, and calculations provided to the staff for review is provided in Appendix 3A, "Computational Modeling Software."

Figure 4-1 presents an overview of the evaluation process and can be used as a guide to assist in coordinating with other review disciplines.

Figure 4-1 Overview of the Thermal Evaluation

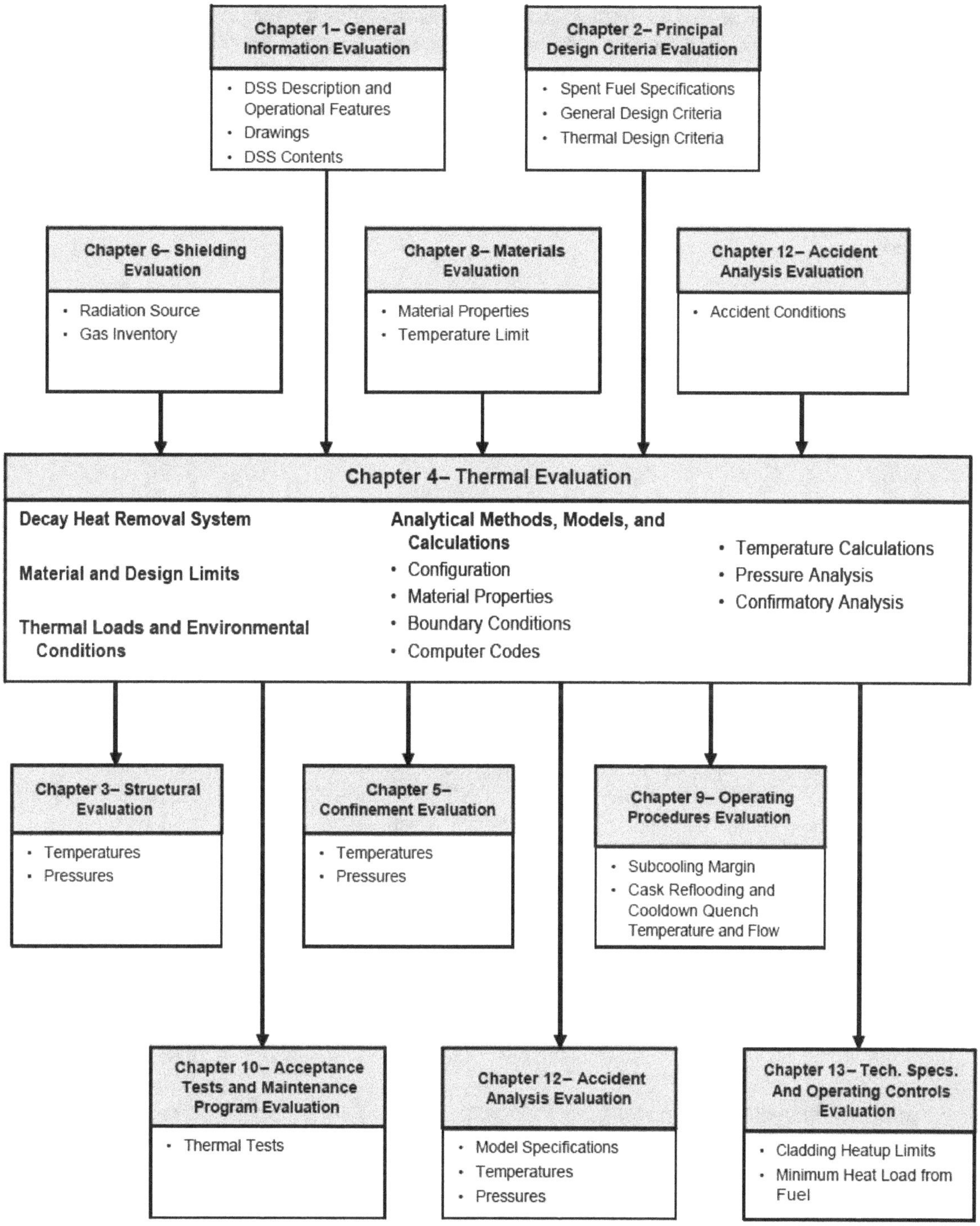

Chapter 1– General Information Evaluation
- DSS Description and Operational Features
- Drawings
- DSS Contents

Chapter 2– Principal Design Criteria Evaluation
- Spent Fuel Specifications
- General Design Criteria
- Thermal Design Criteria

Chapter 6– Shielding Evaluation
- Radiation Source
- Gas Inventory

Chapter 8– Materials Evaluation
- Material Properties
- Temperature Limit

Chapter 12– Accident Analysis Evaluation
- Accident Conditions

Chapter 4– Thermal Evaluation

Decay Heat Removal System

Material and Design Limits

Thermal Loads and Environmental Conditions

Analytical Methods, Models, and Calculations
- Configuration
- Material Properties
- Boundary Conditions
- Computer Codes

- Temperature Calculations
- Pressure Analysis
- Confirmatory Analysis

Chapter 3– Structural Evaluation
- Temperatures
- Pressures

Chapter 5– Confinement Evaluation
- Temperatures
- Pressures

Chapter 9– Operating Procedures Evaluation
- Subcooling Margin
- Cask Reflooding and Cooldown Quench Temperature and Flow

Chapter 10– Acceptance Tests and Maintenance Program Evaluation
- Thermal Tests

Chapter 12– Accident Analysis Evaluation
- Model Specifications
- Temperatures
- Pressures

Chapter 13– Tech. Specs. And Operating Controls Evaluation
- Cladding Heatup Limits
- Minimum Heat Load from Fuel

4.5   Review Procedures

Design features and acceptance criteria, initially presented in SAR Chapter 1, "General Information," and Chapter 2, "Principal Design Criteria," should be reviewed for additional insight about the thermal models that are being presented. Reviewers should examine the appropriateness of the proposed heat loads and environmental conditions. Modeling details such as simulation options, simplifications, and accuracy of results should be assessed. The DSS is to be analyzed under normal, loading, off-normal, and accident scenarios. If necessary, the resulting temperature distributions and internal pressures calculated in the SAR should be confirmed in order to verify compliance with design criteria and regulatory requirements.

One of the most important results of the DSS thermal evaluation is confirmation that the fuel cladding temperature will remain below a specified limit to prevent unacceptable degradation during storage.

Thermal performance of the cask under accident conditions is also evaluated in accordance with Chapter 12, "Accident Analyses Evaluation," of this SRP, as appropriate, in the overall accident analyses presented in the SAR.

In conducting a comprehensive thermal evaluation, reviewers should perform the established review procedures, as applicable, for each of the following areas of review.

4.5.1   Decay Heat Removal System (HIGH Priority)

The reviewer should examine the description of the DSS presented in SAR Chapter 1, "General Information Evaluation" as supplemented by the additional information provided in SAR Chapter 4, "Thermal Evaluation." These two sources of information should be consistent and supplementary. In addition to the material compositions, the dimensions of the cask components and SNF assemblies are to be clearly indicated. All drawings, figures, and tables should be sufficiently detailed to support in-depth staff evaluation.

The applicant's analysis should include the description of the significant thermal design features and operating characteristics of all pertinent DSS components and subsystems. Design features typically include the cask body, thermal fins, shielding materials, fuel baskets, heat transfer disks, confinement seals, drain and vent ports, and external pressure relief devices for the case of transfer casks, among others. The reviewer should verify that the thermal design features will adequately perform their intended safety functions during normal, loading, off-normal, and accident-level conditions. All thermal design features should be passive. Applicants have requested temporary external forced cooling of cask systems during loading operations or as a Technical Specification action statement during transfer operations. Such requests need to be examined by the staff to ensure that they meet the original intent of the regulations; that cask systems remain passively cooled during normal operations.

Any instrumentation used to monitor cask thermal performance should also be described by the applicant in sufficient detail to support in-depth staff evaluation. The monitoring instrumentation components should have a safety classification (presented in SAR Chapter 2, "Principal Design Criteria Evaluation") commensurate with their function and should be fully justified. Applicable operating controls and criteria, such as temperature criteria and surveillance requirements, should be clearly indicated in SAR Chapter 13, "Technical Specifications and Operational Controls and Limits" discussed in the Safety Evaluation Report (SER), and included in the Certificate of Compliance (CoC), as appropriate.

4.5.2 Material and Design Limits (Priority - as indicated)

(MEDIUM Priority) One of the most important results of the thermal evaluation is the confirmation that the fuel cladding temperature is sufficiently low to prevent cladding damage or potential failure during storage. Section 4.4.2, "Material and Design Limits," of this SRP identifies the criteria for cladding temperature limits. The application must clearly agree with these criteria.

(MEDIUM Priority) During licensing reviews, the thermal reviewer should ensure that either of the following criteria are used: (1) the maximum calculated temperatures for normal conditions of storage and for fuel loading operations do not exceed 400°C (752°F), or (2) the maximum calculated temperatures for normal conditions of storage do not exceed 400°C (752°) and that the materials reviewer has verified that the best estimate cladding hoop stress is less than 90 MPa (13.1 ksi) for the maximum allowable temperature specified by the applicant for short-term fuel loading. If the applicants use the latter approach, the thermal reviewer will verify that the materials reviewer has verified that the cladding hoop stresses are less than 90 MPa (13.1 ksi) for each fuel assembly type (e.g., 14x14, 17x17, 9x9, etc.) proposed for storage. Cladding oxide thickness used to compute hoop stress should be evaluated by the materials reviewer. Since the hoop stress is dependent on the rod internal pressure, cladding geometry, and the temperature of the gases inside the rod, the staff will verify that the applicant has calculated the best estimate hoop stress corresponding to the rod internal pressure of the highest burnup fuel assemblies of the specific type of assembly.

(MEDIUM Priority) To limit the amount of SNF that could be released from the cladding under off-normal conditions or accidents, the maximum calculated cladding temperatures should be maintained below 570°C (1058°F).

(MEDIUM - bolted closure/LOW - welded closure) The reviewer should verify that temperature restrictions (upper and lower allowable limits) on all components important to safety (e.g., confinement, shielding, subcriticality, heat removal) during normal, loading, off-normal, and accident scenarios are clearly identified in the application and that the predicted thermal behavior of the entire DSS is indeed within the specified limits. The thermal reviewer should confirm with the assigned materials reviewer the acceptability of all proposed temperature limits.

(LOW Priority) The maximum internal pressure of the fuel container should remain within its design limits for normal, off-normal, and accident-level conditions assuming rupture of 1 percent, 10 percent, and 100 percent of the fuel rods, respectively. The thermal reviewer should confirm with the assigned structural reviewer the acceptability of the proposed design pressure limits.

(HIGH Priority) Any operating scenario (loading or unloading) that results on a time-dependent limiting condition (e.g., number of hours allowed for vacuum drying before fuel cladding temperature reaches its allowable limit) should also be addressed in Chapter 13, "Technical Specifications and Operating Controls and Limits Evaluation," of the SRP and should be included as a limiting condition for operation (e.g., technical specification) in the CoC, as appropriate.

4.5.3 Thermal Loads and Environmental Conditions (Priority - as indicated)

(LOW Priority) The reviewer should examine the specification for the design-basis fuel decay heat presented in SAR Chapter 2, "Principal Design Criteria Evaluation" and ensure that this decay heat is consistent with the specified fuel types, burnups, enrichments and cooling times, if included. Some applications, however, may provide a bounding decay heat load (kW/assembly) without specifying details about the SNF (design, enrichment, cooling time).

(LOW Priority) The axial distribution for the decay heat sources should also be discussed by the applicant with clear justification for a bounding approach. The reviewer should expect a somewhat flat-at-the center axial distribution with a peak-to-average value in the range of 1.1 to 1.2, tapering towards both ends.

(MEDIUM Priority) In general, the NRC staff accepts insolation values presented in 10 CFR Part 71 for 10 CFR Part 72 applications. Because of the large thermal inertia of a storage cask, the insolation values listed in 10 CFR Part 71.71 may be averaged over a 24-hour day assuming steady-state conditions.

(MEDIUM Priority) The reviewer should verify that the ambient temperatures used for normal and off-normal condition evaluations do indeed bound the available historical temperature data for any suggested storage site (current or future). The National Oceanic Atmospheric Administration (NOAA) National Climatic Data Center provides temperature statistics for many American cities and regions. (http://www.ncdc.noaa.gov/oa/ncdc.html).

(MEDIUM Priority) Loading and unloading evaluations should be established on the basis of the SNF pool's technical specification maximum temperature limit (typically 46°C (115°F)).

4.5.4 Analytical Methods, Models, and Calculations (MEDIUM Priority)

For cask system components in which material properties and performance vary with temperature, the reviewer should examine the assumptions used in determining temperature maxima, minima, gradients, and differences for the cask system, as well as review the assumptions used to determine fuel cladding temperatures. The assumed temperature changes over time should result in the bounding conditions for the structural analysis. The calculated temperatures in the various cask system components should be compared to the limiting temperature criteria for the appropriate materials. Ferritic materials are subject to failure by brittle fracture at low temperatures. The reviewer should verify the assumed low temperatures for cask system handling operations for consistency with material properties. Ambient temperature restrictions may be appropriate for cask handling operations. Any limiting conditions regarding ambient temperatures should be addressed in SAR Chapter 13, as well as SER Chapter 13, "Technical Specifications and Operating Controls and Limits Evaluation," and should be included as a limiting condition for operation (e.g., technical specification) in the CoC, as appropriate.

Analysis for accident-level ("design-basis") temperatures should not be considered to envelop the analysis of normal or off-normal temperatures. The acceptance criteria for normal and off-normal temperature demands for structural capacity will differ. Therefore, all three conditions should be analyzed. In addition, the duration over which accident temperature conditions may exist should be evaluated.

4.5.4.1   Configuration (HIGH Priority)

The reviewer should verify that any model used in the thermal evaluation is clearly described. Separate models and submodels may be used for the evaluation of different conditions (normal storage, loading, off-normal situations, and accidents). Coordination with the structural review is necessary to evaluate any damage that may result from accidents or natural phenomena events. All models should be shown as conservative.

Examination by the reviewer of the sketches or figures of all models ensures their proper use in the thermal calculations and verifies that the dimensions and materials are consistent with those in the drawings of the actual cask, as presented in SAR Chapter 1, "General Information Evaluation". If possible, the reviewer should examine the computer input files to verify consistency with the model sketches and engineering drawings. Differences between the actual cask configuration and the model should be identified, and the model should be shown to be conservative.

Particular attention during the review should be paid to gaps between cask components. Tolerances should be considered so that the thermal resistance of each gap is treated conservatively. Gases (e.g., air, helium) assumed to be present in the gap shall be described and justified. If a specific gas other than air in the cask cavity or gaps between cask components is relied upon for heat removal, the reviewer should verify that the applicant shows that the gas is retained *and* that the gas is not diluted by other gases having lower thermal conductivities during the entire storage period. For cask components that are important to heat removal, manufacturing techniques for joining components, surface roughness, contact pressures, and gap conductance values should be adequately described and justified.

The reviewer should verify that decay heat generated in the SNF is limited to the active fuel region of the assemblies. The model should specifically account for the peaking in the central region or provide another conservative approach. Heat from any other stored component (e.g., control rods), if applicable, should also be distributed appropriately. In addition, the positions of heat sources relative to other cask components should be identified.

The application should address the thermal interaction among casks in an array by using a view factor less than unity. Generally, this will result in an operating control and limit in SAR Chapter 13 that imposes a minimum spacing between storage casks.

Coordination with the structural reviewer is necessary to ensure that the applicant has analyzed situations that may produce the worst-case cask loads. The greatest gradients and loadings caused by thermal expansion may occur with casks in alternative storage or in temporary handling positions.

The heat transfer processes used in the analysis should be examined. Conduction and radiation are typically defined as the primary heat transfer mechanisms within the cask itself. In narrow regions of any orientation, little or no convective heat transfer will occur, and only conduction through the gas filled void spaces is assumed. Larger gas volume regions can experience a significant level of convective heat transfer. The staff suggests that the applicant demonstrate the existence of convection in the larger gas regions and quantify the contribution of convection heat transfer to the overall removal of heat from the package. Traditionally, the staff has maintained that natural convection in enclosed cavities should be validated through sufficient CFD calculations or physical experiments.

4.5.4.1.1   General Guidance on Computational Fluid Dynamics Analyses (HIGH Priority)

Since the computational resources necessary to fully resolve flow between individual fuel pins in a cask model with numerous fuel assemblies would be enormous, one acceptable approach would be to treat fuel assemblies as a porous media for applications seeking to credit heat removal from fuel via internal convection. The reviewer should verify that any CFD approach utilizes realistic or bounding flow friction factors in the porous media representation of the fuel, and that friction factors are obtained for each of the limiting fuel assembly types sought as approved contents for the cask.

An acceptable approach to calculate the friction factors would be to perform a computational fluid dynamics (CFD) analysis for each type of fuel assembly for the expected operating conditions (pressure and average gas temperature). From the detailed CFD analysis of a single fuel assembly, wall shear stresses should be obtained separately for bare fuel rods and for fuel rods and associated grid straps. The friction factor shall be calculated based on the wall shear stress method.

The reviewer should evaluate the method used to obtain the friction factors and ensure that the obtained values are realistic or bounding for the intended fuel assembly types. Also, since the friction factor is generally very sensitive to the geometric information (dimensions) and fuel assembly configuration, the reviewer should verify this information by reviewing the fuel assembly design drawings provided by the applicant.

For ventilated spent fuel storage systems (a canister containing the fuel within an outer overpack), the mesh spacing (computational cell size) and density between an overpack liner and canister outer shell wall play an important role when selecting a turbulence model for the air flow through this annular gap.

The near-wall modeling significantly impacts the fidelity of numerical solutions, inasmuch as walls are the main source of flow mean vorticity and turbulence. After all, it is in the near-wall region that the solution variables have large gradients, and the transport of momentum and other scalar variables occurs more vigorously. Therefore accurate representation of the flow in the near-wall region determines a successful prediction of wall-bounded turbulent flows. When dealing with wall effects on the flow usually two modeling options are available to the analyst. The first one is the use of the semi-empirical formulas called "standard wall functions" which are used to bridge the viscosity-affected region between the wall and the fully-turbulent core region. Generally a uniform mesh would be used when these wall functions are invoked. The use of wall functions obviates the need to modify the turbulence models to account for the presence of the wall. This modeling approach is usually applicable to flows with high Reynolds number. In the second approach, the viscosity-affected region is resolved with a mesh all the way to the wall, including the viscous sublayer. This type of approach is referred to as "near wall modeling" approach. The dimensionless distance between the wall and the cell center near the wall (y+) for the mesh used for this case should generally be around 1. Guidance on how to apply any of these modeling approaches should be provided in the CFD program documentation used in the application. Any modeling approach taken should be fully justified and validated.

To properly characterize the flow (internal, external, annular, etc.), Reynolds number estimates shall be made using velocities from initial runs for the cooling air in the annulus and helium fill inside the canister. Reynolds number above 3000 based on the channel hydraulic diameter are above the critical Reynolds number of 2300 for internal flows, characterizing the flow in the transitional range between the laminar and turbulent zone. Since these are buoyancy driven flows, both the Grashof (Gr) number based on the hydraulic diameter of the channel and the modified Grashof number defined as Graetz number (Gz = Gr * W/H), where W and H are the width and height of the air channel,should also be calculated to properly characterize the annular flow. On the other hand, buoyancy driven helium flow, cooling the inside of the canister, generally would be laminar based on both the Grashof and the Reynolds numbers due to higher kinematic viscosities, and low achieved velocities within the canister.

Actual SNF properties and uncertainties (e.g., friction factors, crud and oxide buildup, eccentricities, non-uniform axial and radial decay heat profiles) should also be addressed. Applicants must avoid using an effective thermal conductivity for the cover gas (e.g., helium) in lieu of a specific convection model.

If applicable, the applicant should evaluate the added heat from components stored with the SNF assemblies (e.g., control rods, fuel channels, etc.). This would ultimately affect the maximum predicted cladding temperature.

4.5.4.1.2   General Guidance on Application of Effective Conductivity Models (MEDIUM Priority)

In addition to a CFD method utilizing a porous media, fuel assemblies may be modeled as a homogenous region using an effective thermal conductivity (this is a typical approach when utilizing a finite element analysis approach). The manner in which effective conductivity is determined for each fuel assembly should be examined by the reviewer. Guidance on effective thermal conductivity of the fuel is presented in Section 4.5.4.2, "Material Properties."

Use of effective thermal conductivity coefficients for regions within the confinement cask other than the fuel (e.g., gaps) may overestimate heat transfer. If effective thermal conductivity is used in this manner, the reviewer should verify that the same values have been determined from test data, or CFD submodels, or other appropriate sources that are representative of similar geometry, materials, temperatures, and heat fluxes used in current application. The reviewer should pay particular attention to the effective thermal conductivity of neutron shield regions, such as those embedded within thermal fins. Voids or gaps typically exist as a result of either tolerances or shrinkage, and should be considered in calculating effective thermal conductivity. Also, the applicant should pay particular attention to the values assumed for surface emissivities and view factors, as well as the manner used to account for radiation heat transfer in determining the effective thermal conductivities.

4.5.4.2   Material Properties (MEDIUM Priority)

The reviewer should coordinate with the materials discipline to verify that the material compositions and thermal properties are provided for all components used in the calculational model that the thermal properties used in the safety analysis are appropriate, and that potential degradation of materials over their service life has been evaluated. Temperature and anisotropic dependencies of thermal properties should be considered. If regional thermal properties are determined from a combination of individual materials, the manner in which these effective properties are calculated should be fully described and justified.

If the thermal model is axisymmetric or three-dimensional, the longitudinal thermal conductivity should generally be limited to the conductivity of the cladding (weighted by its fractional area) within the fuel assembly. Gaps between fuel pellets and cracks in the pellets themselves can result in a considerable uncertainty regarding the contribution of the fuel to longitudinal heat transfer. High-burnup effects should also be considered in determining the fuel region effective thermal conductivity.

4.5.4.3 Boundary Conditions (Priority - as indicated)

(MEDIUM Priority) The reviewer should verify that the applicant identifies boundary conditions for normal, loading, off-normal, and accident conditions. The required boundary conditions include the decay heat rate from each fuel assembly and the external conditions on the cask surface. The peak power factor for a fuel assembly should be specified and the peak linear power ("peaking factor") of a fuel assembly should be stated for a given active fuel length.

(MEDIUM Priority) The boundary conditions on the cask surface depend on the environment surrounding the cask. Consequently, the temperature of the environment should be specified for all simulated conditions, as should the incident and absorbed insolation. The mechanisms and models for dissipating the absorbed insolation and decay heat from the surface of the cask to the environment should also be identified and described. The mechanisms for transferring heat from the cask surface usually consist of natural (free) convection and thermal radiation. A heat balance on the surface of the cask should be conducted and the results presented in the applicant's SAR.

(LOW Priority) The initial temperature distribution of the storage cask system before a fire accident should be established on the basis of the hottest temperature distribution during normal or off-normal storage conditions. The duration and flame temperature of the fire should be specified, as should gas velocities and flame emissivity. The flame and cask surface emissivities specified in 10 CFR 71.73(c)(4) for a hypothetical accident test of transportation packages are satisfactory for use with regard to a fire accident involving a storage cask.

(LOW Priority) The applicant should identify and describe the mechanisms and models for coupling the fire energy to the cask surface. These mechanisms include forced convection in relation to the flame velocity (5 to 15 m/s, or about 16 to 49 ft/s) as well as thermal radiation. In addition, justification of the convection coefficients during the fire should be provided. Natural convection coefficients are not appropriate; as such coefficients imply downward gas flow adjacent to relatively cool cask walls. In general, for the fire condition, buoyant, upward flow, driven by hot gasses, will dominate. The orientation of the cask should also be considered.

(LOW Priority) Following the fire, the cask is subject to insolation and content decay heat while being cooled by natural convection and thermal radiation to the environment. The applicant should identify the post-fire conditions of the cask, including any changes in surface conditions and/or geometry that may affect radiation and convection heat losses. Identification and description of the models used for the analysis of the post-fire processes should also be provided by the applicant.

4.5.4.4 Computer Codes (HIGH Priority)

The reviewer should verify that the applicant has provided information on any computer-based modeling as described in Appendix A to Chapter 3.0, "Structural Evaluation," and review the thermal analysis submitted by the applicant in accordance with the Appendix.

4.5.4.5   Temperature Calculations (Priority – as indicated)

(MEDIUM - bolted closure/LOW - welded closure) The application should include a table that lists the maximum and minimum temperatures of all components important to safety under normal, loading, off-normal, and accident-level conditions. This table should specify the operating temperature range for each component. The reviewer should verify that temperatures have been calculated for key components and that they do not exceed the allowable range for each. Justification shall be provided in the application for any material important to safety that exceeds acceptable temperature ranges. If compliance with minimum temperature criteria relies on a specific minimum heat load from the fuel, such heat load shall be quantified and included as an operating control and a technical specification criterion in SAR Chapter 13.

(MEDIUM Priority) The reviewer should pay particular attention to the maximum temperature of the cladding. These temperature limits are discussed in Section 4.4.2, "Material and Design Limits," with review guidance presented in Section 4.5.2, "Material and Design Limits."

(MEDIUM Priority) Some storage systems rely upon natural circulation of air through internal passages to remove heat from the stored confinement canister. For storage systems with internal air flow passages, blockage of inlet and/or outlet flow is an accident situation that should be evaluated. Total blockage of all inlets and outlets may result in fuel heatup, which has been assumed to approach adiabatic conditions. To ensure that blockages do not go undetected for significant periods, the NRC has required objective evidence that inlet and outlet flows are not obstructed. Consequently, for these types of storage systems, the NRC has accepted periodic visual inspection of the vents coupled with temperature measurements to verify proper thermal performance and detect flow blockages. The inspections should take place within an interval that will allow sufficient time for corrective actions to be taken before the accident temperature is reached. The inspection interval should be more frequent than the time interval required for the fuel to heatup to the established accident temperature criteria, assuming a total blockage of all inlets and outlets.

(MEDIUM Priority) The review of the heatup calculations should specifically address any assumptions regarding limiting components and quasi-steady state responses. The initial ambient temperature for the heatup calculations should bound the maximum "normal condition" temperature. The resulting heatup time history should be included in the SAR documentation, and should support the proposed inspection and monitoring intervals. This information is also useful in developing contingency operation procedures, since it indicates the available time in which to take corrective actions before the fuel accident temperature criteria may be exceeded.

(HIGH Priority) Some storage systems may use a transfer cask to move the loaded confinement canister from the fuel handling building to the independent spent fuel storage installation (ISFSI) site. When the canister is within the transfer cask, the rate of cooling is typically less than for normal operation. Therefore, fuel cladding temperatures are expected to be higher than for normal storage conditions.

(HIGH Priority) The reviewer should examine the temperature distribution calculations for the canister inside the transfer cask and verify that heat transfer through gap regions has been treated in a conservative manner, and that material properties and dimensions of the transfer cask are consistent with the design data defined in the SAR documentation. The initial ambient temperature should be the maximum "normal condition" temperature. Cask preparation for storage or unloading operations may include situations in which the canister is evacuated while it is in the transfer cask. If the fuel cladding temperature calculation is based on heatup over a limited time period for cask drying operations, the reviewer should verify that limiting conditions for the operations have been imposed in the technical specifications. Such limiting conditions should ensure that the temperature will remain acceptable during the operations, and that normal cooling will begin before the temperature criterion is exceeded.

(HIGH Priority) During wet fuel transfer operations, the liquid in the fuel canister should not be permitted to boil. This practice avoids uncontrolled pressures on the canister and the connected dewatering, purging, and recharging system(s), unacceptable discharge of liquids which may be providing radiation shielding, and a potentially unacceptable reduction in the safety margin. The reviewer should ensure that to prevent any of the above conditions, an adequate subcooling margin is identified in both the SAR and corresponding operating procedures to prevent boiling. This margin may be cask-specific, depending on the design of the fuel basket and key assumptions used in the criticality analysis. The reviewer should ensure that the applicant reviews the heatup and time-to-boil calculations and assesses whether any technical specification or limiting conditions for operation are needed. Heatup calculations should be established on the basis of the SNF pool's technical specification maximum temperature limit (typically 46°C (115°F)).

(HIGH Priority) For unloading operations, the thermal reviewer should ensure that the applicant evaluates temperature and pressure calculations supporting procedural steps presented in SAR Chapter 9, "Operating Procedures Evaluation," for cask cooldown and reflooding of the cask internals. To ensure that the cask does not overpressurize and that the fuel assemblies are not subjected to excess thermal stresses, the applicant's analysis should specify and justify the appropriate temperature and flow rate of the quench fluid, assuming maximum fuel cladding temperatures in the unloading configuration. This analysis should also be referenced in Chapter 12, "Accident Analyses Evaluation," of the SAR as having been considered in the development of thermal models for the unloading procedures, and be included, as appropriate, in the technical specifications The thermal reviewer should provide thermal profiles to the materials reviewer so that latter can determine if the applicant has adequately addressed the issue of fuel rod response to a reflood incident in Chapter 8, "Materials Evaluation".

(LOW Priority) The most extreme thermal conditions may result from credible ambient temperatures, temperature-time histories, an adjacent fire, or any off-normal or design-basis event (DBE) resulting in blockage of ventilation passages. The worst-case structural loads may occur at temperatures lower than those of design-basis accidents (DBAs) or natural phenomena since load combination expressions effectively require greater safety factors for normal and off-normal analyses than for any DBE. Typically, fire has been the worst-case accident thermal condition for storage systems without internal air flow passages.

(LOW Priority) The burning of fuel and other combustibles associated with vehicles involved in transfer operations should, at a minimum, be presumed to be a DBE with the cask in the most exposed situation during transfer or loading into storage. Fire parameters included in 10 CFR 71.73 have been accepted for characterizing the heat transfer during the in-storage fire. However, a bounding analysis that limits the fuel source thus limits the length of the fire (e.g., by limiting the source to the fuel in the transporter) has also been accepted.

(LOW Priority) Some structures, systems, and components (SSC) may experience the most severe conditions if exposure to high temperatures is followed by dousing with water (such as rain or fire suppression activities). A small amount of exterior concrete spalling may result from a fire, the application of fire suppression water, rain on heated surfaces or other high-temperature condition. The damage from these events is readily detectable and appropriate recovery or corrective measures may be presumed. Therefore, the loss of such a small amount of shielding material is not expected to cause a storage system to exceed the regulatory requirements in 10 CFR 72.106 and need not be estimated or evaluated in the SAR. The NRC accepts that concrete temperatures may exceed the temperature criteria of American Concrete Institute (ACI) 349 for accidents if the temperatures result from a fire. In that case, corrective action may be required for continued safe storage.

(LOW Priority) The methods that are acceptable for analyzing and reviewing the consequences of a fire depend upon the duration of the fire and the margin between the predicted temperatures and the actual thermal limits of the components. A fire of sufficient duration, or one in which material temperatures are close to the criteria of their acceptable operational range, will require a detailed model of the cask and its contents. Cask system components (e.g., the neutron shield) may be assumed to be intact at the start of the fire.

(LOW Priority) If a cask tipover is a credible accident, the reviewer should verify that the applicant has evaluated the effect on cask and fuel temperatures in the new configuration. An analysis may be warranted when a significant portion of heat removal capability is attributed to internal convection if a change in orientation of that cask may have a significant effect.

4.5.4.6   Pressure Analysis (LOW Priority)

Pressure calculations should be performed using the ideal gas law (i.e., $PV = nRT$ where P is pressure, V is volume, n is the number of moles of a gas, R is a constant for a given gas, and T is the absolute temperature) and summing the partial pressures of each of the gas constituents in the cask cavity. The application should identify the method and all assumptions used in the pressure analysis, including the determination of the fission gas inventory.

It is necessary to consider the temperature distribution of all components within the cask cavity and the cavity walls in calculating the gas pressure in the cavity. For the fire accident analysis, the application should identify the maximum gas temperature reached during the post-fire accident phase, explain the method used to determine the average gas temperature, and specify the time in the accident at which the peak gas temperature is attained.

This pressure also depends on the free volume in the cask cavity, the amount (moles) of cover gas (helium) in the cavity, and the amount of gases released from ruptured fuel pins. The free volume calculation should be reviewed to determine if all components internal to the cask cavity (e.g., fuel assemblies, basket, structural supports, spacer disks, reactor control components) have been properly considered.

The NRC accepts that normal conditions occur with less than 1 percent of the fuel rods failed, off-normal conditions occur with up to 10 percent of the fuel rods ruptured, and 100 percent of the fuel rods will have ruptured following a DBE. The NRC also accepts that a minimum of 100 percent of the fill gas and 30 percent of the significant radioactive gases (e.g., ^{3}H, Kr, and Xe) within a ruptured fuel rod is available for release into the cask cavity.

Under the conditions where any of the cask component temperatures are close (within 5 percent) to their limiting values during an accident or the Maximum Normal Operating Pressure (MNOP) is within 10 percent of its design basis pressure, or any other special conditions, the applicant should consider, by analysis, the potential impact of the fission gas in the canister to the cask component temperature limits and the cask internal pressurization.

The reviewer should coordinate with the structural reviewer to verify that the confinement pressure of the cask is within its design limits for normal and accident conditions.

4.5.4.7   Confirmatory Analysis (HIGH Priority)

Reviewers may need to perform a confirmatory analysis of the thermal performance of the cask SSCs identified as important to safety. Confirmatory analyses are recommended where margins between the calculated temperatures and prescribed component temperature limits are small, where particularly complex thermal analyses are submitted by applicants, or where the applicant is submitting a new thermal methodology or analysis approach.

The application should be reviewed to ensure that the applicant made the correct assumptions and provided the correct input, and that the output is consistent with established physical (thermal) behavior. These results should specifically include steady-state temperature distributions, local heat balances, temperatures reached and temperature distributions within any reinforced concrete SSCs, and cask cavity pressures for the bounding ambient temperatures.

To provide the most reliable confirmation, confirmatory analysis should, to the degree possible, use a different thermal analysis method than that used by the applicant. The code used for the confirmatory analysis may be the same as or different from that used by the applicant. Regardless, a review of the applicant's analytical approach and analysis models should be considered part of the overall confirmatory analysis. Similar confirmation of accident temperatures (e.g., during a fire) should be performed, as applicable to the SAR analysis.

If a full confirmatory analysis is not deemed necessary, the minimum confirmatory review should include verifying that key design parameters have been appropriately determined and correctly expressed as input into the computer program(s) used for the thermal analysis. Key parameters include proper dimensions, material properties (including surface emissivities and view factors for radiation), and definition of heat sources. A heat balance at the outer surface of the cask should be performed to verify that the heat from the SNF and insolance, balance that removed by convection and radiation. Correlations for the heat transfer coefficient should then be assessed to confirm that they are appropriate for the existing storage conditions. The temperature of the cask's inner surface should be estimated by calculating the temperature distribution across the cask body with simple heat balance approximations. Finally, the difference between the cask's inner surface temperature and the maximum cladding temperature should be compared with that of similar casks and baskets reviewed in previous SARs.

As discussed above, a more detailed confirmatory analysis may be required, and could include a model of a portion of the cask or basket to ensure that the SAR results are realistic and conservative. A more extensive confirmatory analysis may involve the full geometry of the cask, with relevant component details, to determine temperature distributions in the cask system.

Additional guidance on review of analytical models and conduct of confirmatory analyses can be found in Appendix 3A, "Computational Modeling Software."

As an alternative to a confirmatory analysis, the applicant may be required to perform design-verification testing of an as-built cask or properly scaled mock-up system (when applicable) to confirm the thermal analyses presented in the SAR. Such testing may include verifying gap conductance values assumed in modeling thermal resistance. The test conditions, configuration, and type and location of instrumentation used, if any, should be sufficiently described in SAR Chapter 10, "Acceptance Criteria and Maintenance."

4.6   Evaluation Findings

The reviewer should review the 10 CFR Part 72 acceptance criteria and provide a summary statement for each. These statements should be similar to the following model:

F4.1   Structures, systems, and components (SSCs) important to safety are described in sufficient detail in Chapters _____ of the SAR to enable an evaluation of their thermal effectiveness. Cask SSCs important to safety remain within their operating temperature ranges.

F4.2   The [cask designation] is designed with a heat-removal capability having verifiability and reliability consistent with its importance to safety. The cask is designed to provide adequate heat removal capacity without active cooling systems.

F4.3   The spent fuel cladding is protected against degradation leading to gross ruptures by maintaining the cladding temperature for _____ -year cooled fuel below _____°C (___ °F) in an [applicable gas] environment. Protection of the cladding against degradation is expected to allow ready retrieval of spent fuel for further processing or disposal.

The reviewer should provide a summary statement similar to the following:

"The staff concludes that the thermal design of the [cask designation] is in compliance with 10 CFR Part 72, and that the applicable design and acceptance criteria have been satisfied. The evaluation of the thermal design provides reasonable assurance that the [cask designation] will allow safe storage of spent fuel for a licensed (certified) life of years. This finding is reached on the basis of a review that considered the regulation itself, appropriate regulatory guides, applicable codes and standards, and accepted engineering practices."

5 CONFINEMENT EVALUATION

5.1 Review Objective

In this portion of the dry storage system (DSS) review, the U.S. Nuclear Regulatory Commission (NRC) evaluates the confinement features and capabilities of the proposed cask system. In conducting this evaluation, the NRC staff seeks to ensure that radiological releases to the environment will be within the limits established by the regulations and that the spent fuel cladding and fuel assemblies will be sufficiently protected during storage against degradation that might otherwise lead to gross ruptures.

5.2 Areas of Review

This chapter of the DSS Standard Review Plan (SRP) provides guidance for use in evaluating the design and analysis of the proposed cask confinement system for normal, off-normal, and accident conditions. This evaluation includes a more detailed assessment of the confinement-related design features and criteria initially presented in Chapters 1, "General Information Evaluation" and 2, "Principal Design Criteria Evaluation" of the applicant's Safety Analysis Report (SAR), as well as the proposed confinement monitoring capability, if applicable. In addition, the NRC staff assesses the potential releases of radionuclides associated with spent fuel by independently estimating their potential leakage to the environment and the subsequent impact on a hypothetical individual located at or beyond the controlled area boundary.

As prescribed in U.S. Code of Federal Regulations (CFR) Part 72, "Licensing Requirements for the Independent Storage of Spent Nuclear Fuel and High-Level Radioactive Waste," Title 10, "Energy" (10 CFR Part 72), the regulatory requirements for doses at and beyond the controlled area boundary include both the direct dose and that from an estimated release of radionuclides to the atmosphere (based on the tested leak tightness of the confinement). Thus, an overall assessment of the compliance of the proposed DSS with these regulatory limits is deferred to Chapter 11, "Radiation Protection Evaluation," of this SRP. In addition, the performance of the cask confinement system under accident-level conditions, as evaluated in this chapter, may also be addressed in the overall accident analyses as discussed in Chapter 12, "Accident Analyses Evaluation," of this SRP.

As described in SRP Section 5.5, "Review Procedures," a comprehensive confinement evaluation should encompass the following areas of review:

> **Confinement Design Characteristics**
> Design Criteria
> Design Features
>
> **Confinement Monitoring Capability**
>
> **Nuclides with Potential for Release**
>
> **Confinement Analyses**
> Normal Conditions
> Off-Normal Conditions (Anticipated Occurrences)
> Design Basis Accident Conditions (Including Natural Phenomenon Events)
>
> **Supplemental Information**

5.3   Regulatory Requirements

This section presents a summary matrix of the portions of 10 CFR Part 72 that are relevant to the review areas addressed by this chapter. The NRC staff reviewer should read the exact referenced regulatory language. Table 5-1 matches the relevant regulatory requirements associated with this chapter to the areas of review.

Table 5-1 Relationship of Regulations and Areas of Review			
Areas of Review	**10 CFR Part 72 Regulations**		
	72.104 (a)	72.122(a), (b)(1), (h)(1), (4), (i)	72.236 (d), (e), (i), (j), (l)
Confinement Design Characteristics		•	•
Confinement Monitoring Capability		•	
Nuclides with Potential for Release	•		•
Confinement Analyses	•	•	•

5.4   Acceptance Criteria

In general, the DSS confinement evaluation seeks to ensure that the proposed design fulfills the following acceptance criteria that the NRC staff considers to be minimally acceptable to meet the confinement requirements of 10 CFR Part 72.

5.4.1   Confinement Design Characteristics

The design should provide redundant sealing of the confinement boundary (10 CFR 72.236(e)). Typically, this means that field closures of the confinement boundary should either have two seal welds or two metallic O-ring seals.

The confinement design should be consistent with the regulatory requirements as well as the applicant's "General Design Criteria" reviewed in Chapter 2, "Principal Design Criteria Evaluation," of this SRP. The NRC staff has previously accepted construction of the primary confinement barrier in conformance with the American Society of Mechanical Engineers (ASME) Boiler and Pressure Vessel (B&PV) Code, Section III, "Rules for Construction of Nuclear Facility Components," Division 1, Subsections NB or NC. This code defines the standards for all aspects of construction including materials, design, fabrication, examination, testing, inspection, and certification required in the manufacture and installation of components. In such instances, the staff has relied upon Section III to define the minimum acceptable margin of safety. Therefore, the applicant must fully document and completely justify any deviations from the specifications of Section III. In some cases, after careful and deliberate consideration, the staff has made exceptions to this requirement. In addition, the ASME has published in 2005 Division

3 to Section III which is written specifically for Containments for the Transportation and Storage of Spent Nuclear Fuel and is considered to be the governing code for this component, but has not yet been reviewed and endorsed by the NRC.

The design must provide a nonreactive environment to protect fuel assemblies against fuel cladding degradation, which might otherwise lead to gross rupture (PNL, 1987). Measures for providing a nonreactive environment within the confinement cask typically include drying and backfilling with a nonreactive cover gas (such as helium). Experimental data have not demonstrated an acceptably low oxidation rate for UO_2 spent fuel over the 20-year licensing period to permit safe storage in an air atmosphere during dry storage. Therefore, to reduce the potential for fuel oxidation and subsequent cladding failure, an inert atmosphere (e.g., helium cover gas) has been used for storing UO_2 spent fuel in a dry environment. See Chapter 9, "Operating Procedures Evaluation," of this SRP for more detailed information on the cover gas filling process. Note that other fuel types, such as graphite fuels for the high-temperature gas-cooled reactors (HTGRs), may not exhibit the same oxidation reactions as UO_2 fuels and, therefore, may not require an inert atmosphere. Applicants proposing to use atmospheres other than inert gas should discuss how the fuel and cladding will be protected from oxidation.

5.4.2　　　　Confinement Monitoring Capability

The reviewer should ensure the application describes the proposed monitoring capability and/or surveillance plans for mechanical closure seals. In instances involving welded closures, the staff has previously accepted that no closure monitoring system is required. This practice is consistent with the fact that other welded joints in the confinement system are not monitored, since the initial staff review ensures the integrity of the confinement boundary for the licensing period. Typical surveillances include checking for blockage of the air vents or temperature monitoring.

To show compliance with the requirement for continuous monitoring, 10 CFR Part 72.122(h)(4), cask vendors have proposed, and the staff has accepted, routine surveillance programs and active instrumentation to meet the continuous monitoring requirements.

5.4.3　　　　Nuclides with Potential for Release

The applicant must estimate the maximum credible quantity of radionuclides with potential for release to the environment. The radionuclides potentially available for release to the environment are based on the radiological source term evaluation presented in Chapter 6, "Shielding Evaluation," of this SRP.

5.4.4　　　　Confinement Analyses

The application should specify the maximum allowed leakage rates for the total primary confinement boundary and redundant seals. Applicants frequently display this information in tabular form including the leakage rate of each seal. The maximum allowed leakage rate is the "as tested" leak rate measured by the leak test performed on the cask field closure. Generally, as discussed below, the allowable leakage rate must be evaluated for its radiological consequences and its effect on maintaining an inert atmosphere within the cask. However, the analyses discussed below are unnecessary[1] for storage casks including its closure lid that are

---

[1]　For casks that are demonstrated to be leak tight, the review procedures discussed in Sections 5.5.3 and 5.5.4 are not applicable.

designed and tested to be "leak tight" as defined in the American National Standards Institute (ANSI), Institute for Nuclear Materials Management's "American National Standard for Leakage Tests on Packages for Shipment of Radioactive Materials" (ANSI N14.5-1997).

- The analysis of potential releases should be consistent with the methods described in ANSI N14.5-1997 (ANSI, 1997).

- During normal operations and anticipated occurrences, dose calculations based on the allowable leakage rate must demonstrate that the annual dose equivalent to any real individual who is located beyond the controlled area does not exceed the limits given in 10 CFR 72.104(a).

- For any design-basis accident, dose calculations based on the allowable leakage rate must demonstrate that an individual at the boundary or beyond the nearest boundary of the controlled area does not receive a dose that exceeds the limits given in 10 CFR 72.106(b)-(discussed further in Chapter 12, "Accident Analyses Evaluation")

- The analysis of potential releases must demonstrate that an inert atmosphere will be maintained within the cask during the storage lifetime.

- For casks that employ a pressurized inert gas to facilitate internal natural convection heat transfer, the analysis of potential releases must demonstrate that the pressurized atmosphere will be maintained within the cask during the storage lifetime.

5.4.5   Supplemental Information

The reviewer should ensure all supportive information or documentation that justifies assumptions or analytical procedures is provided in the application.

5.5   Review Procedures

Figure 5-1 presents an overview of the evaluation process for coordination with other review disciplines.

5.5.1   Confinement Design Characteristics (MEDIUM Priority)

5.5.1.1   Design Criteria

The reviewer should examine the principal design criteria presented in SAR Chapter 2 as well as any additional detail provided in SAR Chapter 5, "Confinement."

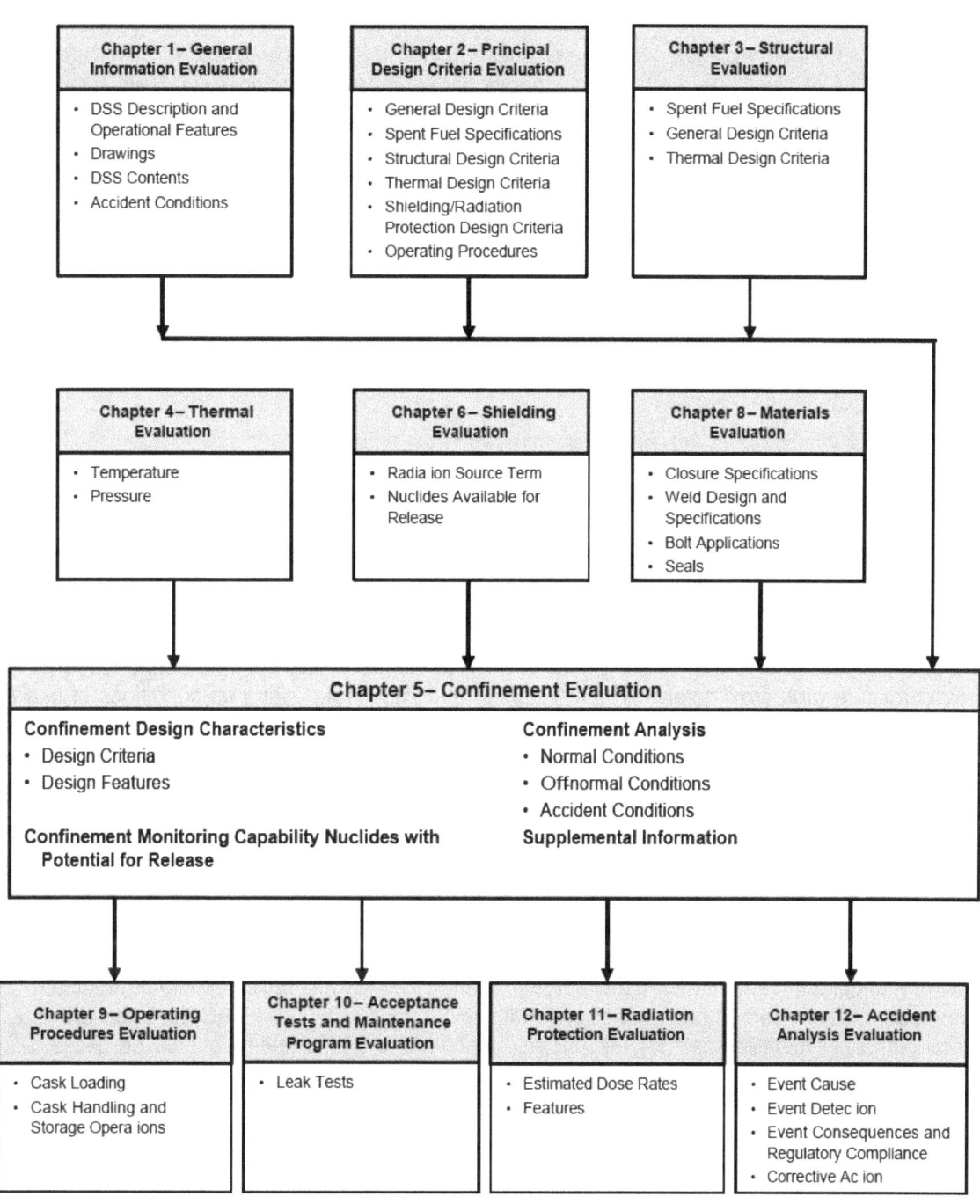

Figure 5-1 Overview of the Confinement Evaluation

5.5.1.2 Design Features

The reviewer should examine the general description of the cask presented in SAR Chapter 1, "General Description," as well as any additional information provided in SAR Chapter 5, "Confinement Evaluation". All drawings, figures, and tables describing confinement features should be sufficiently detailed to stand alone.

The reviewer should verify that the applicant has clearly identified the confinement boundaries. This identification should include the confinement vessel, its penetrations, valves, seals, welds, and closure devices, and corresponding information concerning the redundant sealing.

The reviewer should verify that the design and procedures provide for drying and evacuation of the cask interior as part of the loading operations. Also, the reviewer should verify that the confinement design is acceptable for the pressures that may be experienced during normal, off-normal and accident conditions.

The reviewer should verify that, on completion of cask loading, the gas fill of the cask interior is at a pressure level that is expected to maintain a nonreactive environment and heat transfer capabilities for at least the 20-year storage life of the cask interior under both normal and off-normal conditions and events. This verification can include pressure testing, seal monitoring, and maintenance for casks with seals that are not welded if these are included in Chapter 13, "Technical Specifications and Operating Controls and Limits Evaluation," of this SRP as conditions of use. Acceptance tests for pressure testing are described in Section 10.5.1.1, "Structural/Pressure Tests," of this SRP. The NRC has previously accepted specification of an overpressure of approximately 14 kPa (~2 psig) and cask leak testing as conditions of use for satisfying this requirement. However, this general rule is not applicable to those designs that employ a pressurized content (i.e., to several atmospheres) to facilitate natural circulation cooling within the canister

The reviewer should coordinate with the structural and materials disciplines respectively reviewing Chapter 3, "Structural Evaluation," and Chapter 8, "Materials Evaluation," of this SRP to ensure that the applicant has provided proper specifications for all welds and, if applicable, that the bolt torque for closure devices is adequate and properly specified. If applicable, the reviewer should verify the capability of the seal to maintain long-term closure. Because of the performance requirements over the 20-year license period, the reviewer should evaluate the potential for seal deterioration associated with bolted closures. The NRC staff has previously accepted only metallic seals for the primary confinement. This review should be coordinated with the thermal discipline to ensure that the operational temperature range for the seals (specified by the manufacturer) will not be exceeded.

The staff has concluded that welded canisters can be used as a confinement system provided that the following design/qualification guidance is met:

- The canister is constructed from austenitic stainless steel.

- The canister closure welds meet the guidance of Section 8.5.2.3, "Weld Design and Specifications," of this SRP.

- The canister maintains its confinement integrity during normal conditions, anticipated occurrences, and credible accidents and natural phenomena as required in 10 CFR Part 72.

- The canister shell has been helium leak tested prior its loading as required by 10 CFR 72.236(i). This test demonstrates that the canister is free of defects that could lead to a leakage rate greater than the design basis leakage rate which could result in doses at the control area boundary in excess of the regulatory limits.

- Records documenting the fabrication and closure welding of canisters shall comply with the provisions of 10 CFR Part 72.174, "Quality Assurance Records" and SRP Section 8.5.2.3. Records storage should comply with ANSI N45.2.9, "Requirements for Collection, Storage, and Maintenance of Quality Assurance Records for Nuclear Power Plants."

- Activities related to inspection, evaluation, documentation of fabrication, and closure welding of canisters shall be performed in accordance with a NRC-approved quality assurance program as required in 10 CFR Part 72, Subpart G, "Quality Assurance."

The qualification standards discussed above provide a sufficient alternative to the fabrication, periodic, and pre-shipment leak-testing requirements of ANSI 14.5 for the final closure welds.

5.5.2   Confinement Monitoring Capability (LOW Priority)

The NRC staff has found that casks closed entirely by welding do not require seal monitoring. However, for casks with bolted closures, the staff has found that a seal monitoring system is required to adequately demonstrate that seals can function to limit releases and maintain a helium atmosphere in the cask for the term of the 10 CFR Part 72 general license. A seal monitoring system, combined with periodic surveillance, enables the licensee to determine when to take corrective action to maintain safe storage conditions.

Although the details of the monitoring system may vary, the general design approach has been to pressurize the region between the redundant seals with a nonreactive gas to a pressure greater than that of the cask cavity and the atmosphere. The monitoring system is leak tested to the same leak rate as the confinement boundary. Installed instrumentation is routinely checked per surveillance requirements. A decrease in pressure between these seals indicates that the nonreactive gas is leaking either into the cask cavity or out to the atmosphere. For normal operations, radioactive material should not be able to leak to the atmosphere; hence, this design allows for detecting a faulty seal without radiological consequence. Note that the volume between the redundant seals should be pressurized using a nonreactive gas, thereby preventing contamination of the interior cover gas.

The staff has accepted monitoring systems as not important to safety and classified them as Category B under the guidelines of NUREG/CR-6407, "Classification of Transportation Packaging and Dry Spent Fuel Storage System Components According to Importance to Safety (INEL-95/0551)." Although its function is to monitor confinement seal integrity, the failure of the monitoring system alone does not result in a gross release of radioactive material. It is termed as not important to safety since most of the associated hardware have not met the important to safety programmatic controls, like design, or procurement. Consequently, the monitoring system for bolted closures need not be designed to the same requirements as the confinement boundary (i.e., ASME Section III).

Dependant on the monitoring system design, there could be a lag time before the monitoring system indicates a postulated degraded seal leakage condition. Degraded seal leakage is leakage greater than the tested rate that is not identified within a few monitoring system surveillance cycles. The occurrence of a degraded seal without detection is considered a "latent" condition and should be presumed to exist concurrently with other off-normal and design-basis events (see Section 2.5.2.2, "External Conditions," of this SRP). Note that once the degraded seal condition is detected, the cask user will initiate corrective actions.

For the "latent" condition, the monitoring system boundary would remain intact and this condition would be bounded by the off-normal analysis. If the monitoring system would not maintain integrity under design-basis accident conditions, additional safety analysis may be necessary. The staff recognizes that the possibility of a degraded seal condition is small and that the possibility of a degraded seal condition concurrent with a design-basis event that breaches the monitoring system pressure boundary is very remote. However, these probabilities have not been quantified. To address this concern, the staff accepts a demonstration that the dose consequences of this event are within the limits of 10 CFR 72.106(b).

The reviewer should examine the specified pressure of the gas in the monitored region to verify that it is higher than both the cask cavity and the atmosphere. The reviewer should coordinate with the structural and thermal reviewers associated with Chapters 3 and 4 of this SRP to verify the pressure in the cask cavity.

The reviewer should examine the applicant's analysis to verify that the total volume of gas in the cavity is such that normal seal leakage will not cause all of the gas to escape over the lifetime of the cask. The proposed maximum leakage rate should be based on the confinement evaluation described in Sections 5.5.3 and 5.5.4 of this SRP. The maximum allowable leakage rate should be specified as a minimum acceptance test criterion in SAR Chapter 9, "Acceptance Criteria and Maintenance Program," and Chapter 13, "Technical Specifications and Operating Controls and Limits Evaluation," even though the actual leakage rate of the seals is expected to be significantly lower.

For redundant welded closures, the reviewer should ensure that the applicant has provided adequate justification that the welds have been sufficiently designed, fabricated, tested and examined to ensure that the weld will behave similarly to the adjacent parent material of the cask.

The reviewer should verify that any leakage test, monitoring, or surveillance conditions are appropriately specified in SAR Chapter 10 "Acceptance Tests and Maintenance Program Evaluation"; Chapter 12, "Accident Analyses"; Chapter 13, "Technical Specifications and Operational Controls and Limits Evaluation" ; and/or the Certificate of Compliance (CoC).

5.5.3   Nuclides with Potential for Release (LOW Priority)

The quantities of radioactive nuclides are often presented in the SAR Chapter 6, "Shielding Evaluation," since they are generally determined during the evaluation of gamma and neutron source terms in the shielding analysis. The reviewer should coordinate with the shielding discipline to verify that the applicant has adequately developed the source term.

For determination of the radionuclide inventory available for release, the NRC staff has accepted, as a minimum for the analysis, the activity from the ^{60}Co in the crud, the activity from iodine, fission products that contribute greater than 0.1 percent of design basis fuel activity, and actinide activity that contributes greater than 0.01 percent of the design basis activity. In some cases, the applicant may have to consider additional radioactive nuclides, depending upon the specific analysis. The total activity of the design basis fuel should be based on the cask design loading that yields the bounding radionuclide inventory (considering initial enrichment, burnup, and cool time).

The staff has determined that, as a minimum, the fractions of radioactive materials available for release from spent nuclear fuel (SNF), provided in Table 5-2 for pressurized-water reactor (PWR) fuel and boiling-water reactor (BWR) fuel for normal, anticipated occurrences (off-normal) and accident-level conditions, should be used in the confinement analysis to demonstrate compliance with 10 CFR Part 72. These fractions account for radionuclides trapped in the fuel matrix and radionuclides that exist in a chemical or physical form that is not releasable to the environment under credible normal, off-normal, and accident-level conditions. Other release fractions may be used in the analysis provided the applicant properly justifies the basis for their usage. For example, the staff has accepted, with adequate justification, reduction of the mass fraction of fuel fines that can be released from the cask. Also, for the applicant to utilize the release fractions in Table 5-2, the reviewer should ensure that the condition of the fuel described in the SAR is bounded by the experimental data presented in NUREG/CR-6487. Specifically, this experimental data is based on the release from a single breach of one fuel rod and this data should not be used for spent fuel described as damaged.

Fuel rods that are classified as damaged due to a preloading cladding breach may not have a driving force for the release of particulate from the rod under normal or off-normal conditions, providing the canister is not pressurized. However, under an impact accident damaged fuel rods might release additional fuel fines to the fracture of the fuel, especially the rim region in high burnup fuel. In addition, some canisters may be pressurized to several atmospheres and cask blowdown could also affect releases. Each applicant should establish release fractions for damaged fuel based on applicable physical data and other analyses appropriate for the specific type of fuel, accident impacts, and damaged condition of DSS. Alternatively, a leak-tight confinement boundary may be specified to preclude the release analyses of damaged fuel.

The staff has accepted the rod breakage fractions in Section 4.5.4.6, "Pressure Analysis," of this SRP for the confinement evaluations. It is important to recognize that confinement boundary failure under design basis normal or accident-level conditions is not acceptable. Confinement boundary structural integrity during design basis conditions is confirmed by the structural analysis. The confinement analyses demonstrate that, at the measured leakage rates and assumed nominal meteorological conditions, the requirements of 10 CFR 72.104(a) and 10 CFR 72.106(b) can be met. Each independent spent fuel storage installation (ISFSI), whether it is a site-specific or general license, is also required to have a site-specific confinement analysis and dose assessment to demonstrate compliance with these regulations.

Table 5-2 Fractions of Radioactive Materials Available for Release from Spent Fuel[a]	
Variable	**Fractions Available for Release[b]**

	PWR and BWR Fuel	
	Normal and Off-normal Conditions	Design Basis Accident Conditions
Fraction of Fuel Rods Assumed to Fail	0.01 (normal) 0.10 (off-normal)	1
Fraction of Gases Released Due to a Cladding Breach, $f_G{}^c$	0.3	0.3
Fraction of Volatiles Released Due to a Cladding Breach, $f_V{}^c$	2×10^{-4}	2×10^{-4}
Mass Fraction of Fuel Released as Fines Due to Cladding Breach, f_F	3×10^{-5}	3×10^{-5}
Fraction of Crud that Spalls Off Cladding, f_C	0.15^d	1.0^d

a   Values in this table are taken from NUREG/CR-6487.

b   Except for Co-60, only failed fuel rods contribute significantly to the release. Total fraction of radionuclides available for release should be multiplied by the fraction of fuel rods assumed to have failed.

c   In accordance with NUREG/CR-6487, gases species include H-3, I-129, Kr-81, Kr-85, and Xe-127; volatile species include Cs-134, Cs-135, Cs-137, Ru-103, Ru-106, Sr-89, and Sr-90.

d   The source of radioactivity in crud is Co-60 on fuel rods. At the time of discharge from the reactor, the specific activity, S_c, is estimated to be 140 $\mu Ci/cm^2$ for PWRs and 1254 $\mu Ci/cm^2$ for BWRs. Total Co-60 activity is this estimate times the total surface area of all rods in the cask (Sandoval, et al., 1991). Decay of Co-60 to determine activity at the minimum time before loading is acceptable.

5.5.4   Confinement Analyses (MEDIUM Priority)

The reviewer should examine the applicant's confinement analysis and the resulting doses for the normal, off-normal, and accident conditions at the controlled area boundary.

The analysis typically includes the following common elements:

- Calculation of the specific activity (Ci/cm^3) for each radioactive isotope in the cask cavity based on rod breakage fractions, release fractions, isotopic inventory, and cavity free volume.

- Using the <u>tested</u> leak rate and conditions during testing as input parameters, calculation of the adjusted maximum seal leakage rates (cm^3/s) under normal, off-normal, and accident conditions (e.g., temperatures and pressures).

- Calculation of isotope specific leak rates (Q_i-Ci/s) by multiplying the isotope specific activity by the maximum seal leakage rates for normal, off-normal, and accident conditions.

- Determination of doses to the whole body, thyroid, other critical organs, lens of the eye, and skin from inhalation and immersion exposures at the controlled area boundary (considering atmospheric dispersion factors -χ/Q).

The application should specify maximum allowable "as tested" seal leakage rates as a Technical Specification, as discussed in SRP Chapter 13. Guidance on the calculations of the specific activity for each isotope in the cask and the maximum allowable helium seal leakage rates for normal, off-normal, and accident-level conditions can be found in NUREG/CR-6487, "Containment Analysis for Type B Packages Used to Transport Various Contents" (Anderson, 1996), and ANSI N14.5-1997. The minimum distance between the casks and the distance to the controlled area boundary is generally also a design criterion; however, 10 CFR 72.106(b) requires this distance to be at least 100m from the ISFSI.

For the dose calculations, the NRC staff has accepted the use of either an adult breathing rate (BR) of 2.5×10^{-4} m^3/s (8.8×10^{-3} ft^3/s), as specified in Regulatory Guide (RG) 1.109, "Calculations of Annual Doses to Man from Routine Releases of Reactor Effluents for the Purpose of Evaluating Compliance with 10 CFR Part 50 Appendix I," or a worker breathing rate of 3.3×10^{-4} m^3/s (1.2×10^{-2} ft^3/s), as specified in the U.S. Environmental Protection Agency (EPA) Guidance Report No. 11, "Limiting Values of Radionuclide Intake and Air Concentration and Dose Conversion Factors for Inhalation, Submersion, and Ingestion" (EPA, 1988). The dose conversion factors (DCF) in EPA Guidance Report No. 11 for the whole body, critical organs, and thyroid doses from inhalation should be used in the calculation. The bounding DCFs from EPA Report No. 11 should be used for each isotope unless the applicant justifies an alternate value. The staff is not accepting weighting or normalization of the dose conversion factors. For each isotope, the committed effective dose equivalent (CEDE$_i$ - for the internal whole body dose) or the committed dose equivalent (CDE$_i$ - for the internal organ dose) are calculated as follows:

> CEDE$_i$ or CDE$_i$ (in mrem per year for normal/off-normal or mrem per accident)
> $= Q_i * DCF_i * \chi/Q * BR * $ Duration $*$ conversion factor (The conversion factor, if required, converts the input units into the desired form [CEDE$_i$ or CDE$_i$] in mrem per year for normal/off-normal or mrem per accident).

For the contributions to the whole body, thyroid, critical organs, and skin doses from immersion (external) exposure, the DCFs in EPA Guidance Report No. 12, "External Exposure to Radionuclides in Air, Water, and Soil" (EPA, 1993), should be used. Again, the NRC staff is not accepting weighting or normalization of the dose conversion factors.

The deep dose equivalent (DDE$_i$ - for the external whole body) and the shallow dose equivalent (SDE$_i$ - for the skin dose) are calculated as follows:

> DDE$_i$ or SDE$_i$ (in mrem per year for normal/off-normal or mrem per accident)
> $= Q_i * DCF_i * \chi/Q * $ Duration $*$ conversion factor[2]

The total effective dose equivalent, TEDE = Σ CEDE$_i$ + Σ DDE$_i$
For a given organ, the total organ dose equivalent, TODE = Σ CDE$_i$ + Σ DDE$_i$
The total skin dose equivalent SDE = Σ SDE$_i$

Compliance with the lens dose equivalent (LDE) limit is achieved if the sum of the SDE and the TEDE does not exceed 0.15 Sv (15 rem). This approach is consistent with guidance in the Publication 26 of International Commission on Radiological Protection (ICRP), "Statement from the 1980 Meeting of the ICRP" (ICRP, 1980) and as specified in SRP Chapter 11, "Radiation Protection Evaluation."

In general, the staff evaluates analyses for normal, off-normal, and accident-level conditions.

---

[2] The conversion factor, if required, converts the input units into the desired form, e.g., mrem/year.

5.5.4.1   Normal Conditions

For normal conditions, a bounding exposure duration assumes that an individual is present at the controlled area boundary for one full year (8,760 hours). An alternative exposure duration may be considered by the staff if the applicant provides justification.

Because any potential release resulting from seal leakage would typically occur over a substantial period of time, the staff accepts (for applications for certificates) calculation of the atmospheric dispersion factors (χ/Q) according to RG 1.145, "Atmospheric Dispersion Models for Potential Accident Consequence Assessments at Nuclear Power Plants," assuming D-stability diffusion and a wind speed of 5 m/s (16 ft/s).

For the likely case of an ISFSI with multiple casks, the doses need to be assessed for a hypothetical array of casks during normal conditions according to Section 2.5.3.4, "Shielding/Confinement/Radiation Protection," of this SRP. Therefore, the staff anticipates that the resulting doses from a single cask will be a small fraction of the limits prescribed in 10 CFR 72.104(a) to accommodate the array and the external direct dose.

Note: If the region between redundant, confinement boundary, mechanical seals is maintained at a pressure greater than the cask cavity, the monitoring system boundaries are tested to a leakage rate equal to the confinement boundary, the pressure is routinely checked, and the instrumentation is verified to be operable in accordance with a Technical Specification Surveillance Requirement, the NRC staff has accepted that no discernible leakage is credible. Therefore, calculations of dose to the whole body, thyroid, and critical organs at the controlled area boundary from atmospheric releases during normal conditions would not be required.

5.5.4.2   Off-Normal Conditions (Anticipated Occurrences)

For off-normal conditions, the bounding exposure duration and atmospheric dispersion factors (χ/Q) are the same as those discussed above for normal conditions.

To demonstrate compliance with 10 CFR 72.104(a), the staff accepts whole body, thyroid, and critical organ dose calculations for releases from a single cask. However, the dose contribution from cask leakage should also be a fraction of the limits specified in 10 CFR 72.104(a) since the doses from other radiation sources are added to this contribution.

5.5.4.3   Design-Basis Accident Conditions (Including Natural Phenomenon Events)

For accident-level conditions, the duration of the release is assumed to be 30 days (720 hours). A bounding exposure duration assumes that an individual is also present at the controlled area boundary for 30 days. This time period is the same as that used to demonstrate compliance for reactor facilities licensed per 10 CFR 50 and provides good defense in depth since recovery actions to limit releases are not expected to exceed 30 days.

For accident-level conditions, the staff has accepted calculation of the atmospheric dispersion factors (χ/Q) of RG 1.145 or RG 1.25, "Assumptions Used for Evaluating the Potential Radiological Consequences of a Fuel Handling Accident in the Fuel Handling and Storage Facility for Boiling and Pressurized Water Reactors," on the basis of F-stability diffusion and a wind speed of 1 m/s (3.3 ft/s).

To demonstrate compliance with 10 CFR 72.106(b), the staff accepts whole body, thyroid, critical organ, and skin dose calculations for releases of radionuclides from a single cask.

5.5.5 Supplemental Information

The reviewer should ensure that all supportive information or documentation has been provided or is readily available. This includes, but is not limited to, justification of assumptions or analytical procedures, test results, photographs, computer program descriptions, input and output, and applicable pages from referenced documents. Reviewers should request any additional information needed to complete the review.

5.6 Evaluation Findings

The reviewer should examine the 10 CFR Part 72 acceptance criteria and provide a summary statement for each. These statements should be similar to the following model:

F5.1   Chapter(s) _____ of the SAR describe(s) confinement structures, systems, and components (SSCs) important to safety in sufficient detail in to permit evaluation of their effectiveness.

F5.2   The design of the (cask designation) adequately protects the spent fuel cladding against degradation that might otherwise lead to gross ruptures. Chapter 4, Thermal Evaluation" of the safety evaluation report (SER) discusses the relevant temperature considerations.

F5.3   The design of the (cask designation) provides redundant sealing of the confinement system closure joints by _____.

F5.4   The confinement system is monitored with a _____ monitoring system as discussed above (if applicable). No instrumentation is required to remain operational under accident conditions.

F5.5   The quantity of radioactive nuclides postulated to be released to the environment has been assessed as discussed above. In Chapter 11, "Radiation Protection Evaluation" of the SER, the dose from these releases will be added to the direct dose to show that the (cask designation) satisfies the regulatory requirements of 10 CFR 72.104(a) and 10 CFR 72.106(b).

F5.6   The cask confinement system has been evaluated (by appropriate tests or by other means acceptable to the NRC) to demonstrate that it will reasonably maintain confinement of radioactive material under normal, off-normal, and credible accident conditions.

A summary statement similar to the following should be made:

"The staff concludes that the design of the confinement system of the (cask designation) is in compliance with 10 CFR Part 72 and that the applicable design and acceptance criteria have been satisfied. The evaluation of the confinement system design provides reasonable assurance that the (cask designation) will allow safe storage of spent fuel. This finding is reached on the basis of a review that considered the regulation itself, appropriate regulatory guides, applicable codes and standards, the applicant's analysis and the staff's confirmatory analysis, and accepted engineering practices."

6 SHIELDING EVALUATION

6.1   Objective

The shielding review evaluates the ability of the proposed shielding features to provide adequate protection against direct radiation from the dry storage system (DSS) contents. The shielding features should limit the dose to the operating staff and members of the public so that the dose remains within regulatory requirements during normal operating, off-normal, and design-basis accident (DBA) conditions. The review seeks to ensure that the shielding design is sufficient and reasonably capable of meeting the operational dose requirements of 10 CFR 72.104 and 72.106 in accordance with 10 CFR 72.236(d).

6.2   Areas of Review

This chapter of the DSS Standard Review Plan (SRP) provides guidance for use in evaluating the shielding features of the proposed cask system. As defined in Section 6.5, "Review Procedures," the shielding evaluation may encompass the following areas of review:

Shielding Design Description
    Design Criteria
    Design Features

Radiation Source Definition
    Gamma Source
    Neutron Source

Shielding Model Specification
    Configuration of Shielding and Source
    Material Properties

Shielding Analyses
    Computer Codes
    Flux-to-Dose-Rate Conversion
    Dose Rates
    Confirmatory Analysis

Supplementary Information
    Shielding model description
    Computer model input and output

As prescribed in 10 CFR Part 72, the regulatory requirements for doses at and beyond the controlled area boundary include both direct radiation and radionuclides in effluents. An overall assessment of the compliance of the proposed DSS with these regulatory limits is contained in Chapter 11, "Radiation Protection Evaluation," of this SRP.

In order to ensure that the shielding design of the DSS meets the regulatory requirements as defined in 10 CFR Part 72, the applicant should also include information in the SAR regarding the technical specifications which are necessary for the DSS system to meet the dose limits at the controlled area boundary (See Chapter 13).

In addition, the applicant should demonstrate that the system design, uses, and operating procedures follow the ALARA Principle.

6.3   Regulatory Requirements

10 CFR Part 72 requires that spent fuel storage and handling systems be designed with adequate shielding to provide sufficient radiation protection under normal, off-normal, and accident-level conditions.   The DSS application should describe the design principle and functional features of the shielding structures, systems, and components (SSCs) important to safety in sufficient detail to allow the U.S. Nuclear Regulatory Commission (NRC) staff to thoroughly evaluate their effectiveness.   It is the responsibility of the vendor and the facility owner to analyze such SSCs with the objective of assessing the impact of direct radiation doses and effluent releases to the environment on public health and safety.   The reviewers should verify the applicant's evaluations through review of the applicant's model, or confirmatory analyses or independent modeling analysis.   In addition, SSCs important to safety should be designed to withstand the effects of both credible accidents and severe natural phenomena without impairing their capability to perform their safety functions.

This section presents a summary matrix of the portions of 10 CFR Part 72 that are relevant to the review areas addressed by this chapter.   The NRC staff reviewer should read the exact referenced regulatory language.   The NRC staff reviewer should verify the association of regulatory requirements with the areas of review presented in the matrix to ensure that no requirements are overlooked as a result of unique design features.   Table 6-1 matches the regulatory requirements associated with this chapter to the areas of review.

Table 6-1 Relationship of Regulations and Areas of Review				
Areas of Review	**10 CFR Part 72 Regulations**			
	72.104	72.106(b)	72.122(b), (c)	72.236(d)
Shielding Design Description			•	•
Radiation Source Definition	•	•	•	•
Shielding Model Specification	•	•	•	•
Shielding Analyses	•	•	•	•

6.4   Acceptance Criteria

Several technical and licensing factors should be considered during the shielding evaluation of the proposed DSS. First, 10 CFR Part 72 states regulatory dose limits in terms of annual site-specific doses for normal conditions and total absorbed dose from accident conditions. Because the regulations do not specify cask dose rates (such as package dose rates in 10 CFR Part 71), site-specific factors will have to be considered at each ISFSI when determining compliance with the dose limits in 10 CFR 72.104 and 10 CFR 72.106. These site-specific factors include the geometric arrangement of storage cask arrays, topography, distances to dose receptors, exposure times of dose receptors, actual spent nuclear fuel (SNF) loading patterns in each storage cask, and dose contributions from other surrounding fuel cycle facilities. Because all of these potential site-specific factors at various sites cannot be fully considered in the safety analysis report (SAR) for a DSS design, the regulations in 10 CFR 72.236(d) only require that a demonstration of the shielding design is sufficient to satisfy 10 CFR 72.104 and 72.106. The general licensee is required by 10 CFR 72.212 to consider its site-specific factors and ultimately demonstrate compliance with 10 CFR 72.104. Therefore, the acceptance criteria for DSS shielding seek to define standard analyses for single casks, and a generic array of casks, to demonstrate a sufficient shielding design. In addition, the acceptance criteria seek to establish acceptable dose rate levels surrounding each DSS and acceptable dose calculation methodologies for further use by general licensees.

In general, the DSS shielding evaluation should provide reasonable assurance that the proposed design fulfills the following acceptance criteria:

1. The radiation shielding features of the proposed DSS are sufficient for it to meet the radiation dose requirements in 10 CFR 72.104 and 72.106(b). The applicant demonstrates this with:

   a. A shielding analysis of the surrounding dose rates that contribute to occupational exposure and off-site doses at large distances (for a single storage and transfer cask with bounding fuel source terms at various cask locations), and

   b. A shielding analysis of a single cask and a generic array of casks at large distances.

2. The shielding features of and the radiations emitted by the cask, in conjunction with its proposed operating procedures presented in Chapter 9, "Operating Procedures," of the SAR, are consistent with a well-established "as low as is reasonably achievable" (ALARA) program for activities in and around the storage site.

3. Radiation shielding and confinement features must be sufficient to meet the requirements in 10 CFR 72.106. 10 CFR 72.106(b) states: "Any individual located on or beyond the nearest boundary of the controlled area may not receive from any design basis accident the more limiting of a total effective dose equivalent [TEDE] of 0.05 Sv (5 rem), or the sum of the deep dose equivalent [DDE] and the committed dose equivalent [CDE] to any individual organ or tissue (other than the lens of the eye) of 0.5 Sv (50 rem). The lens dose equivalent [LDE] may not exceed 0.15 Sv (15 rem) and the shallow dose equivalent [SDE] to skin or any extremity shall not exceed 0.5 Sv (50 rem)."

4. The proposed shielding features should demonstrate that the DSS is capable of meeting the regulatory requirements prescribed in 10 CFR Part 20.

The following sections provide additional guidance on acceptance criteria for each area of review for acceptability of SAR informational content and the details and method of evaluation of the proposed shielding features.

6.4.1   Shielding Design Description

6.4.1.1   Design Criteria

The requirements of 10 CFR 72.104 provide dose criteria for the members of the public. Chapter 2, "Principal Design Criteria," of the SAR should specify the criteria that have been used as a basis for protection against direct radiation. Design criteria should include the identification of maximum dose rates and should also be specified for occupancy areas and correlated with occupancy duration and distance to radiation sources. An estimate of collective doses (person-rem per year) should be provided for each occupancy area under various operations (see Chapter 11, "Radiation Protection Evaluation" of this SRP).

The design should consider the ALARA principle. The reviewer should note that it is the responsibility of the general licensee using the DSS design to develop detailed procedures that incorporate the ALARA objectives of its site-specific radiation protection program. Further information on ALARA considerations is contained in the Radiation Protection Chapter.

6.4.1.2   Design Features

The SAR should describe the use of shielding to reduce direct radiation dose rates, and may consider the following:

- Self-shielding provided by the radioactive material being stored;

- Gamma and neutron shielding provided by the structural and nonstructural materials forming the walls and ends of the cask;

- Neutron capture provided by borated materials incorporated into the cask;

- Shielding provided by the temporary placement of water into the cask system during loading and unloading procedures; and

- Shielding provided by temporary placement of equipment and portable shields on and around the cask during loading and unloading procedures.

6.4.2   Radiation Source Definition

The SAR should describe each type of contained radiation source used as a basis for shield design calculations. For spent nuclear fuels, the source terms in particles/s or MeV/s should be described in form of either group structure or a continuous function of energy. For non-fuel hardware, source in Curies or Becquerel is acceptable. For contents other than fuel or non-fuel hardware components, isotopic composition and photon yields for each constituent should be specified. For confinement evaluation purposes, the physical and chemical form, source geometry, radionuclide content, and estimated radiation source strength should be described.

The energy group structure from the source term calculation should correspond to that of the cross-section set of the shielding calculation. The computer methodology or database application used to compute source term strength should be specified.

6.4.2.1   Gamma Sources

The SAR should specify gamma source terms for both spent fuel and activated materials. For spent nuclear fuels, the source terms should be described in a format that is compatible with shielding calculation input, typically in the form of photons/s or MeV/s per energy bin. For assembly hardware and non-fuel hardware, source terms specified by an amount of ^{60}Co activity (in Curies or Becquerel) are acceptable. For contents other than fuel or non-fuel hardware components, isotopic composition and photon yields for each constituent should be specified. A tabulated form of the radiological characteristics is acceptable.

The SAR should include a discussion of energetic radiations created by nuclear reactions such as (n,γ) in the packaging materials and the contents. The SAR should also provide source term descriptions for induced radioactivity and the bases (assumptions and analytical methods) used for their estimation. Alternatively, the SAR may describe the bases for excluding induced radioactivity source terms.

6.4.2.2   Neutron Sources

The SAR should describe the neutron source terms and tabulate the neutron yield by energy group and the bases used to determine the source terms.

6.4.3   Shielding Model Specification

The application should include information in the SAR relative to materials and arrangements of all SSCs important to safety.

6.4.3.1   Configuration of Shielding and Source

The SAR should describe the geometric arrangement of shielding and include illustrations that identify the spatial relationships among sources, shielding, and design dose rate locations. The SAR should clearly indicate the physical dimensions of sources and shielding materials. The SAR should also identify penetrations, voids, or irregular geometries that provide potential paths for gamma or neutron streaming. These potential streaming paths should be clearly identifiable on submitted drawings. The SAR should describe design features used to minimize streaming through these penetrations.

The SAR should clearly state any differences between shielding features during normal or off-normal conditions and accident-level conditions.

6.4.3.2   Material Properties

The shielding reviewer should consult with the materials reviewer to assure that the SAR adequately describes the composition and geometry of the shielding materials.

6.4.4   Shielding Analyses

The SAR should describe the computer codes, version, computational models, data, and assumptions with their bases used in evaluating shielding effectiveness, and should provide dose rate estimates for areas of concern. The reviewer should perform confirmatory calculations, as necessary, to verify the results of the applicant's shielding analyses.

6.4.4.1   Computer Codes

The SAR should identify the computer codes and models used in evaluating shielding for each significant radiation source identified in Section 6.4.2, "Radiation Source Definition," and reference the appropriate documentation. For each computer code used, test problem solutions that demonstrate substantial similarity to solutions from other sources (hand calculations, published literature results, etc.) should be provided. A summary should be provided in the SAR that compares the test problem solutions in either graphical or numeric form. These solutions may be referenced and need not be submitted in the SAR if the references are widely available or have been previously submitted to the NRC for the same computer code and version.

The SAR should clearly present the data used as input for computational purposes and identify any differences between actual material properties or physical dimensions and those used in the analytical method (e.g., for simplifying the computational process). The applicant should defend any simplifications and assumptions by showing that the approach used will result in conservative (bounding) estimates.

The SAR should address calculational error in computer codes for both radiological and thermal source terms. Because validation data are relatively limited for burnups above 45 GWd/MTU (i.e., high burnup fuel), the SAR should numerically specify source term uncertainties for high burnup fuels.

The SAR should determine whether source term values with uncertainties should be applied to the shielding, thermal, and confinement analyses, instead of nominal calculated values. In this determination, the SAR may consider: (1) other conservative assumptions and design margins in the respective analyses; (2) the maximum fuel assembly heat loads; (3) the maximum gamma and neutron dose rates; and (4) any measurable temperature or dose rate limitations proposed in the technical specifications.

A representative computer code input file used in the shielding computation performed for the DSS should be included in the SAR.

6.4.4.2   Flux-to-Dose-Rate Conversion

The basis for the flux-to-dose-rate conversion in the shielding analysis should be stated in the SAR, including conversions that are done by a computer code using its own data library. The SAR should include a table that shows the one to one conversion factor for each energy group of the cask specific source term spectrum. The NRC accepts flux-to-dose-rate conversion factors in American National Standards Institute/American Nuclear Society Standard 6.1.1-1977 (ANSI/ANS-6.1.1-1977).

6.4.4.3   Dose Rates

The SAR evaluation of shielding effectiveness should include calculated or estimated dose rates in representative areas around the DSS. The dose rate calculations should account for such factors as a minimum distance no less than100m (328 ft.), contributions from radionuclide releases, and other significant factors. These criteria are identified and evaluated in the radiation protection evaluation described in Chapter 11 of this SRP. The criteria below relate primarily to the completeness of information provided in the SAR.

The SAR should clearly indicate the physical locations on and around the casks for which dose rate calculations have been performed. These locations should include points on or in the immediate vicinity of cask surfaces where workers will perform operations during loading, retrieval, handling, and any projected maintenance and surveillance. For storage casks with internal labyrinthine air flow passages, the SAR should include dose rate estimates for the air inlets and air outlets using a computer code appropriate for streaming calculations. The SAR should identify points that have the highest calculated dose rates.

The SAR should include dose rate estimates for all onsite areas at which workers will be exposed to elevated dose rates. Dose rates within restricted areas should be calculated in enough detail to estimate doses received by workers performing ISFSI operations and off-site doses at large distances. This should be demonstrated with a standard dose-versus-distance curve or table for a single cask and for a generic DSS array.

The SAR should calculate the dose rate from the cask surface for off-normal events and DBA conditions to ensure compliance with 10 CFR 72.104(a) and 72.106(b), respectively. The computational model used for these calculations should be consistent with the expected condition of the cask after the event.

6.5   Review Procedures

Figure 6-1 presents an overview of the evaluation process and can be used as a guide to assist in coordinating with other review disciplines.

6.5.1   Shielding Design Description

6.5.1.1   Design Criteria (MEDIUM Priority)

Dose rates at the cask surface and in the vicinity of a loaded cask may vary during storage, transfer, and in-storage activities. While 10 CFR Part 72 establishes dose requirements for the ISFSI and its operation, it does not impose specific dose rate limits on the individual casks. Storage cask dose rates from 20 to 400 mrem/hour have been accepted in previous Part 72 evaluations. Acceptable dose rates depend on a number of factors such as the geometry of the storage array, the time workers will routinely spend in the storage array for activities like monitoring or maintenance, the proximity to other areas frequently occupied by workers, and the proximity to the controlled area boundary or other public access areas. The dose requirements are based on 10 CFR 20.1201 for the total expected exposure to workers during anticipated DSS operations, and 10 CFR 72.104 for members of the public who are located beyond the controlled area (i.e., assumed to be at the closest boundary but, in accordance with 10CFR 72.106(b), at least 100m from the storage cask).

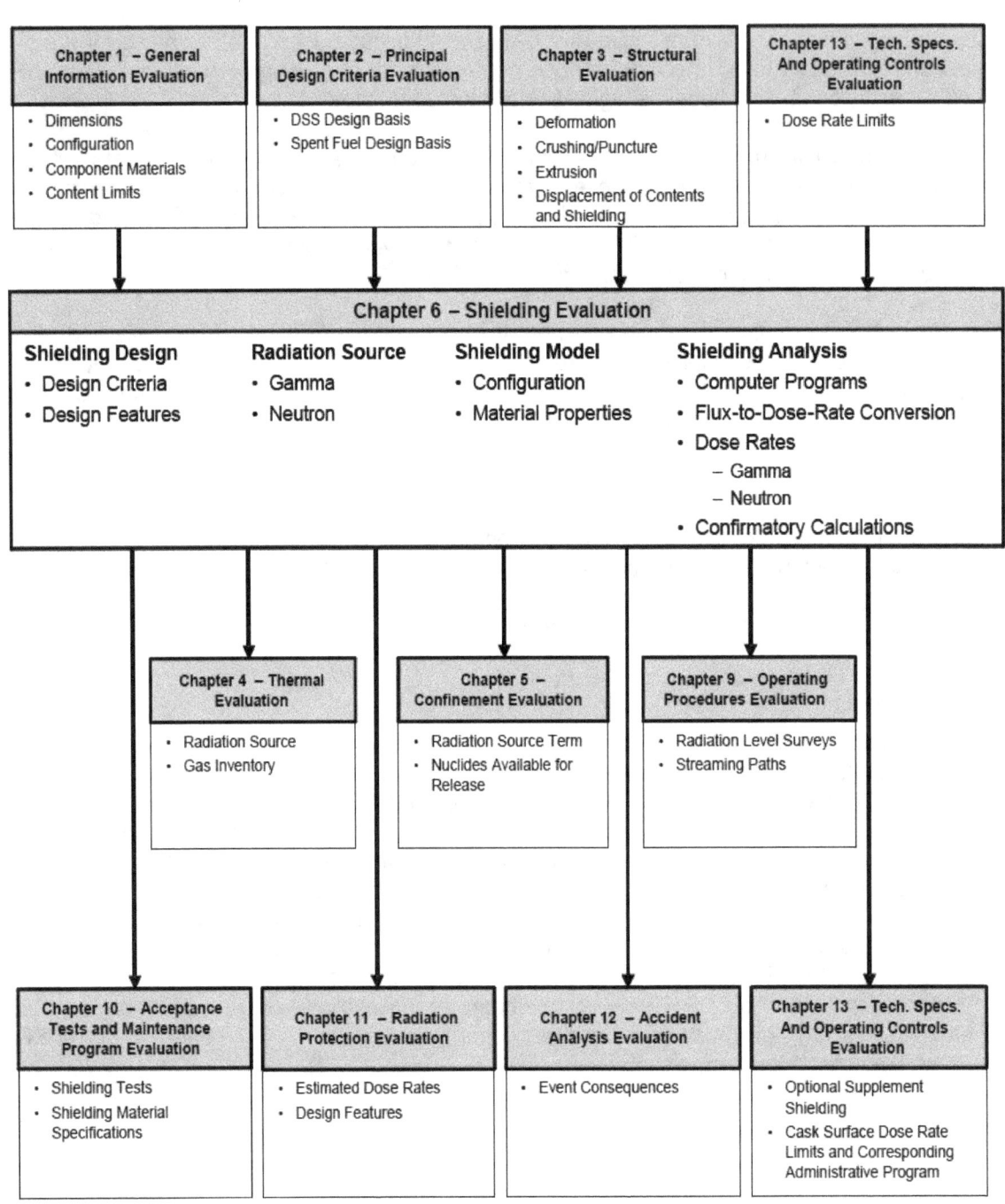

Figure 6-1 Overview of the Shielding Evaluation

The shielding reviewer should coordinate with the review of SRP Chapter 2, "Principal Design Criteria Evaluation," as well as review any additional shielding-related criteria. The reviewer should also refer to SRP Chapter 9, "Operating Procedures Evaluation," to consider any expected operating procedures that would require close proximity to the cask such as cask equipment that should be monitored or serviced frequently. However, the evaluated dose rates at the side of the same cask should be reviewed to ensure that the ALARA principles are either engineered into the design or evoked by specific operating procedures in Chapter 9, "Operating Procedures Evaluation" of the SAR.

6.5.1.2 Design Features (HIGH Priority)

The reviewer should be familiar with the general description of the DSS presented in Chapter 1, "General Description," of the SAR, as well as any additional information provided in Chapter 6, "Shielding Evaluation," of the SAR. All drawings, figures, and tables describing shielding features should be sufficiently detailed to allow the staff to perform an in-depth evaluation.

6.5.2 Radiation Source Definition (HIGH Priority)

Burnup, cooling time, initial uranium loading, and initial enrichment are parameters that affect the total source term of SNF. The reviewer should examine the description of the design-basis fuel in Chapter 2, "Principal Design Criteria" of the SAR to verify that the applicant calculated the bounding source term. The review confirms that the applicant examined all fuel designs and burnup conditions for which the cask system is to be certified, to ensure that the bounding fuel type and values are used. Particular attention should be devoted to the combined effects of gamma and neutron source terms as a function of fuel burnup, cooling times, and enrichment. In many cases, there is no single specific enrichment-burnup combination and cooling time that bounds all potential cask loadings (see the analysis presented in NUREG/CR-6716). Variations in fuel assembly type play a secondary role for pressurized-water reactor (PWR) fuel. For boiling-water reactor (BWR) fuel, void fractions and channel sizes may affect the strengths of neutron and gamma sources. For a cask that contains spent fuel assemblies with irradiated burnable poison rod assemblies (BPRAs), a potential large effect is from activated component hardware (mainly activated cobalt in steel). Again, NUREG/CR-6716 demonstrates that for BPRA designs containing stainless steel, the impact on the gamma dose rate can be large.

The design-basis radiation source term should be based on a saturation value for activation of cobalt impurities or on cobalt activation from a specified maximum burnup and minimum cool time. The reviewer should consider other activation products, as appropriate. These values should be bounded by those listed in the Technical Specifications.

6.5.2.1 Initial Enrichment

The specifications in Chapter 2, "Principal Design Criteria" of the SAR should indicate the maximum fuel enrichment used in the criticality analysis. For shielding evaluations, however, the neutron source term increases considerably with lower initial enrichment for a given burnup. As present in Section 3.4.1.2 of NUREG/CR-6716, as the initial enrichment decreases, the fuel is exposed to a larger neutron fluence to achieve the same burnup. The larger neutron fluence generates larger actinide content which results in larger neutron source term and secondary gamma source term as illustrated in NUREG/CR-6716, Section 3.4.1.2. Consequently, the SAR should specify the minimum initial enrichment as an operating control and limit for cask use, or justify the use of a neutron source term, in the shielding analysis, that specifically bounds the neutron sources for fuel assemblies to be placed in the cask. Because average initial enrichments typically increase with increasing burnup within the spent fuel population, the latter option may be used if the applicant uses low enrichments that bound the historical enrichments for fuels at the proposed burnups. However, the staff should not attempt to use specific source terms as bases for establishing operating controls and limits for cask use because these are not readily inspectable parameters. The fuel assembly initial enrichment, burnup, and cooling time are more appropriate for use as loading controls and limits.

6.5.2.2   Computer Codes for Radiation Source Definition

The reviewer should verify that the applicant determines the source terms using a computer code, such as ORIGEN-S (e.g., as a SAS2 sequence of Oak Ridge National Laboratory's [ORNL] "SCALE" computer code package) that is well benchmarked and recognized and widely used by the industry. If a vendor proprietary code is used, the reviewer should check the code validation and verification records and procedures, preferably with sample testing problems.

The reviewer should ensure that appropriate descriptive information, including validation and verification status, and reference documentation has been provided. The reviewer also should determine if the computer code is suitable for determining the source terms and it has been correctly used. Area of Applicability (AOA) is an important aspect. The reviewer should pay particular attention to AOA to verify if the application falls into the parameter ranges that the code is validated. The reviewer should determine whether the computer code is appropriately applied and the SAR includes verification that the chosen cross-section library is appropriate for the fuel specifications being considered. Many libraries are not appropriate for a burnup exceeding 45,000 MWd/MTU because validation data are limited at high burnups.

The reviewer should verify that the applicant has adequately addressed calculational error and uncertainties of the computer codes used to determine source terms for the thermal, shielding, and confinement analyses. The reviewer should consider: (1) other conservative assumptions and design margins in the analyses; (2) the maximum fuel assembly heat loads; (3) the maximum gamma and neutron dose rates (including relative contributions to total); and (4) any measurable temperature or dose rate limits proposed for the technical specifications.

When reviewing the source term calculations, the reviewer should also consider the factor that nuclide importance changes in high burnup fuels as a function of burnup and validation data. The data for benchmarking the calculations and computer codes is limited at high burnups. Additional data and information on high burnup source term issues are provided in several NRC-sponsored studies (DeHart, 1996; Hermann, 1998; NUREG/CR-6700, NUREG/CR-6701, NUREG/CR-6798.)

6.5.2.3   Gamma Source

The reviewer should verify that the applicant specified gamma source terms as a function of energy for both the spent fuel and activated hardware. If the energy group structure from the source term calculation differs from that of the cross-section set of the shielding calculation, the applicant may need to regroup the photons. Regrouping can be accomplished by using the nuclide activities from the source term calculation as input to a simple decay computer code with a variable group structure. Some applicants will convert from one structure to another using simple interpolation. In general, only gammas with energies from approximately 0.8 to 2.5 MeV will contribute significantly to the dose rate through typical types of shielding; thus, regrouping outside this range is of a lesser importance. The reviewer should determine whether the source terms are specified per assembly, per total assemblies, or per metric ton, and ensure that the total source is correctly used in the shielding evaluation.

Determining source terms for fuel assembly hardware is generally not as straightforward as for the SNF due to cobalt contained in the fuel assembly hardware. The potential impact on the gamma dose rate could be very large during the cooling times in which ^{60}Co is the dominant gamma ray source (up to about 50 years) (NUREG/CR-6716). In particular, steel clad fuel potentially increases the cask dose rate by more than an order of magnitude over that from conventional Zircaloy clad fuel. The stainless steel in the BPRAs was assumed to have a nominal cobalt impurity level of 800 ppm, a value associated with older assembly designs. As presented in NUREG/CR-6716, the largest potential effect from assemblies residing in a cask that contains irradiated BPRAs is from activated component hardware (mainly activated cobalt in steel). For BPRA designs containing stainless steel, the impact on the gamma dose rate can be large. The effort devoted to reviewing this calculation should be based on the contribution of these terms to the dose rates presented in the shielding evaluation. Also, it should be noted whether or not the cask is intended to contain special hardware, such as control assemblies or shrouds, and ensured that source terms from these components are included, if applicable. The reviewer should confer with the Chapter 2, "Principal Design Criteria Evaluation" review team to make this determination.

Depending on the cask design, neutron interactions may result in the production of high energy gammas near the cask surface. If this source term is not treated by the shielding analysis computer code, the reviewer should verify that it is determined by other appropriate means.

As part of the source term determination, the reviewer should verify that the applicant calculates the quantities of certain nuclides (e.g., ^{85}Kr, ^{3}H, and ^{129}I) for use in analyzing doses from the release of radioactive material during postulated accidents in later sections of the SAR. These calculations are typically presented in Chapter 5, "Confinement," of the SAR with the shielding reviewer, in coordination with the confinement reviewer, verifying the information.

6.5.2.4   Neutron Source

The reviewer should verify that the neutron source term is expressed as a function of energy. The neutron source will generally result from both spontaneous fission and alpha-n reactions in the fuel. Depending on the method used to determine these source terms, the applicant may need to independently determine in the SAR, the energy group structure. This analysis is often accomplished by selecting the nuclide with the largest contribution to spontaneous fission (e.g., ^{244}Cm) and using that spectrum for all neutrons, since the contribution from alpha-neutron reactions is generally small. For SNF with cooling times less than 5 years, the analysis should address the spectra of ^{242}Cm and ^{252}Cf.

The specification of a minimum initial enrichment may be a necessary basis for defining the allowed contents. The reviewer should verify that the assumed minimum enrichments bounds all assemblies proposed for the casks in the application. Specific limits are needed for inclusion in the Certificate of Compliance (CoC). Lower enriched fuel, irradiated to the same burnup as higher enriched fuel, produces a higher neutron source. Consequently, the reviewer should verify that Chapter 13, "Technical Specifications and Operational Controls and Limits Evaluation," of the SAR specifies the minimum initial enrichment as an operating control and limit for cask use. Alternately, the applicant should specifically justify the use of a neutron source term, in the shielding analysis, that bounds the neutron sources for fuel assemblies to be placed in the cask. An applicant may demonstrate that the assumed enrichment(s) bound the proposed fuel population except for possible outliers in the SNF population. This is acceptable if the SAR specifically requires each user to verify minimum enrichment with the Final SAR values, and if there are specific dose rate limits in the technical specifications. The applicant and the staff should not attempt to establish specific source terms as the operating controls and limits for cask use.

6.5.2.5 Other Parameters Affecting the Source Term

The reviewer should ensure the SAR contains specific information concerning reactor operations that affects the source term. Several NRC technical reports (specifically, NUREG/CR-6716, but also NUREG/CR-6700, NUREG/CR-6701, and NUREG/CR-6798) discuss the potential affects of other parameters not typically included as a shielding technical specification (e.g., moderator soluble boron concentrations, maximum poison loading, minimum moderator density (for BWR fuels), and maximum specific power). For example, the net impact of moderator density on cask dose rates is expected to be low for PWR fuels. However, the reviewer should be aware that the axial variation in moderator density in BWR cores can have a measurable effect on the axial dose rate profile of a BWR spent fuel assembly. The dose rate may increase near the top of the assemblies where the moderator density was the lowest. This is particularly important for neutron sources because reduced moderator density will harden neutron spectrum and hence induce more actinide production.

6.5.3 Shielding Model Specification (HIGH Priority)

The reviewer should verify that the applicant adequately describes the models that were used in the shielding evaluation for storage under normal, off-normal, and accident-level conditions. For example, if the cask has an external neutron shield, it should be determined whether the cask would be damaged by a tipover accident or degraded in a fire. Applicants should assume liquid, polyesters, or other resin neutron shields are not present after an accident, unless justification is made that they remain intact. The reviewer should confirm this analysis with the structural and thermal evaluation reviews of Chapter 3, "Structural Evaluation," and Chapter 4, "Thermal Evaluation," of the SAR, as appropriate. The reviewer should also confirm that the shielding assumptions made in dose rate calculations, for both occupational workers and the public, are consistent with the design criteria and design drawings.

6.5.3.1 Configuration of the Shielding and Source

The reviewer should examine the sketches or figures that indicate how the shielding design of the canister, storage overpack, and transfer cask is modeled. The reviewer should verify that the model dimensions and materials are consistent with those specified in the cask drawings presented in Chapter 1, "General Information Evaluation" of the SAR. Voids, streaming paths, and irregular geometries should be accounted for or otherwise treated in a conservative manner. In addition, the reviewer should verify that the applicant clearly states the differences, if any, between normal, off-normal, and accident-level conditions.

The reviewer should verify that the applicant properly modeled the source term locations for both spent fuel and structural support regions (i.e., fuel assembly hardware). In some cases, the fuel and basket materials may be homogenized within the fuel region to facilitate the shielding calculations. The reviewer should watch for cases when homogenization may not be appropriate. For example, homogenization should not be used in neutron dose calculations when significant neutron multiplication can result from moderated neutrons (i.e., when significant amounts of moderating materials are present such as when the cask is flooded).

Similarly, homogenization should not be used in configurations where significant radiation streaming can occur between the basket components.

If the applicant has requested storage of damaged fuel assemblies, ensure that the applicant has adequately described the proposed damage assemblies. If the fuel assemblies are damaged to the extent that reconfiguration of the fuel into a geometry different from intact fuel assemblies can occur, ensure that the applicant provides appropriate close assessments for normal, off-normal and accident conditions.

SNF typically has a cosine shape burnup profile along its axial length. If axial peaking appears to be significant, the reviewer should verify that the applicant has appropriately accounted for the condition. Typically, fuel gamma source terms vary proportionally with axial burnup. Fuel neutron source terms vary exponentially by a power of 4.0 to 4.2 (NUREG/CR-6802, "Recommendations for Shielding Evaluations for Transport & Storage Packages") with axial burnup (NUREG/CR-6801, "Recommendations for Addressing Axial Burnup in PWR Burnup Credit Analyses"). In addition, the structural support regions (e.g., top and bottom end pieces and plenum) of the assembly should be correctly positioned relative to the SNF. These support regions may be individually homogenized with the basket materials when particle streaming through the gaps between basket components is not an issue. Generally, however, at least three source regions (i.e., fuel and top/bottom assembly hardware) are necessary. Some canisters may also employ fuel spacers to center the SNF inside the canister.

The reviewer should verify that the SAR shows or adequately describes the locations selected for the various dose calculations. The reviewer should ensure that these dose points are representative of all locations relevant to radiation protection issues. The reviewer should pay particular attention to dose rates from streaming paths to which occupational workers would be exposed (e.g., at vent/drain port covers, lid bolts, air vents, etc.). The shielding end points should be noted as well (such as lead in the cask wall in relation to the assembly hardware and use of fuel spacers to center the fuel). See Section 6.5.4.3 for additional information regarding the selection of locations for dose calculations.

6.5.3.2   Material Properties

The reviewer should verify that the SAR provides information concerning compositions and densities for all materials used in the calculation model. For nonstandard materials, such as neutron shields, Chapter 10 of the SAR, "Acceptance Tests and Maintenance Program Evaluation," should also reference the source of the data and indicate validation criteria. Many shielding computer codes allow the densities to be input directly in g/cm^3. If input is required in atoms/barn-cm the reviewer should pay particular attention to the conversion.

The shielding reviewer should ensure that the elemental composition and density of shielding materials are conservatively adjusted in the shielding analyses to account for any degradation from aging, high temperature, accumulated radiation exposure, and manufacturing tolerances. The shielding reviewer should coordinate with the materials reviewer to obtain reasonable assurance that any degradation that may occur will not impact the safe performance of the shielding materials for the term proposed in the CoC application.

6.5.4   Shielding Analyses

6.5.4.1   Computer Codes (MEDIUM Priority)

The reviewer should evaluate the computer codes or programs used for the shielding analysis. There are several recognized computer codes widely used for shielding analysis. These include computer codes that use Monte Carlo, deterministic transport, and point-kernel techniques for problem solution. The point-kernel technique is generally appropriate only for gammas since casks typically do not contain sufficient hydrogenous material to apply removal cross-sections for neutrons. It is also important for the reviewer to assess whether the number of dimensions of the computer code being applied for the shielding analysis is appropriate for the dose rates being calculated. Typically, NRC staff does not accept the use of one-dimensional codes for calculations other than shielding designs with simple cylindrical geometries. At the least, a two-dimensional calculation is generally necessary. One-dimensional computer codes provide little information about off-axis locations and streaming paths that may be significant to determining occupational exposure. Even a two-dimensional calculation may not be adequate for determining any streaming paths if the modeled configuration is not properly established. These considerations in applying a particular computer code also apply to the computation of dose rates at the end of storage confinement casks. In some cases, the applicant will use the flux output from a deep-penetration shielding code as input to a large distance, skyshine code. The reviewer should verify that the use and interface of these codes are appropriate.

The reviewer should be aware that the applicants often use transport or point-kernel methods to calculate neutron and/or gamma importance functions (unit of $(mrem/hr)/(particle/s\text{-}cm^2)$). Multiplying the importance functions by a neutron and gamma source term-per-unit length yields dose rates on the surface of the cask. Using the neutron and gamma importance functions, the applicant could determine the minimum cooling time required to meet both a decay heat limit and any technical specification at the maximum dose rate limit on the side of the cask. The reviewer, however, should pay close attention to the applicability of the importance function to the actual cask content, and geometry of contents and shielding.

A valuable primer on shielding computer codes and analysis techniques has been published by EPRI (Broadhead, 1995).

The computer codes given below have been previously applied for DSS source and shielding analysis in applications reviewed by the NRC. However, their previous use does not constitute generic NRC approval and, as presented above, the reviewer is cautioned that these computer codes can produce errors when used incorrectly. Specifically, care should be taken to ensure any streaming paths in the cask are appropriately determined with multi-dimensional computer codes under normal, off-normal, and accident-level conditions. The reviewer should also determine that the SAR has specified design control measures that will ensure the quality of computer codes used for shield analysis.

The source of the computer codes given below vary from government sources, such as the Radiation Safety Information Computational Center[3] (RSICC) and other U.S. Department of Energy (DOE) national laboratories, to commercial shielding computer codes. It is also important for the reviewer to be aware that due to proliferation and security concerns, access to specific U.S. government-sponsored computer code packages may be restricted and special permission may be required when granting their use to the applicant. The applicant should use a computer code version that is demonstrated to be adequate for the analysis and is valid for the particular computational platform used to perform the analysis. Computer codes are periodically updated to be compatible with the latest operating system, correct errors found in

_____

[3] Radiation Safety Information Computational Center, Oak Ridge National Laboratory, P.O. Box 2008, Oak Ridge, Tennessee, 37831-6362 and on the Internet at <http://www-rsicc.ornl.gov>.

prior versions, or incorporate updated methodologies. The reviewer should also consider whether additional confirmatory assessments and review are needed to validate the shielding predictions by an applicant that uses older or unsupported codes, especially in cases where NRC may have updated codes and no longer have the capability to directly examine unsupported code models from the applicant.

The computer codes previously applied for DSS source and shielding analyses include:

- MicroSkyshine (air-scattering computer code);

- MORSE (Monte Carlo multigroup three-dimensional neutron and gamma transport computer code);

- MCBEND (Monte Carlo multigroup three-dimensional neutron and gamma transport computer code similar to MORSE developed by the United Kingdom (UK) National Radiation Protection Board (NRPB));

- MCNP (Monte Carlo n-particle transport computer code maintained by Los Alamos National Laboratory (LANL));

- RANKERN (three-dimensional point kernel gamma transport shielding computer code similar to QAD-CGGP);

- SCALE (a modular computer code system for performing standardized computer analyses for licensing evaluation maintained for the NRC by ORNL);

- SKYSHINE-II (air-scattering computer code); and

- STREAMING (computer code for calculation of attenuation of a gamma flux incident on a variety of shielding penetrations, such as ducts and voids).

Some other shielding computer code packages available through RSICC which have potential application to DSS sources include:

- DOORS3.2 (one-, two-, and three-dimensional discrete ordinates neutron/photon transport code system that includes ANISN for one-dimensional, DORT for two-dimensional, and TORT for three-dimensional analysis maintained by ORNL).

- DANTSYS (a code system maintained by the Los Alamos National Laboratory (LANL) that provides discrete ordinates solutions to the neutral particle transport equation that include ONEDANT for one-dimensional, TWODANT for two-dimensional, and THREEDANT for three-dimensional multigroup discrete-ordinate transport analysis.

Some of the above computer codes have been modified or improved to perform adjoint calculations. Examples of the computer codes with adjoint capability are as follows:

- DORT (part of the DOORS3.2 computer code package),

- A^3MCNP (Automated Adjoint Accelerated MCNP),

•   MCBEND.

The reviewer should verify that the SAR describes each of the numerical models of the computer codes used in the shielding evaluation. For each computer code used, the reviewer should ensure that an approved, validated, and verified version of the computer code is being applied by verifying that the following information has been provided in the SAR:

•   The author, source, and dated version;

•   A description of the numerical model applied in the computer code and the extent and limitation of its application; and

•   Either (1) the evaluation of computer code solutions to a series of test problems, demonstrating substantial similarity to solutions obtained from hand calculations, analytical results published in the literature, acceptable experimental tests, a similar computer code, or benchmark problems; or (2) the specification of publically available references for commonly used and well-established codes (e.g. SCALE and MCNP) that demonstrate validation..

The reviewer should examine the solution comparisons provided by the SAR and determine whether satisfactory agreement of computer and test solutions (or resolution of deviations) is evident. Ideally (though not a requirement), the computer code used for evaluation of shielded storage containers should have been validated with actual dose rate measurements from similar or prototypical SNF or high-level waste storage systems.

6.5.4.2   Flux-to-Dose-Rate Conversion (MEDIUM Priority)

The shielding analysis computer code may perform flux-to-dose-rate conversion using its own data library. For the conversions, the NRC accepts the use of ANSI/ANS 6.1.1-1977. While this standard was revised in 1991, the NRC has not adopted the methodology given in ANSI/ANS 6.1.1-1991 principally for two reasons. First, the 10 CFR Part 20 radiation protection requirements are based on fluence-to-dose conversions that are essentially the same as those defined by ANSI/ANS 6.1.1-1977, and are conservative relative to those of ANSI/ANS 6.1.1-1991. Second, neutron dose rates determined on the basis of conversions performed according to ANSI/ANS 6.1.1-1991 may be significantly lower than those determined on the basis of 10 CFR Part 20 or ANSI/ANS 6.1.1-1977.

6.5.4.3   Dose Rates (MEDIUM Priority)

On the basis of experience, comparison to similar systems, or scoping calculations, the reviewer should make an initial assessment of whether the dose rates appear reasonable and whether their variation with location is consistent with the geometry and shielding characteristics of the cask system. The following guidance pertains to the selection of points at which the dose rates should be calculated.

For normal and off-normal conditions, the applicant should indicate the dose rate at all locations accessible to occupational personnel during cask loading, transport to the ISFSI, and maintenance and surveillance operations. Generally, these locations include points at or near various cask components and in the immediate vicinity of the cask. Example of locations include vent areas, trunnion areas, peak side of the cask, peak top of the cask, the canister-gap region, and the bottom of the transfer cask. The applicant should also calculate the dose rates at a distance of 1m from these locations because they typically contribute to occupational exposures.

The application for a cask design is required by 10 CFR 72.236(d) to demonstrate that the shielding and confinement features of the cask are sufficient to meet the requirements in 10 CFR 72.104 for any real individual. The real individual is an individual at or beyond the controlled area, For example, a real individual may be anyone living, working, or recreating close to the facility for a significant portion of the year. The dose to any real individual must not exceed the limits specified in 10 CFR 72.104 from both the storage facility and other surrounding fuel cycle activities.

However, for approval of a cask design, the applicant should evaluate the shielding and confinement features of a single cask and a theoretical array of casks, assuming design-basis source terms and full-time occupancy. The applicant should also provide analyses to facilitate future site-specific evaluations for each general ISFSI licensee. The single cask analysis should identify the minimum distance that is required to meet the dose in 10 CFR 72.104. Past applications have shown this distance to be typically within 200m (656 ft.) of the cask. The applicant should include a dose or dose rate versus distance curve for a single cask to facilitate a site-specific evaluation for general licensees. To satisfy 10 CFR 72.106(b), dose evaluations should be determined at a minimum of 100m (328 ft.) distance to the closest boundary of the controlled area. However, the applicant may use a longer distance, provided that the longer distance is made a condition of use.

The applicant should also include a dose rate-versus-distance curve for a theoretical cask array. The theoretical cask array should consist of at least 20 storage casks (typically in a 2x10 array), and may account for shadowing effect among casks.

It is important to note that the general ISFSI licensee is permitted to use distance or additional engineering features, such as berms, or both, to mitigate doses to real individuals near the site. If such features are used in the cask SAR evaluations, they should be included in the system and described in the CoC. In addition, the SAR should determine the degree to which the normal condition dose rates could change for the identified off-normal conditions.

As required by 10 CFR 72.212(b)(2)(i)(C), a general licensee must perform a written evaluation to demonstrate that the dose limits in 10 CFR 72.104 are met. An evaluation similar to that for a site-specific ISFSI should be performed. The licensee may use information provided in the cask SAR, as well as site specific information to perform the evaluation. Evaluations performed by the general ISFSI licensee are not reviewed for approval by NRC; however, they are subject to NRC inspection and must be recorded and maintained by the general licensee.

The general licensee should establish measures in the radiological protection program, environmental monitoring program, and/or operating procedures to identify and reevaluate potential increases in exposure to the real individuals. Compliance with the dose limits in 10 CFR 72.104 will be verified by the environmental monitoring program with direct radiation measurements and/or effluent measurements, as appropriate.

The reviewer should review the technical specifications of Chapter 13 of this SRP to ensure appropriate requirements are addressed in the technical specifications of the cask. In addition, the degree to which the normal condition dose rates could change for the identified off-normal conditions should be verified. The need for additional calculations should be indicated in the Safety Evaluation Report (SER) and in the conditions set forth in the CoC.

If the above dose rate criteria are satisfied, NRC accepts that the direct-dose regulatory requirements can also be satisfied, although the exact details needed to comply with these limitations will vary from ISFSI site to site. Therefore, the SAR needs to address such requirements only in general terms. Detailed calculations need not be presented if Chapter 13 of the SAR, "Technical Specifications and Operational Controls and Limits Evaluation," assigns ultimate compliance responsibilities to the ISFSI site licensee.

In addition, the applicant should calculate the dose rate at 100m (328 ft.) from the cask surface for accident-level conditions to assist in demonstrating the design is sufficient to meet the requirements of 10 CFR 72.106. The model used for these calculations should be consistent with the expected condition of the cask after an accident or natural event.

The potential reconfiguration of damaged fuel within the damaged-fuel can, if applicable, must be analyzed to demonstrate that the cask/fuel meet the dose limits of normal and design basis events of storage. The shielding analysis should assume a worst case or bounding configuration of the canned fuel.

6.5.4.4 Confirmatory Calculations (HIGH Priority)

The reviewer should independently evaluate the dose rates in the vicinity of the cask for normal, off-normal, and accident-level conditions. In determining the level of effort appropriate for these calculations, the reviewer should consider the following factors:

- the degree of sophistication in the SAR analysis;

- a comparison of SAR dose rates with those of similar casks that have previously been reviewed, if applicable;

- the typical variation in dose rates expected between different computer codes and cross-section sets;

- the fact that actual dose rates will be monitored and limited by the requirements of 10 CFR Part 20;

- the restrictions to be placed on the DSS operations or the limits to be placed on dose rates, as documented in the CoC and/or technical specifications.

- the applicant's experience in using the methods and computer codes in previous submittals;

- the use of new, or previously reviewed, computational methods or computer codes; and,

- the inclusion in the design of any significant departures from previous cask system designs (e.g., unusual shield geometry, new types of materials, or different source terms).

At a minimum, the review should include examination of the applicant's input to the computer code used for the shielding analysis. The reviewer should verify use of proper dimensions, material properties, and an appropriate cross-section set. In addition, the reviewer should independently evaluate the use of gamma and neutron source terms.

If a more detailed review is required (e.g., a new and not previously reviewed shielding computer code), the reviewer should independently confirm the dose rates to ensure that the SAR results are reasonable and conservative. As previously noted, the use of a simple computer code for neutron calculations often does not provide results with sufficient accuracy and confidence. An extensive and more detailed evaluation may be necessary if large uncertainties are suspected. To the degree possible, the use of a different shielding computer code with a different analytical technique and cross-section set from that used in the SAR analysis will usually provide a more independent evaluation.

A good reference regarding the treatment of uncertainty in thick-shielded cask analyses is the Electric Power Research Institute's "Evaluation of Shielding Analysis Methods in Spent Fuel Cask Environments," published in 1995 (Broadhead, 1995).

6.5.5   Supplemental Information

Supplemental information can include copies of applicable references (especially if a reference is not generally available to the reviewer), computer code descriptions, input and output files, and any other information that the applicant deems necessary. Likewise, the reviewer should request any additional information needed to complete the review process.

6.6   Evaluation Findings

The reviewer should review the 10 CFR Part 72 acceptance criteria and provide a summary statement for each. These statements should be similar to the following model:

F6.1   Section(s) _____ of the SAR describe(s) shielding structures, systems, and components (SSCs) important to safety in sufficient detail to allow evaluation of their effectiveness. The reviewer should cite specific drawings that are used to define the SSCs for shielding.

F6.2   Section(s) _____ of the SAR provide reasonable assurance that the radiation shielding features are sufficient to meet the radiation protection requirements of 10 CFR Part 20, 10 CFR 72.104 and 10 CFR 72.106.

F6.3   Operational restrictions to meet dose and ALARA requirements in 10 CFR Part 20, 10 CFR 72.104, and 10 CFR 72.106 are the responsibility of the site licensee. The [cask designation] shielding features are designed to assist in meeting these requirements.

A summary statement similar to the following should be made:

"Based upon its review, the staff has reasonable assurance that the design of the shielding system of the [cask designation] is in compliance with 10 CFR Part 72 and that the applicable design and acceptance criteria have been satisfied. The evaluation of the shielding system design provides reasonable assurance that the [cask designation] will allow safe storage of spent fuel in accordance with 10 CFR 72.236(d). This finding is reached on the basis of a review that considered the regulation itself, appropriate regulatory guides, applicable codes and standards, accepted engineering practices, and the statements and representations in the application.

7 CRITICALITY EVALUATION

7.1 Review Objective

The criticality review and evaluation ensures that spent nuclear fuel (SNF) to be placed into the dry storage system (DSS) remains subcritical under normal, off-normal, and accident conditions involving handling, packaging, transfer, and storage. The criticality review is designed to fulfill the strategic outcome of no inadvertent criticality events, part of the strategic goal of safety described in the agency's strategic plan (NUREG-1614).

7.2 Areas of Review

This portion of the DSS review evaluates the criticality design and analysis related to SNF handling, packaging, transfer, and storage procedures for normal, off-normal, and accident conditions. Consequently, this chapter of the DSS Standard Review Plan (SRP) provides guidance for use in conducting a comprehensive criticality evaluation that may encompass any or all of the following areas of review:

Criticality Design Criteria and Features

Fuel Specification
        Non-Fuel Hardware
        Fuel Condition

Model Specification
        Configuration
        Material Properties

Criticality Analysis
        Computer Codes
        Multiplication Factor
        Benchmark Comparisons

Burnup Credit
        Limits for the Licensing Basis
        Code Validation
        Licensing-Basis Model Assumptions
        Loading Curve
        Assigned Burnup Loading Value
        Estimate of Additional Reactivity Margin

Supplemental Information

7.3 Regulatory Requirements

SNF storage systems must be designed to remain subcritical unless at least two unlikely independent events occur. Moreover, the SNF cask must be designed to remain subcritical under all credible conditions. Regulations specific to nuclear criticality safety of the cask system are specified below. Normal and accident conditions to be considered are also identified in U.S. Code of Federal Regulations (CFR) Part 72, "Licensing Requirements for the Independent Storage of Spent Nuclear Fuel and High-Level Radioactive Waste," Title 10, "Energy" (10 CFR

Part 72). The reviewer should read the exact regulatory language. Table 7-1 matches the relevant regulatory requirements associated with this chapter to the areas of review.

Areas of Review	10 CFR Part 72 Regulations		
	72.124	72.236(a)	72.236(b), (c), (g), (h), (m),
Criticality Design Criteria and Features	●	●	●
Fuel Specification	●	●	
Model Specification	●	●	●
Criticality Analysis	●	●	●
Burnup Credit	●	●	

Table 7-1 Relationship of Regulations and Areas of Review

7.4   Acceptance Criteria

In general, the DSS criticality evaluation seeks to ensure that a subcritical condition is maintained for the given design by fulfilling the following acceptance criteria:

- The effective neutron multiplication factor, k_{eff}, including all biases and uncertainties at a 95-percent confidence level, should not exceed 0.95 under all credible normal, off-normal, and accident-level conditions.

- At least two unlikely, independent, and concurrent or sequential changes to the conditions essential to criticality safety, under normal, off-normal, and accident-level conditions would need to occur before an accidental criticality is deemed to be possible (i.e., double contingency principle).

- When practicable, criticality safety of the design should be established on the basis of favorable geometry, permanently fixed neutron-absorbing materials (poisons), or both. Where solid neutron-absorbing materials are used, the design should provide for a positive means to verify their continued efficacy during the storage period. The neutron-absorbing materials' continued efficacy may be confirmed by a demonstration or analysis before use, showing that significant degradation of these materials cannot occur over the life of the facility.

- Criticality safety of the cask system should not rely on credit for more than 75 percent of the neutron poison material in fixed neutron absorbers when subject to standard acceptance tests. For greater credit allowance, special, comprehensive fabrication tests capable of verifying the presence and uniformity of the neutron absorber are needed.

7.5   Review Procedures

The interrelationship of the criticality evaluation review with other disciplines is shown in Figure 7-1.   The figure shows that this review draws upon information from the general information section as well as information reviewed or developed for the design criteria, structural, and operating procedures evaluations.   Information collected or developed during the review of this chapter is useful in the evaluation of the materials, operating procedures, acceptance tests and maintenance program, accident analysis, and technical specifications and operating controls for the DSS.

The reviewer should examine the criticality design features and criteria in SAR Chapter 1, "General Information," and SAR Chapter 2, "Principal Design Criteria," in addition to SAR Chapter 7, "Criticality Evaluation," for any additional details concerning criticality design features and criteria.   The reviewer should assess the bounding specifications for the SNF and assure consistency with the models used by the applicant in the criticality analyses.   The reviewer should verify that criticality safety considerations under normal, off-normal, and accident-level conditions are addressed by the applicant and that the cask system design complies with 10 CFR Part 72.   In addition, the reviewer should verify that the criticality calculations determine the highest k_{eff} that might occur for all loading states under normal, off-normal, and accident conditions involving handling, packaging, transfer, and storage.   To the extent practicable, the use of independent methods to perform any k_{eff} calculations by the reviewer should be pursued to evaluate the applicant's design.

7.5.1   Criticality Design Criteria and Features (HIGH Priority)

The reviewer should examine the principal criticality design criteria presented in SAR Chapter 2 as well as any related details provided in SAR Chapter 7, "Criticality Evaluation".   The general cask description presented in SAR Chapter 1 should be examined for any relevant information. The information in Chapter 7 of the SAR should be verified to be consistent with the information in SAR Chapters 1 and 2.   The reviewer should verify that all descriptions, drawings, figures, and tables are sufficiently detailed to support an in-depth staff evaluation.

The criticality design of the cask relies on the general dimensions of the cask components and the spacing of the fuel assemblies.   The criticality design also often relies on neutron poisons. These may be in the form of fixed poisons in the basket structure, which may be used together with flux traps, and/or soluble poisons in the water of the SNF pool.   During loading and unloading operations, NRC staff accepts the use of borated water as a means of criticality control if the applicant specifies a minimum boron content and strict controls are established to ensure that the minimum required boron concentration is maintained.   This condition in turn becomes an operating control and limit in SAR Chapter 13, and in the Technical Specification (TS).   The SER should also discuss these operating controls.   Other design features significant to the criticality design, such as important basket dimensions that control the spacing of the fuel assemblies should also be included in the TS.   These dimensions may be a minimum pitch for the basket cells or a minimum flux trap width.

If borated water is used for criticality control during loading and unloading operations, administrative controls and/or design features should be implemented to ensure that accidental flooding with unborated water cannot occur, or the criticality evaluation should consider accidental flooding with unborated water.   If the cask is also intended for transport, borated water should not be relied upon for criticality control.   Borated water and any other liquids are not acceptable as a means of criticality control for a cask in dry storage.

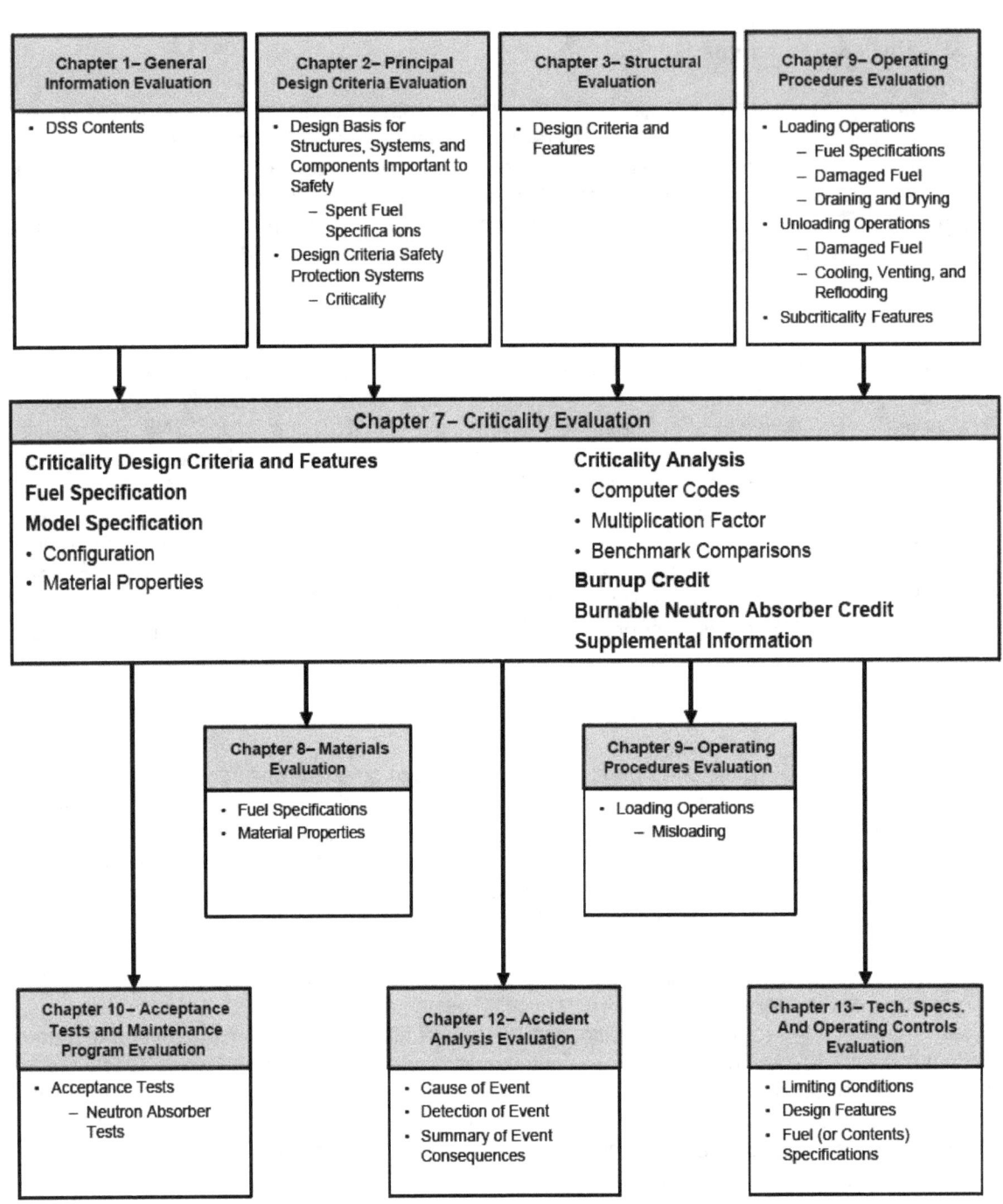

Figure 7-1 Overview of Criticality Evaluation

This includes use of any credit in the criticality analysis for the presence of a liquid that may provide neutron shielding (and is external to the fuel basket); however, its presence and most reactive density should be assumed if it increases k_{eff}. Also, if more than one certified or licensed basket design of the same supplier could fit in the cask; the type of basket to be used with the cask should be stamped in a location on the cask system in a way that allows for easy identification of the basket. Thus, a licensee using the cask system will be able to easily verify the appropriateness of the fuel contents to be loaded in the basket.

7.5.2 Fuel Specification (HIGH Priority)

The reviewer should examine the specifications for the ranges or types of SNF that will be stored in the cask as presented in SAR Chapters 1, "General Information Evaluation" and 2, "Principal Design Criteria Evaluation" as well as any related information provided in SAR Chapter 7,"Criticality Evaluation". The SNF specifications given in Chapter 7 of the SAR should be consistent with, or bound, the specifications given in SAR Chapters 1 and 2 and in the TS. The reviewer should also, keeping in mind that some specifications are more important than others, identify the specifications that are keys to criticality safety and verify that these are appropriately captured in the TS. NUREG-1745 provides a listing of some fuel specifications that may be keys to maintaining the system subcritical.

Of primary interest is the type of fuel assemblies and maximum fuel enrichment that should be specified and used in the criticality calculations. Some boiling-water reactors (BWR) use multiple fuel pin enrichments, in which case the criticality calculations should use the maximum fuel pin enrichment present. Depending upon the fuel design, an applicant may propose use of assembly averaged or lattice averaged enrichments. This may be acceptable if the applicant can demonstrate that the applicant's averaging technique is technically defensible and, for the criticality calculation, produces realistic or conservative results. Because of the natural uranium blankets present in many BWR designs, use of an assembly-averaged enrichment that includes the blankets is not normally considered appropriate or conservative for BWR fuel.

Another parameter of interest is the fuel density assumed in the analysis. The value of the fuel density used in the calculations should be justified to be realistic or conservative.

Although the burnup of the fuel affects its reactivity, many criticality analyses have assumed the cask to be loaded with fresh fuel (the fresh fuel assumption). Alternatively, the NRC staff has provided guidance for limited burnup credit for intact fuel. This guidance is currently limited to burnup credit available from actinide compositions associated with UO_2 fuel of 5.0 wt percent or less enrichment that has been irradiated in a PWR to an assembly-average burnup value not exceeding 50 GWD/MTU and cooled out-of-reactor for a time period between 1 and 40 years. Guidance regarding the review of a criticality analysis that involves burnup credit is provided in Section 7.5.5. Specifications for the fuel that will be stored in the cask, including those important for burnup credit, if applicable, should be included in Chapter 13, "Technical Specifications and Operational Controls and Limits Evaluation" of both the SAR and SER, with those specifications determined to be key to criticality safety also explicitly listed in the Technical Specifications.

For analyses that use the fresh fuel assumption, inadvertent loading of the cask with unirradiated fuel is not a major concern. However, inadvertent loading of the cask with unirradiated fuel is a major concern for casks that rely on criticality analyses that use burnup credit. Therefore, detailed loading procedures for these casks will need to include steps to prevent misloading of unirradiated fuel. Regardless of which analysis is used, detailed loading procedures may need to include steps to prevent misloading if fuel exceeding the design basis for the DSS is present in the pool at the time of loading.

Because casks are typically designed to store many types and configurations of fuel assemblies, the applicant should demonstrate that criticality requirements are satisfied for the most reactive case. A determination of which fuel is bounding in a criticality analysis depends on many factors and usually requires examination of several types of fuel assemblies and compositions. The design-basis fuel has often been the Westinghouse 17x17 optimized fuel assembly (OFA); however, this will not be the case for all cask designs because of cask-specific effects on reactivity. Therefore, the applicant should demonstrate and reviewers should verify that the fuel assembly used as the design basis is the most reactive for the specific cask design. Chapter 1, "General Information Evaluation" of the SAR and Chapter 13, "Technical Specifications and Operation Controls and Limits Evaluation" of the SER should either clearly indicate the design-basis assemblies or reference the SAR chapter in which they are identified.

7.5.2.1   Non-Fuel Hardware

Some fuel assemblies may also have non-fuel components that are positioned or operated within the envelope of the fuel assembly during reactor operation that an applicant may seek to store with the assemblies in the cask. These items include PWR control assemblies such as Rod Cluster Control Assemblies (RCCAs), Control Element Assemblies (CEAs), Burnable Poison Rod Assemblies (BPRAs) and Axial Power Shaping Rods (APSRs). Applicants may also seek approval of storage of fuel assemblies with other items that extend into an assembly's active fuel region, such as stainless steel rod inserts used to displace water in PWR assembly guide tube dashpots. For applications that propose to load assemblies containing non-fuel hardware, ensure that the analysis considers the effects of both inclusion and neglect of non-fuel hardware on system reactivity. If the application relies on the presence of the non-fuel hardware to meet the subcritical criterion, verify that the non-fuel hardware will remain in place under all normal and design basis conditions.

Generally, staff does not allow reliance on, or credit for, fuel-related burnable neutron absorbers. This restriction includes residual neutron-absorbing material remaining in the non-fuel hardware loaded with an assembly. However, credit for any negative reactivity for this latter absorbing material may be accepted if: (1) the remaining absorbing material content is established through physical measurement, where a sufficient margin of safety is included commensurate with the uncertainty in the method of measurement, (2) the axial distribution of the poison depletion is adequately determined with appropriate margin for uncertainties, and (3) adequate structural integrity and placement of the non-fuel hardware under accident conditions is demonstrated. Ensure that the fuel specifications, described in Chapter 13, "Technical Specifications and Operation Controls and Limits Evaluation" of both the SAR and SER, include the important details about the non-fuel hardware to be stored with the fuel assemblies and the associated residual neutron absorbing material, with those details key to criticality safety included in the TS, as appropriate. Also, verify that operating procedures are established that ensure that non-fuel hardware loaded with assemblies meets the approved specifications as well as remains in position.

7.5.2.2   Fuel Condition

Determine if the applicant has included any specifications regarding the fuel condition. To date, a number of applications have requested approval for storage of fuel that is damaged as well as intact, or undamaged. The reviewer should consult the most current staff guidance for detailed descriptions regarding what constitutes damaged, undamaged and intact fuel (e.g., Sections 8.4.17.2 and 8.6 of this SRP or more recent guidance). This guidance gives the applicant the latitude to define fuel with defects (such as missing rods but not loose rods or debris) as undamaged fuel as long as the fuel can meet all the fuel specific or system related functions. For purposes of the criticality function, undamaged fuel is fuel that: (1) is in the form of an assembly, (2) has structural and material properties such that the assembly can withstand normal and design basis events while maintaining its geometric configuration and (3) has had any damaged or missing fuel rods replaced with solid dummy rods that displace an equal amount of water as the original rods. Fuel that cannot meet these criteria is considered to be damaged. However, a fuel assembly with missing fuel rods may be considered undamaged fuel if analyses are performed that show the criterion for subcriticality will be met with the fuel rods missing.

A fuel assembly that is classified as damaged must be placed in a damaged fuel canister, or in an acceptable alternative, for loading into the cask. For a cask that is also intended for transport, it must be kept in mind that the more severe conditions of transport may require re-analysis of assemblies classified as undamaged under storage-only conditions prior to transport. Specifications concerning the condition of the fuel to be stored in the cask and the loading of damaged fuel, as applicable, should be included in Chapter 13, "Technical Specifications and Operation Controls and Limits Evaluation" of both the SAR and SER and in the Certificate of Compliance (in the TS).

The reviewer should verify that the criticality analysis addresses the conditions of the fuel to be stored in the cask system. Analyses for cask systems designed to store damaged fuel should bound the configuration of the damaged fuel assemblies under all credible normal and design basis conditions. For example, some analyses have performed calculations that model the damaged fuel as arrays of bare fuel rods (i.e., the cladding is assumed to be completely removed) having an optimized rod pitch.

7.5.3   Model Specification (HIGH Priority)

Manufacturing and fabrication tolerances should be specified, and the reviewer should verify that the applicant used the most reactive combination of tolerances, within the ranges of their acceptable values, in the cask system model.

7.5.3.1   Configuration

The reviewer should verify that the model used in the criticality evaluation is adequately described for normal, off-normal, and accident conditions. The reviewer should also coordinate with the structural, materials, and thermal reviewers to understand any damage that could result from accident or natural phenomena events.

The reviewer should examine the sketches or figures of the model used for criticality calculations. The reviewer should verify that the dimensions and materials of the model are consistent with the engineering drawings. Differences between the actual cask configuration and the models should be identified, and the models should be shown to be conservative. Substitution of end sections and support structures of the fuel with ordinary water is a common and usually conservative practice in criticality analysis. However, substitution with borated water is typically not conservative. Any such substitutions should be justified.

Tolerances for poison material dimensions and/or concentrations should be defined, and the most reactive conditions should be used in the criticality analysis. In addition, the analysis should identify all important design conditions and then address these conditions for potential variations during normal, off-normal, and accident-level conditions.

The reviewer should verify that the applicant has considered deviations from nominal design configurations. The evaluation of k_{eff} should not be limited to a model in which all of the fuel bundles are neatly centered in each basket compartment with the center line of the basket coincident with the center line of the cask. For example, a cask with steel confinement and lead shielding may have a higher k_{eff} when the basket and fuel assemblies are positioned as close as possible to the lead. However, in some designs, the most reactive configuration may be when all fuel assemblies are shifted toward the center of the basket.

In addition to a fully flooded cask, the SAR should address configurations in which the cask is filled with partial density water or is partially filled with water (borated, if applicable) and the remainder of the cask is filled with steam consisting of ordinary water at partial density. These configurations are considered to be possible during loading and unloading operations. The SAR should also consider the possibility of preferential or uneven flooding within the cask, if such a scenario is credible for the given cask design (e.g., because of blockage in small flow or drain paths). In particular, the reviewer should watch for situations where there is water in the fuel regions but not in the flux traps, if applicable. Cask designs for which this type of flooding is credible are generally unacceptable. The SAR should also consider flooding in the fuel rod pellet-to-clad gap regions with unborated water. Above all, the analysis must demonstrate that the cask remains subcritical for all credible conditions of moderation.

The reviewer should examine whether the applicant has prepared a heterogeneous model of each fuel rod or has homogenized the entire fuel assembly. With current computational capabilities, homogenization is now an uncommon practice and should not be used.

7.5.3.2   Material Properties

The reviewer should verify that the compositions and densities are provided for all materials used in the calculational model. The applicant should also cite, in the SAR Chapter 8, "Materials Evaluation", the source of all materials data, particularly the data for fuel and poison materials. In coordination with the materials reviewer, the criticality reviewer should determine the acceptability of the sources of data that are important to the criticality safety function of the cask. The criticality reviewer should, in coordination with the materials reviewer, ensure that the applicant addressed the validation of the poison concentration in the acceptance testing discussion in SAR Chapter 10, "Acceptance Tests and Maintenance Program Evaluation." Criticality computer codes generally will allow the densities to be input directly in units of g/cm^3 or units of atoms/barn-cm. In either case, the reviewer should pay attention to the final value used directly by the code. Also, the reviewer should confirm that the analysis does not take credit for more than the minimum amount of neutron absorber verified by the acceptance testing, subject to the criteria in Section 7.4.

Among other specifications, 10 CFR Part 72 requires that a positive means to verify the continued efficacy of solid neutron-absorbing materials should be provided when these materials are used. The criticality reviewer should verify that the neutron flux from the irradiated fuel results in a negligible depletion of poison material over the storage period, In coordination with the materials and structural reviewers, the criticality reviewer should ensure that the applicant demonstrates that the required acceptance testing of the poisons during fabrication (specified in SAR Chapter10, "Acceptance Tests and Maintenance Program Evaluation") has been satisfactorily specified, and by analysis or demonstration, the applicant has shown the poison material's durability and resistance to degradation during the storage period.

The neutron flux used for this analysis should be the maximum that may be produced by feasible loadings of irradiated or unirradiated fuel. The reviewer should coordinate review of the applicant's acceptance testing and assessment of the poison material's durability with the materials reviewer to verify that the applicant provides a valid and accurate demonstration of the absorber material's continued efficacy. Consideration should be given to the effects of physical and chemical actions as well as irradiation (gamma and neutron). There may be other ways to provide positive means of verifying the neutron absorber's continued efficacy. For applications that propose an alternative method, the reviewer should verify that the proposed method is reasonable (considering any effects on meeting confinement, shielding, or other system design criteria) and valid and accurate in demonstrating the absorber's continued efficacy.

7.5.4 Criticality Analysis (Priority as indicated)

7.5.4.1 Computer Codes

(MEDIUM Priority) Both Monte Carlo and deterministic computer codes may be used for criticality calculations. Monte Carlo computer codes are better suited to three-dimensional geometry and, therefore, are more widely used to evaluate spent fuel cask designs. The most frequently used Monte Carlo codes are SCALE/KENO (ORNL, 2005), MCNP (MCNP5, 2003), and MONK (AEA Technology, 2001). All three codes permit the use of either multigroup or continuous cross sections. The reviewer should determine that the applicant has used a computer code that is appropriate for the particular application and has used that code correctly.

(LOW Priority) The reviewer should determine whether the applicant has chosen an acceptable set of cross sections. Cross sections may be distributed with the criticality computer codes or developed independently from another source. The applicant should provide or reference the source of cross-section data. For user-generated cross sections, the applicant should specify the method used to obtain the actual data employed in the criticality analysis. For multigroup calculations, the neutron flux spectrum used to construct the group cross sections should be similar to that of the cask. If a multigroup treatment is used, the reviewer should ensure the applicant has appropriately considered the neutron spectrum of the cask. In addition to selecting a cross-section set collapsed with an appropriate flux spectrum, a more detailed processing of the energy-group cross sections is required to properly account for resonance absorption and self-shielding. The use of multigroup KENO as part of the CSAS sequences in SCALE will directly enable appropriate cross-section processing. Some cross-section sets include data for fissile and fertile nuclides (based on a potential scattering cross section, s_p) that can be input by the user. If the applicant has used a stand-alone version of KENO, the reviewer should ensure that potential scattering has been properly considered. Furthermore, information has been published concerning problems with some cross-section libraries once commonly distributed with SCALE/KENO. One library, the "working-format" library, was used for calculations of the code manual's sample problems but is not intended for criticality calculations of actual systems (IN 91-26, 1991). Another library, the SCALE 123-group library, has demonstrated inadequacies for non-thermalized, highly enriched systems (NUREG/CR-6328, 1995).

MEDIUM Priority) The reviewer should pay particular attention to the proper selection of scattering cross section data for important compounds that may be in the system. Use of a free atom cross section for nuclides in a compound may not adequately account for the scattering effects of atoms bound in molecules and lattices. This misrepresentation can cause the underprediction of k_{eff}, particularly in the case of a well moderated system where energetic up scattering plays a significant role in the neutronics of the system.

(MEDIUM Priority) For analyses of a cask model with separate regions of water and steam, the use of a multigroup cross-section set raises additional concerns. The reviewer should verify that the applicant has addressed the differences in the flux spectra in the two regions. If the results of these calculations indicate that k_{eff} is close to 0.95, additional independent calculations using a different code and/or cross-section library (a library derived from a different cross-section database if possible and appropriate) may be helpful. The reviewer should also closely examine the applicant's benchmark analysis to verify the applicability of the critical experiments considered.

7.5.4.2   Multiplication Factor

(MEDIUM Priority) The reviewer should examine the results and discussion of the k_{eff} calculations for the storage cask. The reviewer should verify that the calculations determine the highest k_{eff} that might occur during all operational states under normal, off-normal and accident conditions. Sensitivity parametric analyses may be used to provide the required demonstration that the highest k_{eff} with a confidence level of 95 percent has been determined. Variations in the results caused by differences in the models and sensitivity analyses should be explained and found to be reasonable.

(MEDIUM Priority) For Monte Carlo calculations, the reviewer should assess if the number of neutron histories and convergence criteria are appropriate. As the number of neutron histories increases, the mean value for k_{eff} should approach a fixed value, and the standard deviation associated with each mean value should decrease. Depending on the code used by the applicant, a number of diagnostic calculations are generally available to demonstrate adequate convergence and statistical variation. For deterministic codes, a convergence limit is often prescribed in the input. The selection of a proper convergence limit and the achievement of this limit should be described and demonstrated in either the SAR or supporting criticality calculations. When burnup credit is included in the criticality analysis, the reviewer needs to be sure that proper neutron sampling and convergence have been achieved because the flux will be concentrated in the low burned ends of the fuel assemblies.

(HIGH Priority) Because of the importance and complexity of the criticality evaluation, independent calculations should be performed to ensure that the most reactive conditions have been addressed, the reported k_{eff} is conservative and the applicant has appropriately modeled the storage cask geometry and materials. In deciding the level of effort necessary to perform independent confirmatory calculations, the reviewer should consider the following factors: (1) the calculation method (computer code) used by the applicant, (2) uniqueness and complexity of the design and analysis, (3) the degree of conservatism in the applicant's assumptions and analyses, and (4) the extent of the margin between the calculated result and the acceptance criterion of $k_{eff} \leq 0.95$. As with any design and review, a small margin below the acceptance criterion and/or a small degree of conservatism may necessitate a more extensive staff analysis.

(HIGH Priority) The reviewer should develop a model that is independent of the applicant's model. If the reported k_{eff} for the most reactive case is substantially lower than the acceptance criterion of 0.95, a simple model known to produce very bounding results may be all that is necessary for the independent calculations.

(HIGH Priority) If possible and appropriate, the reviewer should perform the independent calculations with a computer code different from that used by the applicant. Likewise, use of a different cross-section set, derived from a different cross-section database where possible and appropriate (e.g., ENDF/B, JEF, JENDL, UKNDL, etc.), can provide a more independent confirmation. The continuous energy (CE) cross sections created for use with KENO in the SCALE code system are generated by the AMPX processing code rather than the more widely used NJOY code. Even though some cross section libraries may not have fully independent data bases because they are all derived from ENDF/B data, the CE library in SCALE still can provide some level of independence and is useful for checking computations performed with libraries which were generated by using NJOY. The reviewer should describe the staff's independent analysis and the analysis general results and conclusions in the SER.

(HIGH Priority) Although a k_{eff} of 0.95 or lower meets the acceptance criterion, the reviewer should watch for design features or content specifications where small changes could result in large changes in the value of k_{eff}. When the value of k_{eff} is highly sensitive to system parameters that could vary, the acceptable k_{eff} limit may need to be reduced below 0.95. When establishing a k_{eff} limit below 0.95, the reviewer should consider the degree of sensitivity to system parameter changes and the likelihood and extent of potential parameter variations.

7.5.4.3   Benchmark Comparisons (HIGH Priority)

Computer codes for criticality calculations should be benchmarked against critical experiments. A thorough comparison provides justification for the validity of the computer code, its use for a specific hardware configuration, the neutron cross sections used in the analysis, and consistency in modeling by the analyst. Ultimately the benchmarking process establishes a bias and uncertainty for the particular application of the code (using the benchmark results for calculations performed by another analyst does not address this last issue). The calculated k_{eff} of the cask should then be adjusted to include the appropriate biases and uncertainties from the benchmark calculations.

The reviewer should examine the general description of the benchmark comparisons. This examination includes verifying that the analysis of the experiments used the same computer code, computer system, cross-section data, modeling methods, and code options that were used to calculate the cask system k_{eff} values.

The reviewer should also closely examine the applicant's benchmark analysis to determine whether the benchmark experiments are relevant to the actual cask design. No critical benchmark experiment will precisely match the fissile material, moderation, neutron poisoning, and configuration in the actual cask. However, the applicant can perform a proper benchmark analysis by selecting experiments that adequately represent cask and fuel features and parameters that are important to reactivity. Key features and parameters that should be considered in selecting appropriate critical experiments include the type of fuel, enrichment, hydrogen-to-uranium (H/U) ratio (dependent largely on rod diameter and pitch), reflector material, neutron energy spectrum, and poisoning material and placement. The applicant should justify, and the reviewer should verify, the suitability of the critical experiments chosen to benchmark the criticality code and calculations. Techniques such as the sensitivity/uncertainty method developed by Oak Ridge National Laboratory (ORNL/TM-2005/39, 2005) can be helpful when assessing the applicability of the critical experiments used to benchmark the design analysis. UCID-21830 (Lloyd, 1990), the "International Handbook on Evaluated Criticality Safety Benchmark Experiments," (NSC,NEA, 9/2003) and NUREG/CR-6361 provide information on benchmark experiments that may apply to the cask being analyzed.

The reviewer needs to assess whether the applicant analyzed a sufficient number of appropriate benchmark experiments and how the results of these benchmark calculations have been converted to a bias for the cask calculations. Simply averaging the biases from a number of benchmark calculations typically is not sufficient, such as when one benchmark yields results that are significantly different from the others, the number of experiments is limited, or benchmarks that over-predict k_{eff} are included. In addition, benchmark comparisons should be checked for bias trends with respect to parameter variations (such as pitch-to-rod-diameter ratio, assembly separation, reflector material, neutron absorber material, etc.). A Lawrence Livermore National Laboratory (LLNL) (Lloyd, 1990) and NUREG/CR-6361 provide some guidance, but other methods, when adequately explained, have also been considered appropriate.

For Monte Carlo codes, the statistical uncertainties of both benchmark and cask calculations also need to be addressed. The uncertainties should be applied to at least the 95-percent confidence level. As a general rule, if the acceptability of the result depends on these rather small differences, the reviewer should question the overall degree of conservatism of the calculations. Considering the current availability of computer resources, a sufficient number of neutron histories can readily be used so that the treatment of these uncertainties should not significantly affect the results.

The reviewer should verify that only biases that increase k_{eff} have been applied. For example, if the benchmark calculation for a critical experiment results in a neutron multiplication that is greater than unity, it should not be used in a manner that would reduce the k_{eff} calculated for the cask. Only corrections that increase k_{eff} should be applied to preserve conservatism.

The reviewer may have already performed a number of benchmark calculations applicable to storage casks and may have a reasonable estimation of the bias to be applied to the independent calculation of the cask. If such is not the case, or if the acceptability depends on small bias differences, the reviewer again needs to determine whether sufficient conservatism has been applied to the calculations.

7.5.5 Burnup Credit (HIGH Priority)

For guidance regarding the use of burnup credit, see the current revision of ISG-8.

7.5.6 Supplemental Information

The reviewer should ensure that all supportive information or documentation is provided. This may include, but not be limited to, justification of assumptions or analytical procedures, test results, photographs, computer program descriptions, input/output, and applicable pages from referenced documents. In addition, the SAR should include a list of fuel designs with the acceptable parametric limits and the maximum enrichments for which the criticality analysis is valid. The reviewer should request any additional information needed to complete the review.

7.6 Evaluation Findings

The reviewer should review the 10 CFR Part 72 acceptance criteria and provide a summary statement for each. These statements should be substantially as follows:

F7.1   Structures, systems, and components important to criticality safety are described in sufficient detail in Chapters _____ of the SAR to enable an evaluation of their effectiveness.

F7.2   The _____ cask and its spent fuel transfer systems are designed to be subcritical under all credible conditions.

F7.3   The criticality design is based on favorable geometry, fixed neutron poisons, and soluble poisons of the spent fuel pool [as applicable]. An appraisal of the fixed neutron poisons has shown that they will remain effective for the term requested in the CoC application and there is no credible way for the fixed neutron poisons to significantly degrade during the requested term in the CoC application; therefore, there is no need to provide a positive means to verify their continued efficacy as required by 10 CFR 72.124(b).

F7.4   The analysis and evaluation of the criticality design and performance have demonstrated that the cask will enable the storage of spent fuel for the term requested in the CoC application.

The reviewer should provide a summary statement similar to the following:

"The staff concludes that the criticality design features for the [cask designation] are in compliance with 10 CFR Part 72, as exempted [if applicable], and that the applicable design and acceptance criteria have been satisfied. The evaluation of the criticality design provides reasonable assurance that the [cask designation] will allow safe storage of spent fuel. This finding is reached on the basis of a review that considered the regulation itself, appropriate regulatory guides, applicable codes and standards, and accepted engineering practices."

8 MATERIALS EVALUATION

8.1 Review Objective

The materials review ensures adequate material performance of components important to safety of a dry cask storage system (DSS), including the spent fuel canister or cask, under normal, off-normal, and accident-level conditions. To ensure an adequate margin of safety in the design basis of the DSS, the reviewer should obtain reasonable assurance that:

- The physical, chemical, and mechanical properties of materials for components important to safety (ITS) meet their service requirements including normal, off-normal, and accident-level conditions, and that the mechanical properties are Code accepted values.

- Materials for components ITS have sufficient requirements to control the quality of the production, fabrication, and test activities.

- Materials for ITS components are selected to accommodate the effects of, and to be compatible with, the independent spent fuel storage installation (ISFSI) site characteristics, environmental conditions, and duration of the license period.

- The spent nuclear fuel (SNF) cladding is protected from gross rupture and from conditions that could lead to fuel redistribution.

- The DSS is designed to maintain the spent fuel in a readily retrievable condition.

- Other materials which support or protect ITS components (such as coatings) are suitable for the application.

In reviewing the materials, the reviewer should consider the sources of information for the physical and mechanical properties of the materials used in the DSS construction and those materials which are part of the spent fuel payload. These material properties should be considered against both static and dynamic loadings for normal, off-normal, accident conditions, and other phenomena such as corrosion. The material properties and characteristics needed to satisfy these functional safety requirements should be maintained and are applicable over the complete licensing period.

Preferred materials information sources are U.S. industry consensus codes, standards, and specifications. The applicability and acceptability of all other sources, such as manufacturer's test data and handbooks, should be reviewed. The reviewer should also examine published articles, research reports, and texts as sources of information concerning material performance. Foreign standards (and codes) may be acceptable on a case by case basis. The applicant should provide complete documentation supporting the use of the foreign standard and show that the foreign standard is equivalent to a comparable US standard (e.g. ASME, ASTM, etc.), or otherwise sufficient for its intended use. The staff may need to review foreign standards in greater depth, depending on the familiarity with the standard and applicability of the standard to the proposed DSS design

8.2   Areas of Review

The materials evaluation encompasses the following listed areas of review.   The various materials engineering related topics requiring review may be addressed in different chapters of the SAR.   However, the review guidance for all materials engineering related topics are provided in this chapter of the SRP.

Areas for materials review:

General

> Cask Design/Materials
> Environmental Conditions
> Engineering Drawings

Materials Selection

> Applicable Codes and Standards and Alternatives to the Code
> Material Properties
> Alternative or Substitute Materials (ITS components)
> Copper bearing or other weathering steels or other corrosion control measures
> for coastal ISFSI locations
> Weld Design, Inspection
> Bolt Applications
> Coatings
> Neutron Shielding Materials
> Gamma shielding
> Neutron Poison Materials for Criticality Control
> Concrete and Reinforcing Steel
> Seals
> Low Temperature Ductility of Ferritic Steels
> Creep Properties/Analyses

Corrosion

> Corrosion Resistance
> Galvanic/Chemical/Radiolytic Reactions of Fuel with Canister Internals

Cladding Integrity/Fuel

> Fuel Burn-up
> Cladding Temperature Limits
> Damaged Fuel Definition

Operational Issues (see Operating Procedures Chapter of SAR)

> Hydrogen gas monitoring/mitigation
> Preventing oxidation of fuel during loading/unloading operations which can lead to Rod Splitting

Examination and Testing (see Acceptance Test Chapter of SAR)

Helium leakage testing of canister welds
Periodic Inspections

Code Case Acceptability

Refer to Regulatory Guide 1.193

8.3 Regulatory Requirements

This section presents a summary matrix of the portions of U.S. Code of Federal Regulations (CFR) Part 72, "Licensing Requirements for the Independent Storage of Spent Nuclear Fuel and High-Level Radioactive Waste," Title 10, "Energy" (10 CFR Part 72) relevant to the review areas addressed by this chapter. The U.S. Nuclear Regulatory Commission (NRC) staff reviewer should read the exact referenced regulatory language. Table 8-1 matches the relevant regulatory requirements associated with this chapter to the areas of review.

Table 8-1 Relationship of 10 CFR Part 72 Regulations and Areas of Review

Chapter 8 Areas of Review	10 CFR Part 72 Regulations				
	72.104(a)	72.106(b)	72.122 (a), (b), (c)	72.122 (h)(1), (i), (l)	72.124
General					
Materials Selection	●	●	●		●
Corrosive Reactions					
Cladding Integrity				●	

Chapter 8 Areas of Review	10 CFR Part 72 Regulations			
	72.236(g)	72.236(h)	72.236(i)	72.236(m)
General				●
Materials Selection	●		●	●
Corrosive Reactions		●		
Cladding Integrity				●

8.4 Review Procedures and Acceptance Criteria

Metallic materials are primarily assumed in this guidance. The interrelationship of the materials evaluation review with other disciplines is shown in Figure 8-1.

8.4.1 General Review Considerations (HIGH Priority)

The reviewer should survey the SAR and design drawings (generally SAR Chapters 1 and 2) to identify the various materials issues that may be associated with the specific design proposal in the application. The reviewer should also examine the criticality, shielding, confinement, and thermal chapters to identify cross-cutting issues that should be coordinated among the technical disciplines. Note, not all design and license changes in the amendment will necessarily be neither separately identified by the applicant nor obvious.

The reviewer should examine the following Technical Specification (TS) items to verify its proposal by the applicant and understand the specific limits, design requirements, and operating constraints proposed by the applicant:

Maximum fuel burn-up
Maximum cladding temperature
Definition of damaged fuel
Code of record and alternatives to specific Code requirements
Specification/requirements for alternative materials for ITS components
Manufacture and testing of neutron poison material(s) for criticality control
Hydrogen monitoring/mitigation during wet loading/unloading
Helium leakage testing of confinement and cover welds
Maintaining inert atmosphere during canister draining/flooding to prevent oxidation
Use of Code Case N-595 (not acceptable)
Use of copper bearing or weathering steel for structural steel components at coastal marine ISFSI sites (or other corrosion mitigation measures)
Operational controls to maintain cladding temperature limits
Low Temperature Ductility of Ferritic Steels
Damaged fuel definitions
Materials acceptance testing
Design temperature for aluminum components used in the fuel basket or canister interior (creep issues)

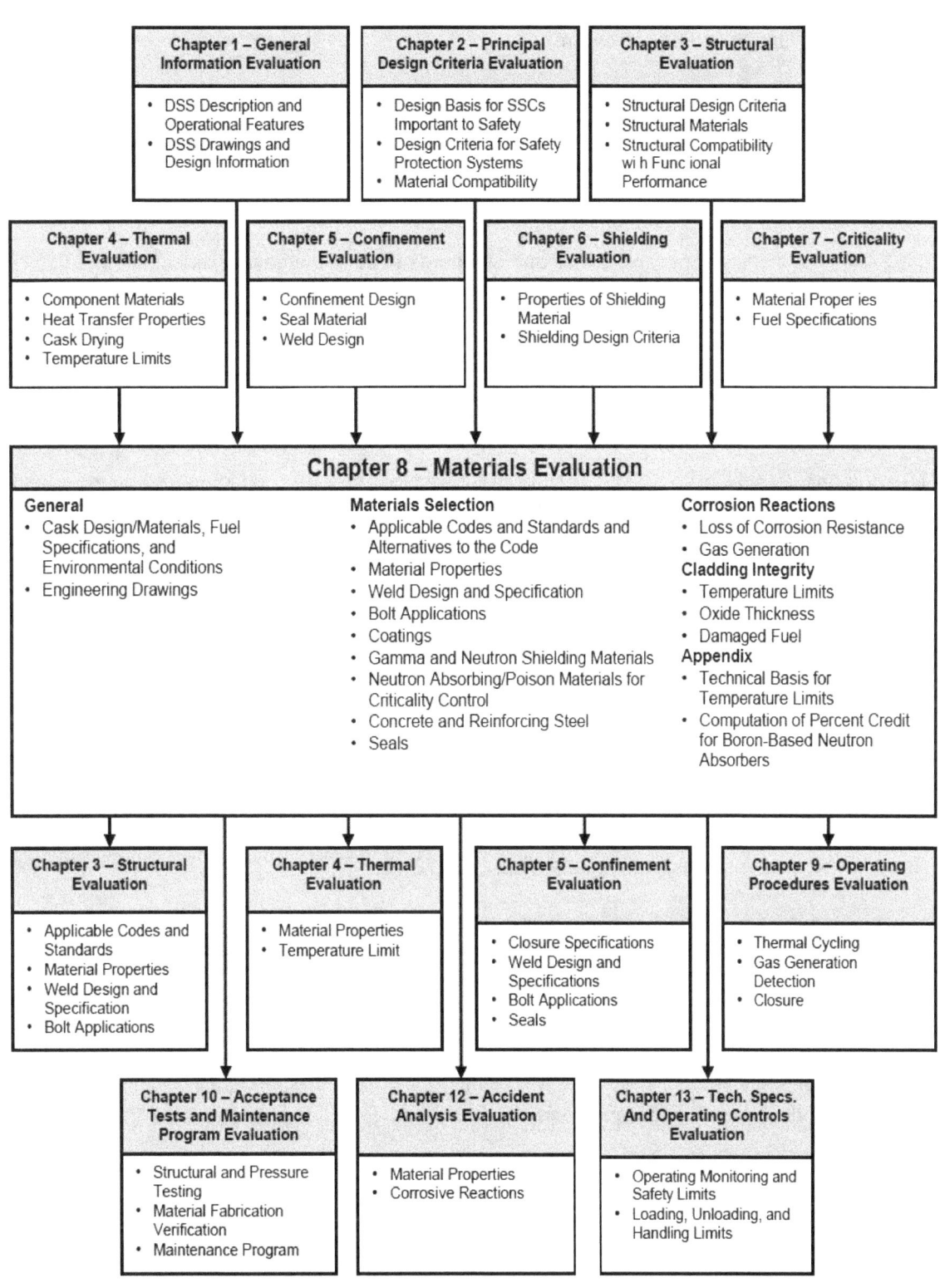

Figure 8-1 Overview of Materials Evaluation

8.4.2 Codes and Standards (HIGH Priority)

8.4.2.1 Usage and Endorsement

Codes (or "construction codes") govern which materials may be used and how they may be employed. Standards detail how a material is produced and establishes chemical and material property requirements. All ASME materials are a subset of AWS and ASTM materials. However, not all ASTM materials are endorsed for use by the ASME or other codes which may be used for canister design.

The SAR must identify applicable codes and standards used in the design, selection, and use of materials. For important-to-safety (ITS) components, U.S. industry consensus codes and standards such as ASME, AWS, ANSI, ACI, and ASTM should be specified.

Foreign codes and standards are generally NOT acceptable for ITS components/materials and would only be approved on a case-by-case basis. However, foreign-produced materials which comply with U.S. codes and standards are acceptable.

ITS components subject to ASME Section III jurisdiction, typically confinement boundary and fuel basket, are normally ASME Section II materials. ITS attachments to the confinement boundary, as well as structural components of the overpack, may be ASME or ASTM materials, depending on the code of record for the component. For non-ASME ITS components, ASTM materials may be used.

Non-ITS items can be specified by generic names such as "stainless steel", "aluminum," "carbon steel," etc., as appropriate for the application.

Proprietary materials which are ITS (specifically neutron poisons) must be described adequately in SAR Chapter 8, "Materials" to permit the staff to make a safety finding. The governing quality assurance and quality control (QA/QC) documents, key manufacturing procedures, and key testing protocols for proprietary materials should be incorporated by reference into the TS. Limited changes to the materials composition, performance, or manufacturing methods may be allowed if the changes satisfy the criteria of 10 CFR 72.48.

Polymeric neutron shielding materials, which are usually proprietary, are considered important-to-safety (ITS) materials in order to meet the regulatory requirements of 72.126(a)(6). NUREG/CR-6407 specifically designates neutron shielding materials as ITS Category B.

8.4.2.2 Code Case Use/Acceptability

Review any referenced ASME Code cases against Regulatory Guide 1.193 for acceptability. Note that Code Case N-595 (any revision) has been found unacceptable to the staff per RG 1.193.

8.4.3 Environment (Priority – as indicated)

(MEDIUM Priority) Generally, the ISFSI site with associated storage canisters are subjected (long-term) to a mild atmospheric environment. Twenty or more years of ISFSI operational experience has verified that no significant corrosion issues generally exist during storage. However, note whether or not the site or potential site is a coastal marine location. Additional corrosion prevention measures may be applied when the ISFSI is located in a coastal marine environment. Detailed review guidance is provided in 8.4.6 Coastal Marine ISFSI Sites–Material Selections.

(LOW Priority) Underground structures require additional consideration due to soil corrosion issues. Additional guidance is provided in 8.4.14.3 Omission of Reinforcement.

(LOW Priority) Fuel loading/unloading conditions assume a borated, demineralized water environment at temperatures up to the boiling point. Experience with the conventional stainless steel and aluminum construction canister internals have verified no significant corrosion of fuel canister ITS components occur during the limited duration of a fuel loading/unloading operation. Pool water is buffered to a pH of about 8.5 to limit corrosion.

8.4.4   Drawings (MEDIUM Priority)

Licensing drawings usually appear in SAR Chapters 1 or 2. Examine the drawings and drawing notes for material specifications and alternatives. Ensure any materials substitutes are adequately specified, either on the drawing or in the SAR. ITS component material substitutes must appear in the TS.

8.4.5   Material Properties (MEDIUM Priority)

8.4.5.1   Structural Properties

The intent of this portion of the materials evaluation is to determine the acceptability of all material properties that have a structural role in confinement system structures and other structures important to safety (e.g., the basket, impact limiters, and shielding) and non-safety. The material properties and characteristics need to be applicable over the term requested in the CoC application. The reviewer should analyze the potential for corrosion and ensure that the applicant established and used appropriate corrosion allowances for the structural analyses. The range of some materials components properties may have to be evaluated over the range of life cycle conditions experienced during cask fabrication, loading, emplacement, storage, transfer, retrieval, unloading, and decontamination.

The information provided on structural materials must be consistent with the application of accepted design criteria, codes, standards, and specifications selected for the storage cask system and as described in this chapter and Chapter 3, "Structural Evaluation" of this SRP. Materials and material properties used for the design and construction of these safety-related structures should comply with the applicable codes and standards identified in Section 3.5.2.2 (i). For example, if the applicant elects to use design criteria from Section III of the ASME B&PV Code, the materials selected for the cask must be consistent with those allowed by the ASME Code subsection related to design. Acceptable requirements include the ASME adopted specifications given in Section II, Part A, "Ferrous Metals;" Part B, "Nonferrous Metals;" Part C, "Welding Rods, Electrodes, and Filler Metals;" and Part D, "Properties." The review of structural materials should be coordinated with the structural discipline.

A list of all materials used and the proposed service conditions for those materials during loading, storage, and unloading is a useful aid during the review. These tables provide various types of information that the reviewer needs from an application to aid in determining the suitability of the materials for the structural evaluation. The tables include the name and safety classification of each component part of the DSS and, where applicable, the function, the material specification(s) to which it is produced, and the nominal values for structural parameters. The tabulation should include all materials used for components with an important-to-safety function (e.g., confinement, transfer, criticality control, shielding). Information in this table can aid the reviewer to formulate the types of performance-related questions that are important for each component of a storage system.

The SAR documentation should fully define the structural materials used for components important to safety. The reviewer may find it useful to tabulate the major structural materials to facilitate the review. The following information could be tabulated: specification number, grade, type, and class of the material, nominal composition, product form, yield strength, tensile strength, and notes about the materials, etc. The SAR should identify properties related to structural performance and resistance or response to thermal, radiation, or other applicable environments that may impact structural performance. The structural and material disciplines should coordinate their reviews as appropriate for these components.

The completeness, accuracy, and acceptability of the identification and stated properties of the safety-related materials should be reviewed. In reviewing the structural materials, the reviewer should consider the sources of information; properties used in the structural evaluation and suitability for term requested in the CoC application. The reviewer should verify that the SAR clearly references acceptable sources of all material properties.

Examine the SAR adopted material properties for ITS component materials and ensure ASME Section II, Part D, properties and stresses are employed. The longstanding staff position (developed by NRR) regarding material properties is that ASME Code values must be used. Use of certified material test report (CMTR) values of UTS, yield, etc., is generally not permissible. Use of CMTR values is at risk of being non-conservative because samples may be taken at a portion of the ingot, billet, or forging that have optimum materials properties during certification.

8.4.5.2   Thermal Materials

The materials reviewer should coordinate with the thermal reviewer to determine the materials properties of the materials important to the thermal analysis. The material compositions and thermal properties such as thermal conductivity, thermal expansion, specific heat, and heat capacity should be verified as a function of the temperature over the range the components are to operate, for all components used in the safety analysis. Verify the change in these material properties due to potential degradation of materials over their service life has been evaluated by the applicant. Temperature and anisotropic dependencies of thermal properties should be considered.

8.4.6   Coastal Marine ISFSI Sites–Material Selections (MEDIUM Priority)

At coastal marine locations, the heavy salt drift can significantly accelerate the normally slight atmospheric corrosion rate to unacceptable values of some canister storage module designs, such as those that employ carbon steel structural elements inside the canister storage module. Experience has shown ordinary grades of structural steel (such as A-36) withstand the nominally dry interior environment of the canister overpack very well over a 20 year operational period.

For such cases, the reviewer must verify that the corrosion allowance specified is adequate for the 20 to 40 year CoC period of the canister. Corrosion rates for carbon steel in air may be found in corrosion references such as Corrosion Engineering by Fontana and Greene,

Corrosion Data Survey by the National Association of Corrosion Engineers (NACE), Corrosion and Corrosion Control by Uhlig, and the publications of the NASA Kennedy Space Center Corrosion Technology Laboratory. For exposures to coastal marine atmospheres, the corrosion rate data from the Kennedy Space Center Corrosion Technology Laboratory appears to be bounding for any location in the continental United States.

To address the increased atmospheric corrosion rates found at coastal marine (salt water) sites, some applicants have specified the use of 0.20%, minimum, copper-bearing steels, or, "weathering steels" such as Cor-Ten. The Kennedy Space Flight Center has collected data which has demonstrated the benefit of copper-bearing and weathering steels for significantly reducing corrosion at coastal marine sites. Therefore, for coastal marine ISFSI sites, the use of copper-bearing steels (containing a minimum of 0.20 percent copper), or weathering steels, may be necessary. Such steels are covered by ASTM A-242 and A-588, and supplemental requirements to ASTM A-36, and/or other specifications.

Other corrosion control measures may be employed, provided adequate documentation is supplied to demonstrate efficacy.

Coatings may be specified to alleviate the coastal atmospheric corrosion issue. However, unless supporting data is available to demonstrate the predicted coating life, the coating must be periodically inspected and maintained.

8.4.7   Weld Design/Inspection (MEDIUM Priority)

8.4.7.1   Welding Codes–Background Discussion

The nationally recognized codes which have been used for spent fuel canister construction include:

- ASME B&PV Code, Section III, "Rules for Construction of Nuclear Facility Components," Division 1.

- AWS D1.1 (current edition), "Structural Welding Code-Steel."

- AWS D1.6 (current edition), "Structural Welding Code-Stainless Steel."

The ASME B&PV Code Section III contains the design requirements for nuclear systems at a commercial nuclear power plant. It contains sections governing the design of welded nuclear components in the plant.

AWS D1.1 is the structural welding code for carbon steel structures such as bridges and steel-framed buildings.

The NRC staff accepts the use of the ASME B&PV Code, Section III, as the preferred construction code for storage casks. Some older cask designs used the AWS D1.1 Code. Note, the various construction codes (e.g., ASME Sections I, III, or VIII, and AWS D1.1) differ from one another in their requirements for materials and welding procedures, because each code is specialized with a particular application in mind.

The ASME construction codes are supplemented by "supporting codes" which detail how special processes such as welding and nondestructive examination (NDE) are to be qualified and executed. ASME B&PV Code Section IX, "Welding and Brazing Qualifications" details the requirements for specifying and qualifying a welding procedure and for testing and qualifying welders. ASME B&PV Code Section V, "Nondestructive Examination," supports the various ASME construction codes by detailing the required qualifications for NDE examiners and the requirements and methods for performing the types of NDE specified by the various construction codes.

Standard welding and NDE symbols may be found in AWS A2.4 (latest edition), "Symbols for Welding, Brazing, and Nondestructive Testing," to aid interpretation of such symbols found on the drawings submitted with the SAR.

Technical specification items related to the welds and testing are discussed separately.

8.4.7.2   Weld Design and Testing

Verify that the canister confinement welds are full penetration welds. Inspection of these welds must follow the ASME Code requirements of full volumetric examination [radiographic testing (RT) or ultrasonic testing (UT)] and a surface examination [liquid penetrant testing (PT), for austenitic stainless steel canisters]. A hydrostatic or pneumatic test is also required by the Code.

Stainless steel fillet welds can only be inspected by PT. Volumetric inspection of fillet welds is not feasible.

Due to the relatively benign operating conditions in storage, imposition of specific weld filler metals, or use/prohibition of certain welding processes is not presently necessary. Sensitization of the stainless steel is not an issue. Hence, solution annealing is unnecessary.

A shop helium leakage test, using ANSI N14.5 testing standards, must be performed to demonstrate that the entire canister or cask confinement body is free of defects that could lead to a leakage rate greater than the allowable design basis leakage rate specified in the confinement analyses. The requirements for the helium leakage test should be specified in the CoC to meet the requirements of 10 CFR 72.236(j) and (l). For bolted closure casks the entire confinement boundary should be similarly helium leak tested and pressure tested. The confinement boundary should be tested at the fabrication shop, with only a leakage test performed on the bolted lid closure seals (including drain and vent port seals) tested in-field by the cask user. The lid-to-shell welds and vent ports should be fabricated and helium leakage tested in accordance with the guidance of Section 8.4.20, as applicable. The staff should note that only lid-to-shell welds are within the scope of leak testing exceptions specified in 8.4.20.

The entire confinement boundary should be pressure tested hydrostatically or pneumatically to 125 or 110 percent of the design pressure, respectively. The test pressure should be maintained for a minimum of 10 minutes prior to initiation of a visual examination for leakage, per the ASME Code.

Following the application of the test pressure for the required time, all joints, connections, and regions of high stress, such as regions around openings and thickness transition sections, should be visually examined for leakage. This visual examination shall be performed in accordance with ASME Code requirements and shall be performed at a pressure equal to or greater than the design pressure or three-fourths of the test pressure. This pressure test and visual examination applies to both the canister body constructed at a fabrication facility and the lid-to-shell welds fabricated and closed in the field by a Part 72 licensee.

If pressure testing is performed only in the field, the visual examination of the portions of the canister shell may be impractical due to its inaccessibility inside the transfer cask. The application should discuss the proposed operations and reasons for inaccessibility for visual examination. Due to the inability to perform the visual examination of inaccessible portions of the canister welds during the field ASME Code hydrostatic test, staff has accepted the results from the shop helium leakage test applied under ANSI-N14.5 standards. The exception and basis should be listed in the table of ASME code exceptions in the Certificate of Compliance (CoC).

After the canister is loaded and lids welded, the confinement welds are pressure tested and helium leakage rate tested as further detailed in section 8.4.20.

8.4.7.3   Lid Welds and Closure Welds

The staff should verify the cask design is in compliance with Section 8.9 of this SRP or as follows:

- This guidance only applies to canisters of all-welded construction, fabricated from austenitic stainless steel, and employing redundant welds for the confinement closure.

- The welded canister (i.e., the confinement boundary) must be leak tested in accordance with ANSI N14.5-1997, except as specified by this guidance. The exemption for leak testing only applies to the closure welds that are typically made in the field and all other welds should be leak tested.

- "Structures, systems, and components important to safety must be designed to withstand postulated accidents" (10 CFR 72.122(b)).

- Records documenting the lid welds shall comply with the provisions of 10 CFR Part 72.174, "Quality Assurance Records" or with NQA-1, "Quality Assurance Requirements for Nuclear Facility Applications," depending upon the standard in effect at the time of licensing.

- Activities related to inspection, evaluation, documentation of fabrication, and lid welding shall be performed in accordance with an NRC-approved quality assurance program as required in 10 CFR Part 72, Subpart G, "Quality Assurance."

A redundant sealing of the canister is required by 10 CFR 72.236(e). One of the redundant seals in a welded canister design will involve a structural weld. The structural lid weld joint will be a full or partial penetration groove weld.

Carbon and Alloy Steel Cask Designs

The reviewer should verify the applicant has considered all the closure lid weld material and technique improvements that accrued from previous DSS design and fabrication experience. For example, the reviewer should refer to the technical evaluation in NRC Confirmatory Action

Letter 97-7-001, 1998 (ADAMS ML 060620420). Some of the DSS improvements resulting from that action include:

- Shell plates made from low sulfur, calcium-treated, vacuum-degassed steel.

- Application of minimum 93°C (200°F) preheat.

- Use of low-hydrogen electrodes.

- Low carbon equivalent base metals and weld metals.

- Magnetic particle examination (MT) of the root pass.

- Maintenance of preheat as a postheat treatment for a minimum of one hour.

- Minimum of two-hour delay after postheat before performing final volumetric NDE.

UT examine the structural lid weld in accordance with ASME Section III, D1, NB method and acceptance criteria requirements
Progressive surface examinations, utilizing a PT or magnetic particle testing (MT), are permitted only if unusual design and loading conditions exist. In addition, a stress-reduction factor of 0.8 is imposed on the weld strength of the closure joint to account for imperfections or flaws that may have been missed by progressive surface examinations. The weld design should be approved by the NRC on a case-by-case basis.

8.4.7.4   Austenitic Stainless and Nickel-Base Alloy Steels Cask Design

NDE of the large structural lid-to-shell weld designs fabricated from austenitic materials may be volumetric UT or multi-pass PT examined. A multi-pass PT is defined as performing a PT inspection of every pre-calculated intermediate weld deposit depth (layer) between the root and final weld layers.

Use ASME Section III, Division 1, Subsection NB (Section III, D1, NB) requirements for UT and PT inspection method and acceptance criteria.

A multiple-pass PT examination may be utilized in lieu of UT inspection and is performed as follows: Note: Impose a stress reduction factor of 0.8 for weld strength.

1. Calculate the critical flaw size (depth) assuming a buried flaw. Postulate a full circumferential (360-degree) flaw. Use ASME Section XI, D1, IWB 3600 requirements for alternative flaw acceptance criteria. Use of J-integral or net section stress is acceptable.

2. Establish the maximum allowable intermediate weld deposit depth (layer)/required in-process PT inspection interval by using the critical flaw depth calculated in Step 1. Note: Lessons learned suggest that the critical flaw depth for many structural lid welds is 3/8-inch.

3. PT the root layer, every intermediate layer established in Step 2 and the final weld layer. It is assumed that the root layer is single pass. If the root layer is multi-pass, calculate the critical flaw depth (Step 1) to establish the maximum allowable intermediate weld deposit depth (layer)/required in-process PT interval. Assume a surface connected flaw when calculating the critical flaw depth for a multi-pass root layer.

The applicant's evaluation of the critical flaw size using the above methodology should be reviewed based on service temperature, dynamic fracture toughness and critical design stress parameters as specified in ASME Section XI, D1.

8.4.8   Galvanic/Corrosive Reactions (LOW Priority)

8.4.8.1   Environmental considerations

The reviewer can find operational issues associated with hydrogen generation and guidance for evaluating galvanic or corrosive reactions in NRC Bulletin 96-04 (1996). The reviewer should confirm the DSS will perform adequately under the operating environments expected (e.g., short-term loading/unloading or long-term storage) for the duration of the license period such that no adverse galvanic or corrosive reactions occur between the canister materials, fuel payload, and the operating environments.

8.4.8.2   Canister Contents

The staff has previously reviewed a number of non-fuel hardware components and materials for compliance with 10 CFR 72.120(d), meaning, compatibility with a canister interior composed of stainless steel and aluminum components. These components are various neutron source assemblies, burnable poison rod assemblies (BPRAs), thimble plug devices, and other types of control elements. The staff has found the following materials to be acceptable for storage when the canister is constructed of stainless steel with stainless steel and aluminum basket components:

Neutron source materials composed of stainless steel or zirconium alloy cladding containing: antimony-beryllium, americium-beryllium, plutonium-beryllium, polonium-beryllium, and californium. Exposure of these various contents to the wet loading and dry storage environment was assessed and found to be satisfactory.

Control elements composed of zircaloy or stainless steel cladding containing: boron carbide, borosilicate glass, silver-indium-cadmium alloy, or thorium oxide. Exposure of these various contents to the wet loading and dry storage environment was assessed and found to be satisfactory.

8.4.9   Creep Behavior of Aluminum Components (HIGH Priority)

Aluminum based metal matrix composites and aluminum / boron carbide laminates (e.g. Boral^{TM}) are employed for all presently utilized neutron poison materials. Also, aluminum components are frequently part of the spent fuel basket. More recent designs have specified ever higher design temperatures for the fuel basket components in order to accommodate higher loading densities and higher burn-up fuel. This trend has pushed the various aluminum components well into creep regime operating temperatures.

Review the design maximum temperatures and stress for any aluminum components and verify a creep analysis has been performed if any structural load bearing aluminum components operate at a design temperature above approximately 200°F.
  In the event temperatures exceed the ASME Section II nominal 400°F temperature limit for aluminum, other sources for creep data must be examined. One previously cited reference for this information is: D.W. Wilson, J.W. Freeman and H.R. Voorhees, Creep-Rupture Testing of Aluminum Alloys to 100,000 Hours, First Progress Report, prepared for the Metal Properties Council, New York, November 1969. The staff makes no judgment as to the acceptability of this reference. This is because the designs reviewed through the time of this writing have had design stresses (on the order of tens of PSI) which were substantially below the creep-rupture stresses provided in the referenced report. None-the-less, an assessment of creep deformation over a 20 to 40 year CoC period should be part of the design calculations.

Borated aluminum neutron poison materials must be considered on a case-by-case basis if they are subjected to structural load bearing beyond their own dead-weight loads. This is due to their inherently low ductility and generally unknown creep properties.

8.4.10   Bolt Applications (MEDIUM Priority)

If threaded fasteners are employed for ITS components, verify the bolt material(s) have adequate resistance to corrosion and brittle fracture and a coefficient of thermal expansion similar to the materials being bolted together.

8.4.11   Protective Coatings (LOW Priority)

Coatings in DSSs are used primarily as corrosion barriers or to facilitate decontamination. They may have additional roles, such as improving the heat rejection capability by increasing the emissivity of cask internal components. Protective coatings are occasionally specified for carbon steel components. Coatings are not ITS components. The structures or components that the coatings are applied to are generally ITS component. No coating should be credited for protecting the substrate material or extending the useful life of the substrate material unless a periodic coating inspection and maintenance program is required for the coating.

The staff has established this section to alleviate confusion regarding coatings on cask components. Coatings generally have low safety significance with the exception of coating issues that may result in adverse chemical or galvanic reactions. Typically, the detailed guidance in this section is not generally subject to further confirmation as part of the review. However, there may be instances in which unique or innovative coatings are specified by the applicant to perform a specific function unique to the cask system. In these instances, the reviewer may use discretion in implementing the detailed review guidance in this section. This section outlines methods and procedures for appropriately assessing coatings. Within the assessment several areas are covered in detail including the scope of the coating application, type of coating system, surface preparation methods, applicable coating repair techniques, and coatings qualification testing.

8.4.11.1   Review Guidance

The reviewer should determine the appropriateness of the coating(s) for the intended application by reviewing the coating specification for each protective coating that is applied to an important to safety component. A specification that describes the scope of the work, required materials, the coating's purpose, and key coating procedures, should ensure that the appropriate and compatible coatings have been selected by the DSS designers. A coating specification should include the following:

- Scope of coating application;
- Type of coating system;
- Surface preparation methods;
- Coating application method;
- Applicable coating repair techniques;
- Coatings qualification testing, as applicable.

8.4.11.2   Scope of Coating Application

The coating specification should identify the purpose of the coating, a list of the components to be coated, and a description of the expected environmental conditions (e.g., expected conditions during loading, unloading, and dry storage).

The reviewer should verify that the coatings will not react with the cask internal components and contents and will remain adherent and inert when exposed to the various environments of a SNF cask. The most prevalent, potentially degrading environments include the immersion in borated SNF pool water during loading and unloading operations, and high-temperature and high-radiation (including neutrons) environments encountered during vacuum drying evolutions and long-term storage.

8.4.11.3   Coating Selection

The reviewer should verify that the coating specification identifies the manufacturer's name, the type of primers and topcoat(s) comprising the coating system, and the minimum and maximum dry coating thickness(es). Due to the unique nature of coating properties, and coating application techniques, the manufacturer's literature may be the only source of information on the particular coating.

The reviewer should verify that the coating selected for cask components is capable of withstanding the intended service conditions over the design service life. Failures can be prevented by ensuring that the selection and the application of the coating are controlled by adhering to the coating manufacturer's recommendations.

8.4.11.4   Surface Preparation

The reviewer should verify that the coating specification identifies whether solvent or abrasive cleaning methods should be used to prepare surfaces prior to coating application. This information should ensure that proper surface preparation techniques can be implemented during cask fabrication.

The reviewer should confirm that the specified type and degree of surface cleaning and the required surface profile meet the coating manufacturer's specification. Any deviations from the manufacturer's standards for surface preparation must be supported by appropriate tests that demonstrate acceptable coating performance under all design conditions.

8.4.11.5   Coating Repairs

The reviewer should verify that the coating specification identifies the general requirements for repairing damage to the coating. This information will assist the reviewer in evaluating the effects of repairs on the integrity of the coating and whether the designated repair methods could be implemented during or after cask fabrication.

The reviewer should examine the design to determine whether the structure is assembled before or after its various parts are coated. If a complex structure is to be coated after assembly, it is very important that the consequences of a potential coating failure be analyzed to determine whether other cask functions or component features could be compromised by the failure.

The consequences of coating failure depend on the type of coating and service environment, and may include the following:

- Partial and/or complete coating failure that alters the corrosion resistance of DSS structural and shielding components (primarily during loading/unloading operations).

- Partial and/or complete coating failure that alters the emissivity and heat transfer of basket components.

- Particulates (cloudiness) that form in SNF pool water or cask during loading or unloading that may affect such operations.

- Aggressive or reactive chemical species that form and consequently impact the performance of other cask components during long-term exposure to radiation (e.g., gamma and neutron).

8.4.11.6   Coating Qualification Testing

Coatings used on cask external surfaces may have been selected upon the basis of their performance requirements and exposure conditions. The applicant may have used related industrial conditions as a documented guide or basis for coating selection without performing further laboratory tests.

Any coating (including paints or plating) used inside a DSS must have been tested to demonstrate the coatings performance under all conditions of loading and storage. The conditions evaluated should include exposure to radiation, high temperature during vacuum drying and storage, and immersion during loading, unloading and transfer operations. The coating must be demonstrated to remain intact and inert for the full duration of the DSS design life.

There are a number of standardized ASTM tests for coatings performance. In reviewing ASTM (or other) tests used to qualify coatings for service in storage casks, consideration should be given to the applicability of a test to the service conditions.

Planning, execution, and interpretation of coating qualification tests must be performed by a qualified coatings engineer (e.g., certified by the National Association of Corrosion Engineers). The reviewer should ensure that appropriate, qualified expertise has been employed by the applicant for any coatings qualification program.

The reviewer should verify that the coating specification includes a description of the coating qualifications testing program, as applicable. The following information, which is important to qualifying a coating, includes, but is not limited to:

-   The size and shape of samples used for the coating tests, as well as the type of material(s), and a description and results of any tests conducted on partial or full-size production mock-ups.

-   The test sample surface preparation method(s) and expected or measured surface profile. Sample surface preparation should be performed in accordance with written production procedures, using the same equipment, materials, and qualified personnel as intended for production coating. Inspection methods and acceptance criteria should be included.

-   Application method(s) and measured control parameters, including records of temperature and humidity, cure cycle and times, and any other monitoring or acceptance tests such as dry film thickness, hardness, and adhesion. The methods and parameters should be employed in accordance with written production procedures using the same equipment, methods, materials, and qualified personnel.

-   A test plan description which clearly describes the rationale for and the types and sequences of all coating qualification tests, lab protocols, numbers of samples, inspection methods, and acceptance criteria. Raw test results should be tabulated or otherwise presented. The test plan should include: (1) laboratory coupons for demonstrating coating suitability/qualification; and (2) partial or full size production mock-up tests that demonstrate that the selected coating can be applied successfully to real production parts under production shop conditions to give reasonable assurance that field performance will meet laboratory, test-based expectations.

- An interpretation and discussion of the test program results by a certified coatings engineer. This evaluation should examine, at a minimum, the coating performance against the specific tests and the overall requirements for coating performance. The overall program must be assessed as to whether it is likely to be an effective predictor of actual performance. A recommendation for the use of the coating, with specific restrictions, if any, must be included.

The application should also include general requirements applying to all tests:

- Test durations for immersion must equal or exceed the combined maximum design (or technical specification) durations for loading and vacuum drying.

- An evaluation of any observed gasses, bubbles or other evidence that a gas was produced during the test. Coatings that produce flammable gas require a mitigation program to prevent burnable or explosive gas concentrations during all phases of cask operations.

8.4.12   Neutron Shielding (MEDIUM Priority)

8.4.12.1 Neutron Shielding Materials

Concrete, steel, uranium, and lead typically serve as gamma shields. Boron-filled polymers are sometimes used for neutron shielding materials (as opposed to neutron poisons used to control criticality). Although dose limits are calculated at the site boundary, not the canister surface, these materials are considered ITS, in order to meet the regulatory requirements of 72.126(a)(6). NUREG/CR-6407 specifically designates neutron shielding materials as ITS Category B.

References for all materials used, including nonstandard materials (e.g., proprietary neutron shield material), should be provided for the source of the material composition and density data along with validation of the data. The SAR should also describe the geometry of the shielding materials.

In-service performance monitoring of these materials is performed during the required periodic radiation surveys. Should a decline in the shielding effectiveness be detected, there is ample time and opportunity for engineering evaluation and corrective action. Therefore, the qualification and acceptance testing of neutron shielding materials should not be required in the TS.

The SAR should describe the composition of shielding materials and geometries. References for all materials used, including nonstandard materials (e.g., proprietary neutron shield material), should be provided for the source of the material composition and density data along with validation of the data.

8.4.12.2   Assessing Previously Unreviewed (New) Neutron Shielding Materials

Should a new material be introduced, review may proceed as follows:

The reviewer should confirm that temperature-sensitive (e.g., polymeric) neutron shielding materials will not be subject to temperatures at or above their design limits during normal conditions. The reviewer should determine whether the applicant properly examined the potential for shielding material to experience changes in material densities at temperature extremes. For example, elevated temperatures may reduce hydrogen content through loss of water in concrete or other hydrogenous shielding materials.

With respect to polymeric neutron shields, the reviewer should verify that the application:

- Describes the test(s) demonstrating the neutron-absorbing ability of the shield material.

- Describes the testing program and provides data and evaluations that demonstrate the thermal stability of the resin over its design life while at the upper end of the design temperature range.

- Describes the nature of any temperature-induced degradation and its effect(s) on neutron shield performance.

- Describes what provisions exist in the neutron shield design to assure that excessive neutron streaming will not occur as a result of shrinkage under conditions of extreme cold. This description is required because polymers generally have a relatively large coefficient of thermal expansion when compared to metals.

- Describes any changes or substitutions made to the shield material formulation. For such changes, describes how they were tested and how that data correlated with the original test data regarding neutron absorption, thermal stability, and handling properties during mixing and pouring or casting.

- Describes the acceptance tests conducted to verify any filled channels used on production casks did not have significant voids or defects that could lead to greater than calculated dose rates.

* Describe the materials ability to withstand the combined aging effects of heat and radiation field.

The potential for shielding material to experience changes in material properties at temperature extremes should be described in the SAR. Temperature sensitivities of shielding materials should be referenced. The SAR should also address degradation from aging, accumulated radiation exposure, and manufacturing tolerances. Twenty years of operational experience has not resulted in any noticeable decline in the performance of previously accepted materials, as verified by examination of periodic radiation survey results on the ISFSI pads at Surry and Robinson sites.

8.4.13   Criticality Control (HIGH Priority)

U.S. Nuclear Regulatory Commission (NRC) staff reviewer should read 72.104(a), 72.106(b), 72.124, and 72.236(g).

Qualification testing is conducted to ensure that (1) the material used will have sufficient durability for the application for which it has been designed, (2) the physical characteristics of the components of the absorber materials will meet the design requirements, and (3) the uniformity of the distribution of ^{10}B is sufficient to meet the requirements of the applications for which the absorber materials will be used. Materials that have passed the qualification tests must be acceptance tested (See Chapter 10 of this SRP) for use in systems to be used in storage or transportation of nuclear fuel.

8.4.13.1 Neutron-Absorbing/Poison Materials

Various boron containing materials are used in the nuclear industry as neutron absorbers. Since these materials are used in storage containers for fissile materials, the materials should have excellent physical and chemical stability, including a high resistance to radiation and corrosion. Further, these materials should experience no reduction in effectiveness under normal/off-normal and accident-level conditions of storage. Neutron absorbers can consist of alloys of boron compounds with aluminum or steel in the form of sheets, plates, rods, liners, and pellets. Likewise, neutron absorbers can consist of a core containing mixed aluminum and boron carbide (B_4C) particles, clad on both sides with aluminum (a composite).

The neutron absorber material must be demonstrated to be adequately durable for the service conditions of the application. These assurances are usually obtained during qualification testing of the material. In addition, acceptance tests (see Chapter 10 of this SRP) are performed on samples from each production run of the material. This procedure will ensure the properties for the plates or other shapes produced are in compliance with the specifications and requirements of the application. The uniformity of the distribution of ^{10}B may be addressed in both the qualification and the acceptance tests.

For all boron-containing absorber materials, the reviewer should verify the SAR, with its supporting documentation, describes the absorber material's chemical composition, physical and mechanical properties, fabrication process, and minimum poison content. The manufacturer's data sheet should be submitted to supplement the above information. In the case of absorber plates or sheets, the minimum poison content should be specified as an areal density (e.g., milligrams of ^{10}B per cm^2).

For all boron-containing absorber materials, the reviewer should verify that the SAR, with its supporting documentation, describes the absorber material's chemical composition, physical and mechanical properties, fabrication process, and minimum poison content. If the applicant intends to uses an absorber material with a specific trade name, the manufacturer's data sheet should be submitted to supplement the above information. In the case of absorber plates or sheets, the minimum poison content should be specified as an areal density (e.g., milligrams of ^{10}B per cm^2).

8.4.13.2 Computation of Percent Credit for Boron-Based Neutron Absorbers

This section illustrates one method used by the materials reviewers to compute the level of credit to be allowed for 1/v neutron absorber materials, such as boron or lithium, in the criticality safety analysis of packages for storing fissile materials, including fresh and SNF. The computation of the allowed level of credit uses the results of neutron attenuation measurements performed on samples of the absorber material placed in a beam of thermal neutrons.

Where such validation uncertainties exist, an upper limit of 90 percent credit is applied to boron-based solid absorbers, meaning that the material is computationally modeled as containing only 90 percent of the ^{10}B shown to be present. The staff has concluded that limiting the poison credit to 90 percent adequately accounts for the uncertainties arising in extrapolating the validation for boron-based absorber materials. Other remedies, beyond the scope of this guidance, may be necessary in addressing the potentially more complex neutron-spectral effects and validation uncertainties encountered with materials based on non-1/v-absorbers such as cadmium or gadolinium. The current guidance applies only to 1/v absorbers such as boron or lithium.

Neutron channeling has been shown to occur in a commercial product that uses coarse particles of natural B_4C dispersed in an aluminum matrix. For one material, neutron channeling effects reduced the measured attenuation of thermal neutrons by about 18 percent. Therefore, whenever uncertainty due to these materials factors exists in a product, it may be necessary to measure the neutron attenuation for that product to assess the expected material performance in service. Thus, in addition to the 90-percent limit on poison credit that is used to offset validation uncertainties, an additional penalty must be considered for material heterogeneity effects and uncertainties. In the absence of a fully documented understanding of non-uniformities and channeling effects in a heterogeneous absorber material, the staff recommends that the poison credit should continue to be limited to 75 percent.

A neutron absorber material is formulated to meet or exceed the neutron absorption effect computed to be required for a given service application. This guidance can be used to extend the range of credit for heterogeneous absorber materials from 75 to 90 percent, as follows:

- Material for which data is presented to show the measured attenuation for thermal neutrons to be at or above the acceptance attenuation (A_a), is given the full credit of 90 percent.

- Material for which data is presented to show the measured attenuation for thermal neutrons to be at levels between 75 and 100 percent of the acceptance attenuation (A_a) is given a fraction of the 90 percent credit allowed for fully effective absorber material.

- Material for which data is presented to show the measured attenuation for thermal neutrons to be at or below 75 percent of the acceptance attenuation (A_a) is not approved for use at any level of credit; the process used to produce such material is judged to be unsuitable.

The sampling, testing, and reporting of results shall be conducted according to the specifications given in ASTM standard C1671-7.

The applicable credit can be calculated by the following method. Using the following definitions:

A = neutron attenuation, a measured value taken on a given absorber material in a beam of thermal neutrons with fixed energy spectrum. A is assumed to be normally distributed with mean μ and standard deviation σ.

A_a = A_a = acceptance value of neutron attenuation, based on a qualified homogeneous absorber standard such as ZrB_2, or a heterogeneous calibration standard that is traceable to nationally recognized standards, or calibrated with a monoenergetic neutron beam to the known cross section of boron-10. Calibration standards should be evaluated at 111 percent (i.e., 1/0.90) of the poison density assumed in the criticality computational model.

A_{tl} = attenuation tolerance limit, a statistic of the data

n = number of coupon measures of attenuation

P = probability

μ = true mean of A

x bar = estimate of μ

σ = true standard deviation of A

S = estimate of σ

C_p = exact number of standard deviations required at probability P

K_p = tolerance coefficient that is substituted for C_p when μ and σ are estimated by x bar and S, respectively

γ = confidence level

The attenuation data can be used to bound the probability P that the value of neutron attenuation A at an arbitrary location on the material is greater than the acceptance attenuation A_a. This is done by computing an attenuation tolerance limit, A_{tl}, such that, with 95-percent confidence, the probability is less than 0.05 that A < A_{tl}.

Let P = 0.95 and γ = 0.95. Compute A_{tl} = (X bar - K_p S), where K_p = f(P, n, γ). The value of K_P may be found in a table of one-sided tolerance coefficients for a normal distribution.

If $A_{tl} \geq A_a$, then 90 percent credit is given.

If $A_{tl} < A_a$, then compute the fractional credit from 0.75 to 0.90 as follows:

Fractional Credit = 0.30 + 0.6(A_{tl} / A_a).

If the computed fractional credit is less than 0.75, the process is regarded as unsuitable and should be given no credit.

8.4.13.3   Qualifying the Neutron Absorber Material Fabrication Process

Not including neutron attenuation, in past reviews the staff has accepted the following qualification testing:

1)   Mechanical testing to ensure that the neutron poison material is structurally sound, even if the absorber is not used for structural purposes.

In the past, the staff has accepted ASTM B 557 – 06 tensile testing of samples which demonstrated:

- 0.2% offset yield strength no less than 1.5 ksi
- ultimate strength no less than 5.0 ksi
- elongation no less than 1%

Alternatively, the staff has accepted ASTM E 290 – 97a bend tests, with a 90° bend without failure as the passing criteria.

2) Porosity measurements to ensure that the corrosion resistance (which is directly linked to hydrogen generation in the spent fuel pool) of the neutron poison material is maintained, and that the general structural characteristics of the material are controlled.

   The methodology for porosity is up to the discretion of the applicant. Limits on both the total porosity of the material, and the "open" or "interconnected" porosity of the material should be explicitly stated in the Technical Specifications. Excluding Boral™, the total open porosity of the neutron poison material should be limited to 0.5 volume percent or less.

3) In general the conditions of spent fuel loading, unloading, and storage do not require qualification testing to demonstrate resistance to thermal, radiation, or corrosion induced degradation if the neutron absorber is only made of boron carbide and an aluminum alloy meeting ASTM chemical requirements for the 1000 or 6000 series of aluminum. Other aluminum alloys (particularly those which are not heat-treatable) may also be acceptable to the staff without qualification testing. Porosity measurements on the neutron poison material should not be waived, regardless of the aluminum alloy used in the neutron absorber, however.

4) A sufficient number of samples should be used to measure the thermal conductivity of the neutron poison material at room and elevated temperature. Reviewers should be aware that clad neutron poison materials are thermally anisotropic.

5) For clad materials, a test demonstrating resistance to blistering during the drying process should be included in the qualifying tests. In the past the staff has accepted testing where:

   Samples of clad materials are soaked in either pure or borated water for 24 hours and then insertion into a preheated oven at approximately 825°F for a minimum of 24 hours. The samples are then visually inspected for blistering and delamination before undergoing qualifying mechanical testing.

Significant, additional qualifying tests should be conducted for structural neutron poisons. Mechanical and thermal tests should include, tensile testing, impact testing (or K_{IC} measurements), creep testing, and (if applicable) mechanical testing of weldments.

Samples of neutron poison material should also be examined [i.e., the use of transmission electron microscopy (TEM) or scanning electron microscopy (SEM)] for the following changes:

- Redistribution or loss of boron.

- Dimensional changes (material instability).

- Cracking, spalling, or debonding of the matrix from the boron-containing particles.

- Weight changes caused by leaching, dissolution, corrosion, wear, or off-gassing.

- Embrittlement.

- Chemical changes such as oxidation or hydriding.

- Molecular decomposition of the material as a result of radiation (radiolysis).

Coupons should be taken so as to be representative of the neutron poison material. To the extent practical, test locations on coupons should be stratified to minimize errors due to location or position within the coupon. Some suggested locations should include the ends, corners, centers, and irregular locations. These locations represent the most likely areas to contain variances in thickness. Adequate numbers of samples should be taken from components (i.e., plate, rod, etc.) produced from a lot to obtain a good representation. A lot is defined as all plates from a single billet. Overall, the coupons should be a representative sample of the material.

For containers that will be loaded or unloaded in a SNF pool or similar environment, the reviewer should verify the absorber material has been evaluated or tested for environmental and galvanic interactions and the generation of hydrogen in the pool environment. If environmental testing is employed, the test conditions (time, temperature) should equal or exceed those expected for loading, unloading, and transfer operations. For environmental tests, the absorber materials should be coupled to dissimilar metals, as may be appropriate to the application. The environment may be borated or deionized water, as appropriate. The evaluation should also consider the effects of any residual pool water remaining in the container after removal from the pool.

Generally, for common engineering materials, an evaluation based upon consultation of a corrosion reference (galvanic series) should suffice for pool loading/unloading situations.

The reviewer should note the applicant must take appropriate measures to assess the strength or ductility of the material, depending on the structural requirements of the application.

Acceptance testing of the fabricated materials is discussed in Chapter 10, "Acceptance Tests and Maintenance Program Evaluation," of this SRP.

8.4.14   Concrete and Reinforcing Steel (LOW Priority)

8.4.14.1   Embedment Materials

The materials discipline should review the material to be used for embedments, inserts, conduits, pipes, or other items embedded in the concrete. Embedments must satisfy the requirements of the code used in designing the reinforced concrete structure in which they are embedded (e.g., ACI 359, ACI 349, or ACI 318). Zinc, zinc rich coatings, zinc-clad materials, and aluminum should not be used for any embedded objects in structures designed to ACI 349 or ACI 359 that will be in contact with wet concrete, because of the potential for concrete degradation from an adverse chemical reaction. Embedments and attachments are considered to include components cast or grouted into the reinforced concrete structure, inserts, embedded pipes, conduits, or lightning protection and grounding systems.

Unless otherwise specified in this SRP, steel structural attachments must comply with the appropriate requirements of ACI-349.

8.4.14.2   Concrete Temperature Limits

The NRC accepts the use of ACI 318 for the design and material specifications for reinforced concrete structures subject to NRC approval, but they are not important to safety. If ACI 349 is used for design of such structures, the NRC accepts the use of ACI 318 for construction. The NRC also accepts the following criteria as an alternative to the temperature requirements of ACI 349 Section A.4, but only for the specified use and temperature ranges:

1.   Concrete temperatures in general or local areas are a maximum of 93°C (200°F) in normal or off-normal conditions and/or occurrences, no tests are needed to prove capability for elevated temperatures or reduced concrete strength.

2.   If concrete temperatures in general or local areas exceed 93°C (200°F) but are less than 149°C (300°F), no tests are required to prove capability for elevated temperatures or reduced concrete strength if Type II cement is used and temperature appropriate aggregates are used. The following criteria for fine and coarse aggregates are acceptable:

-    Satisfy ASTM C33 requirements and requirements references in ACI 349 for aggregates, and

-    Have a demonstrated coefficient of thermal expansion (tangent in temperature range of 20-38°C (70-100°F) no greater than 11×10^{-6} mm/mm/°C (6×10^{-6} in./in./°F), or be one of the following materials: limestone, dolomite, marble, basalt, granite, gabbro, or rhyolite.

•   If concrete temperatures in general or local areas under normal or off-normal conditions do not exceed 107°C (225°F), the requirements of 1 and 2 (above) apply to the coarse aggregate. Fine aggregate that meets 1 (above) and is also composed of quartz sands or sandstone sands may be used in place of 2 (above) and be in compliance.

8.4.14.3   Omission of Reinforcement

Frequently, designers specify the omission of reinforcing steel ("rebar") in concrete above-ground structures which have the purpose of gamma shielding only. This is acceptable since it is to avoid the inadvertent formation of voids in the concrete due to the presence of the rebar, which can act to block the aggregate in the concrete from filling all intended areas.

Concrete applied around buried steel structures should be reinforced to alleviate shrinkage crack propagation. Concrete alleviates soil corrosion by creating a beneficial chemical buffering effect (high pH) around the steel. Cracks allow groundwater plus electrolyte intrusion which reduces the effectiveness of the concrete protective barrier.

8.4.15   Seals

Applicants for spent fuel storage canisters with metallic seals generally rely on seal manufacturer's data to determine the maximum service temperatures for seals. Seals that may potentially be exposed to high temperature may not have been tested by independent laboratories (such as NIST and Factory Mutual). Due to the importance of the integrity of the seals, laboratory test results or data sheets that reference independent test results should be included in applications, if available.

8.4.15.1 Metallic Seals (MEDIUM Priority)

Bolted lid canisters employ redundant metallic seals as part of the confinement boundary. These seals are ITS components. The primary materials issue is the temperature resistance of the seal spring material. Generally this is a nickel-base alloy with excellent temperature and creep resistance. The seal cover material may be soft aluminum or silver. Aluminum faced seals have failed in service due to corrosion from inadvertent rainwater intrusion. Substitution of silver alloy faced seals appears to have alleviated the susceptibility of mechanical seals to this corrosion-induced failure mechanism.

8.4.15.2 Elastomeric Seals (LOW Priority)

Bolted lid canister designs may also employ a weather cover to preclude rainwater from the confinement boundary seals. These weather covers may be sealed against the weather with an elastomeric seal such as Viton. As such, these seals may be susceptible to thermally and radiation induced aging (hardening). Consequently, a replacement program may be warranted if the heat or radiation exposure is sufficient. Guidance as to radiation or thermal resistance is usually obtainable from the seal manufacturer. Elastomeric seals have never been ITS components in storage canisters.

Radiation will generally cause polymerization of elastomers to an extent that would adversely affect the performance when the dose reaches 10^5 Gy (10^7 rads). For higher dose rate environments, elastomer O-rings should not be specified. The use of fluorcarbons, which are known to be particularly susceptible to radiation damage, should be restricted if the expected dose exceeds 100 Gy (10^4 rads).

The reviewer should verify O-ring seals do not reach their maximum operating temperature limit during normal and off-normal conditions of storage. The O-ring manufacturer's data sheets specifying temperature and radiation tolerances should be included in the SAR.

The materials discipline should review the applicant's evaluation demonstrating the minimum normal operating temperature (usually -40°F) will neither fail the O-ring seal by brittle fracture nor stiffen the O-ring (loose elasticity) to an extent that prevents the seal from meeting its service requirements.

The reviewer should verify that under the environmental conditions expected in storage service, O-ring seals will not chemically react or decompose in a manner that would significantly affect other components of the DSS.

8.4.16 Low Temperature Ductility and Fracture Control of Ferritic Steels (MEDIUM Priority)

Regulatory Guides 7.11 and 7.12 specify acceptable ferritic steels for low temperature service where good toughness is required. Metals having a face-centered cubic crystal structure such as austenitic stainless steels remain tough and ductile to very low temperatures and are not a concern in this regard. Toughness testing (e.g., Charpy impact) of welds is governed by ASME Section III, as supported by Section IX.

For designs that specify ferritic steels other than those listed in Reg. Guides 7.11 and 7.12, the Reg. Guide specifies the types of tests and data needed to qualify a material. Those tests and data include dynamic fracture toughness and nil-ductility or fracture appearance transition temperature test data. Toughness testing (e.g., Charpy impact) of welds is governed by ASME Section III, as supported by Section IX.

8.4.17   Cladding

(MEDIUM Priority) This guidance will allow all commercial spent fuel that is currently licensed by the Nuclear Regulatory Commission (NRC) for commercial power plant operations to be stored in accordance with the regulations contained in 10 CFR Part 72. However, cask vendors' requests for the storage of spent fuel with burnup levels in excess of those levels licensed by the Office of Nuclear Reactor Regulation (NRR), or for cladding materials not licensed by NRR, may require additional justifications by the applicant.

The most important issues regarding spent fuel and cladding that must be considered are:

- The maximum cladding temperature during loading/unloading operations and normal conditions of storage. For high burn-up fuel, defined as any fuel with a burn-up greater than 45GWd/MTU, the maximum allowable cladding temperature limit is 400°C. For materials analyses, an appropriate maximum fuel burn-up is to be specified as the peak rod average.

- Compatibility of fuel bundle materials and non-fuel component materials such as burnable poison rod assemblies (BPRAs) with the loading/unloading environment and the cask interior components. Refer to the separate discussion of this in Section 8.4.8.1.

- The fuel is maintained in a water or inert environment during loading/unloading operations to prevent excessive oxidation of fuel pellets. This is discussed in more detail in Section 8.7 of this SRP.

- A definition of damaged fuel is adequate for the intended fuel load and fuel with more severe damage (if any) is precluded from loading.

8.4.17.1   Cladding Temperature Limits (MEDIUM Priority)

The requirements of 10 CFR 72.122(h)(1) seek to ensure safe fuel storage and handling and to minimize post-operational safety problems with respect to the removal of the fuel from storage. In accordance with this regulation, the spent fuel cladding must be protected during storage against degradation that leads to gross rupture of the fuel and must be otherwise confined such that degradation of the fuel during storage will not pose operational problems with respect to its removal from storage. Additionally, 10 CFR 72.122(l) and 72.236(m) require that the storage system be designed to allow ready retrieval of the spent fuel from the storage system for further processing or disposal.

Spent fuel storage casks and systems must be designed to meet four safety objectives:

- Ensure doses from the spent fuel in the casks and systems are less than limits prescribed in the regulations.

- Maintain subcriticality under all credible conditions.

- Ensure there is adequate confinement and containment of the spent fuel under all credible conditions of storage.

- Allow the ready retrieval of the spent fuel from the storage systems.

The acceptance criteria below and review procedures are designed to provide reasonable assurance the spent fuel is maintained in the configuration analyzed in the storage SARs. These criteria are applicable to all commercial spent fuel burnup levels and cladding materials. In order to assure integrity of the cladding material, the following criteria should be met:

- For all fuel burnups (low and high), the maximum calculated fuel cladding temperature should not exceed 400°C (752°F) for normal conditions of storage and short-term loading operations (e.g., drying, backfilling with inert gas, and transfer of the cask to the storage pad). However, for low burnup fuel, a higher short-term temperature limit may be used, if the applicant can show by calculation the best estimate cladding hoop stress is equal to or less than 90 MPa (13,053 psi) for the temperature limit proposed.

- During loading operations, repeated thermal cycling (repeated heatup/cooldown cycles) may occur but should be limited to less than 10 cycles, where cladding temperature variations are more than 65°C (117°F) each.

- For off-normal and accident conditions, the maximum cladding temperature should not exceed 570°C (1058°F).

Given the conservatism used in calculating peak clad temperatures for low burnup fuel, the staff has reasonable assurance that storage cask systems which use the 570°C temperature limit for low burnup fuel loading operations will continue to perform as expected when the casks were originally certified. Therefore, there is no need to require the licensees of storage-only or dual-purpose cask systems to repackage spent fuel loaded using the 570°C temperature limit.

The maximum allowable temperature should be based upon the peak rod temperature, not the average rod temperature. By employing the peak rod temperature, only a small fraction of the rods will experience the temperature and stress conditions that could lead to the formation of radial hydrides during normal conditions of storage.

High burnup fuel (i.e., fuel with burnups generally exceeding 45 GWd/MTU) may have cladding walls that have become relatively thin from in-reactor formation of oxides or zirconium hydride. For design basis accidents, where the structural integrity of the cladding is evaluated, the applicant should specify the maximum cladding oxide thickness and the expected thickness of the hydride layer (or rim). Cladding stress calculations should use an effective cladding thickness that is reduced by those amounts. The reviewer should verify that the applicant has used a value of cladding oxide thickness that is justified by the use of oxide thickness measurements, computer codes validated using experimentally measured oxide thickness data, or other means that the staff finds appropriate. Note that oxidation may not be of a uniform thickness along the axial length of the fuel rods.

Since the hoop stress is dependent on the rod internal pressure, cladding geometry, and the temperature of the gases inside the rod, the staff will verify that the applicant has calculated the best estimate hoop stress corresponding to the rod internal pressure of the highest burnup fuel assemblies of the specific type of assembly.

The intent of the thermal cycling acceptance criteria is to prevent licensees from applying cask drying, loading and transfer operations that could inadvertently enhance an undesirable hydride reorientation to form radial hydrides. Accordingly, these criteria pertain only to periods of fuel loading and transfer operations of the casks to the storage pads.

In general, the materials reviewer should coordinate with the structural reviewer to assure the spent fuel is maintained in the configuration analyzed in the Safety Analysis Reports (SARs) in order to meet the objectives described above.

The materials reviewer should coordinate with the thermal reviewer to assure the temperature criteria stated above are met. If higher peak temperatures are proposed by the applicant, additional justification for the higher temperatures must be supplied.

This guidance will allow all commercial spent fuel that is currently licensed by the Nuclear Regulatory Commission (NRC) for commercial power plant operations to be stored in accordance with the regulations contained in 10 CFR Part 72. However, cask vendors' requests for the storage of spent fuel with burnup levels in excess of those levels licensed by the Office of Nuclear Reactor Regulation (NRR), or for cladding materials not licensed by NRR, may require additional justifications by the applicant.

Background justification for these temperature limits can be found in Sec 8.8 of this SRP.

8.4.17.2   Fuel Classification (HIGH Priority)

The staff should verify that the definitions below are used in the SAR, and where appropriate are also included in the CoC.

Spent Nuclear Fuel (SNF) - See 10 CFR Part 72.3 for definition. This term has been used in the nuclear industry, at different times, to mean the fuel pellets, the rod, or entire fuel assembly. Unless specifically modified, the term will refer to both the rods and fuel assembly.

Damaged SNF - Any fuel rod or fuel assembly that cannot fulfill its fuel-specific or system-related functions.

Undamaged SNF - SNF that can meet all fuel-specific and system-related functions. As shown in Figure 8-2, undamaged fuel may be breached. Fuel assembly classified as undamaged SNF may have "assembly defects."

Breached spent fuel rod - Spent fuel rod with cladding defects that permit the release of gas from the interior of the fuel rod. A breached spent fuel rod may also have cladding defects sufficient to permit the release of fuel particulate. A breach may be limited to a pinhole leak, hairline crack, or may be a gross breach.

Pinhole leaks or hairline cracks - Minor cladding defects that will not permit significant release of particulate matter from the spent fuel rod, and therefore present a minimal as low-as-is-reasonably-achievable concern, during fuel handling operations. (See discussion of gross defects for size concerns.)

Grossly breached spent fuel rod - A subset of breached rods. A breach in spent fuel cladding that is larger than a pinhole leak or a hairline crack. An acceptable examination for a gross breach is a visual examination that has the capability to determine the fuel pellet surface may be seen through the breached portion of the cladding. Alternatively, review of reactor operating records may provide evidence of the presence of heavy metal isotopes indicating that a fuel rod is grossly breached. (See discussion for size concerns.)

Intact SNF - Any fuel that can fulfill all fuel-specific and system-related functions and that is not breached. Note that all intact SNF is undamaged, but not all undamaged fuel is intact, since under most situations, breached spent fuel rods that are not grossly breached will be considered undamaged.

Can for Damaged Fuel - A metal enclosure that is sized to confine one damaged spent fuel assembly. A fuel can for damaged spent fuel with damaged spent-fuel assembly contents must satisfy fuel-specific and system-related functions for undamaged SNF required by the applicable regulations.

Assembly Defect - Any change in the physical as-built condition of the assembly with the exception of normal in-reactor changes such as elongation from irradiation growth or assembly bow. Examples of assembly defects: (a) missing rods; (b) broken or missing grids or grid straps (spacers); and (c) missing or broken grid springs, etc. An assembly with a defect is damaged only if it can't meet its fuel-specific and system-related functions required by the applicable regulations.

A fuel-specific regulation - a characteristic or performance requirement of the fuel specifically named in the applicable Code of Federal Regulations (CFR). These are regulations that specify capabilities that the spent nuclear fuel (SNF) must have. Examples include 10 CFR 72.122(h)(1) and 10 CFR 72.122(l).

A system-related regulation - a performance requirement placed on the fuel so that the storage system can meet its regulatory requirements. Examples include 10 CFR 72.122(h)(5) and 10 CFR 72.124(a)..

Previous definitions of damaged fuel have identified specific characteristics of the fuel that classify it as damaged, irrespective of whether the fuel is being stored or transported and independent of the design of the storage or transportation system. The current staff position is that damaged fuel is defined in terms of the characteristics needed to perform the fuel-specific and system-related functions. The materials properties, and possibly the physical condition, of a fuel rod or assembly can be altered during irradiation or storage. If this alteration is large enough to prevent the fuel or assembly from performing its fuel-specific or system-related functions during storage, then the fuel assembly is considered damaged.

To determine whether a fuel assembly is undamaged, the following should be stated in the SAR:

1)   The functions the applicant has imposed on the fuel rods and assembly by either fuel specific or system-related functions to meet a regulatory requirement for the designated phase (storage, transportation, or both);

2)   The mechanisms of change (alteration mechanisms) or the characteristics of the fuel that could potentially cause the fuel to fail to meet its fuel-specific or system-related functions;

3)   An acceptable analysis showing that the fuel with the designated characteristics will meet the fuel-specific and system-related functions when the mechanisms considered in item #2, above, are evaluated; and

4)   The physical characteristics of the fuel, based on item #3, above, that could cause the fuel or assembly to be classified as "damaged."

A "default" definition of damaged SNF, derived from ANSI N14.33-2005, is provided for those that do not want to perform the assessment outlined in item numbers 1 through 4 above. The default definition, however, may not take full advantage of the flexibility of the performance-based definition of damaged fuel provided in this guidance. This default definition may be more restrictive than necessary, depending on the design of the storage or transportation cask. For example, the default definition of damaged SNF indicates that SNF must be classified as damaged if an individual fuel rod is missing from an assembly. However, if an analysis shows that all fuel-specific and system-related functions will be met (e.g., subcriticality will be maintained, that the SNF assembly will be retrievable and that the structural properties of the assembly are not compromised by the missing rod) the assembly may be classified as undamaged. An alternative default definition of damaged Spent Nuclear Fuel (SNF) is: SNF assemblies must be classified as damaged if any one of the following conditions exist:

On removal of SNF selected for dry storage or transport from the spent fuel pool, any of the following apply:

- There is visible deformation of the rods in the SNF assembly. Note: This is not referring to the uniform bowing that occurs in the reactor. This refers to bowing that significantly opens up the lattice spacing.

- Individual fuel rods are missing from the assembly. Note: The assembly may be reclassified as intact if a dummy rod that displaces a volume equal to, or greater than, the original fuel rod, is placed in the empty rod location.

- The SNF assembly has missing, displaced, or damaged structural components such that either:

    a. Radiological and/or criticality safety is adversely affected (e.g., significantly changed rod pitch).

    b. The assembly cannot be handled by normal means (i.e., crane and grapple).

- Reactor operating records (or other records) indicate that the SNF assembly contains fuel rods with gross breaches.

- The SNF assembly is no longer in the form of an intact fuel bundle (e.g., consists of, or contains, debris such, as loose fuel pellets or rod segments).

Additional background and examples of defining damaged fuel can be found in Section 8.6 of this SRP.

8.4.17.3 Reflood Analysis (HIGH Priority)

The NRC accepts that the total stress on the cladding is maintained below the material's minimum yield stress. The total stress includes the thermal stress combined with the cladding hoop stress from internal rod pressure and the rod-gas plenum temperature. The analysis also should account for high burnup effects on the fuel (e.g., waterside corrosion, high internal rod pressure) and minimum manufacturing wall thickness. Other assembly components should also be examined in a similar manner. Engineering judgment, combined with relevant industry operational experience with unloading SNF from transportation and storage casks, may support the basis for limits on quench fluid temperature and flow rate. This review should be coordinated with the thermal reviewer.

8.4.18 Prevention of Oxidation Damage During Loading of Fuel (MEDIUM Priority)

The guidance in this section is only applicable to irradiated LWR fuel or other uranium oxide based fuel. The reviewer should make sure that the oxidation of other types of fuels during loading is evaluated. The information given in this section and Section 8.7 of this SRP may not be applicable to other fuel types. The characteristics of those fuel types must be considered when evaluating their analysis.

Once the fuel rods are placed inside of the storage cask and water is removed to a level that exposes any part of the rods to a gaseous atmosphere, reasonable assurance the spent fuel cladding will be protected against splitting due to fuel oxidation might occur must be demonstrated. If oxidation occurred, it may lead to loss of retrievability, or to a configuration not adequately analyzed for radiation dose rates or criticality safety. Further, the release of fuel fines or grain-sized powder into the inner cask environment from ruptured fuel may be a condition outside the licensing basis for the cask system. Three possible options exist to address the potential for and consequences of fuel oxidation:

1. Maintain the fuel rods in an appropriate environment such as Ar, N2, or He to prevent oxidation.

2. Assure there are not any cladding breaches (including hairline cracks and pinhole leaks) in the fuel pin sections that will be exposed to an oxidizing atmosphere. This can be done by a review of records (for example, sipping records) or 100 percent eddy current inspection of assemblies.

3. Determine the time-at-temperature profile of the rods while they are exposed to an oxidizing atmosphere and calculate the expected oxidation to determine if a gross breach would occur. The analysis should indicate the time required to incubate the splitting process will not be exceeded. Such an analysis would have to address expected differences in characteristics between the fuel to be loaded and the fuel tested to determine the basis for the analysis. Conversely, the maximum allowable temperature of the rods could be limited to the temperature that calculations show cladding splitting will not be expected to occur. Such evaluations must incorporate the effects of uncertainty in the data base. Calculation of the possibility of cladding splitting, is fraught with all the uncertainties discussed above. Lowering the maximum allowable temperature may impose an economic penalty by limiting the heat load in the cask. The

selection of the methodology used to address this issue is up to the applicant. The use of a non-oxidizing atmosphere in the fuel canister to prevent fuel oxidation is one method accepted by the staff to address the issue.

If Option 3 is chosen, the materials reviewer should coordinate with the thermal reviewer to determine that the operating procedures, technical specification, and associated licensing documentation, as submitted by the applicants, provide a supportable analysis of the potential for cladding splitting, should fuel rods be exposed to an oxidizing gaseous atmosphere. For fuel with burnup below ~45 GWd/MTU and Zircaloy cladding, the time-at-temperature (TT) curves developed to date (R.E. Einziger and R.V. Strain, "Oxidation of Spent Fuel at Between 250° and 360°C," EPRI Report NP-4524, 1986, for example) can be used to determine the allowable exposure duration to an oxidizing atmosphere if the fuel temperature is known, or conversely the maximum allowable temperature if the exposure time is known. For example, using Figure 3-9 of the above reference, at 360°C one would expect to incur splitting between 2 and 10 hours. On the other hand, if one expected to stay at temperature for 100 hours then the fuel temperature should be kept below 290°C.

Additional information on oxidation of damaged fuel can be found in Section 8.7 of this SRP. Please refer to this reference for additional detail and background.

8.4.19 Flammable Gas Generation (MEDIUM Priority)

The reviewer should assume the generation of hydrogen or other gases during wet loading/unloading operations occurs. Field experience has amply demonstrated that any canister design employing aluminum components as part of the fuel basket construction will have a propensity to generate hydrogen. Efforts to passivate the aluminum components have proven inadequate to eliminate the generation of hydrogen. The use of zinc, zinc-rich coatings, or zinc-clad materials (e.g., galvanized steel) in particular, is known to generate potentially large quantities of hydrogen gas during wet-loading in SFP.

Consequently, the reviewer should verify the operating procedures contain adequate guidance for detecting the presence of hydrogen and preventing the ignition of combustible gases during cask loading and unloading operations. These procedures must be incorporated by reference into the TS.

8.4.20   Canister Closure Welds Testing (MEDIUM Priority)

Helium leakage testing of the entire confinement boundary is performed to demonstrate compatibility with the design basis leak rate, and ensures that:

- the fuel payload is protected from the deleterious oxidizing effects of moisture by excluding intrusion of such,

- the helium inerting gas will remain in the canister in sufficient amount over the license period, and

- the helium gas heat transfer medium will remain in sufficient quantity over the license period to assure the cladding temperatures are controlled at safe levels.

This guidance addresses all welds associated with the redundant closures of a spent fuel canister and describes how each individual closure weld must be considered from the overall design and testing standpoint. It only applies to canisters of all-welded construction, fabricated from austenitic stainless steel, employing redundant welds for the confinement closure.

The staff should verify that the cask design under review is in compliance with the guidance of this document. In order for any closure weld to be exempt from the helium leak testing to demonstrate compliance with 10 CFR 72.236, the staff should verify all of the following conditions are satisfied:

- The welded canister (i.e., the confinement boundary) must be leak tested in accordance with ANSI N14.5-1997, except as specified by this guidance.

- Closure welds must conform with the guidance of this SRP, as appropriate.

- "Structures, systems, and components important to safety must be designed to withstand postulated accidents." [10 CFR 72.122(b)(1)].

- Records documenting the lid welds shall comply with the provisions of 10 CFR Part 72.174, "Quality Assurance Records." Records storage should comply with ANSI N45.2.9, "Requirements for Collection, Storage, and Maintenance of Quality Assurance Records for Nuclear Power Plants."

- Activities related to inspection, evaluation, documentation of fabrication, and lid welding shall be performed in accordance with an NRC-approved quality assurance program as required in 10 CFR Part 72, Subpart G, "Quality Assurance."

In addition for exemption of large multi-pass welds from helium leak testing the following must be satisfied.

(1) The weld must be multi-pass, with a minimum weld depth comprised of at least 3 distinct weld layers.

(2) Each layer of weld may be composed of one or more adjacent weld beads.

(3) The layer must be complete across the width of the weld joint.

(4) If only 3 weld layers comprise the full thickness of the weld, each layer must be PT examined.

(5) For more than 3 weld layers, not all weld layers need be PT examined. The maximum weld deposit depth allowed before a PT examination is necessary is based upon flaw-tolerance calculations in accordance with Section 8.9 of this SRP. Note: This criterion does not supersede the flaw acceptance criteria of any construction code. Instead, this criterion is used to establish the maximum allowable weld deposit depth before an in-process PT examination is necessary.

(6) Regardless of conditions (4) or (5) above, at least 3 different weld layers must be examined, e.g., the root pass, a mid-layer, and the cover pass.

(7) The weld cannot have been executed under conditions where the root pass might have been subjected to pressurization from the helium fill in the canister

itself. Credit may not be taken for closure valves, quick-disconnects, or similar. It is assumed that mechanical closure devices (e.g., a valve or quick-disconnect) permit helium leaks. Practical experience has shown such leaks occur and have been responsible for causing leak paths through the weld. Consequently, welds potentially subjected to helium pressure (by way of leakage through a mechanical closure device) during the welding process must be subsequently helium leak tested.

Other closure issues the materials reviewer should evaluate are: Hydrostatic Testing, ASME Code Case N-595-4, and the limiting root pass criteria for the weld.

Closure welds must be hydrostatically or pneumatically tested in accordance with ASME Code Section III requirements to the extent practicable. The two designs discussed in Section 8.9 of this SRP meet this criterion.

ASME Code Case N-595-4 is not endorsed by the NRC staff, per Regulatory Guide (RG) 1.193 and consequently is not permitted as an alternative to the Code requirements.

Cask lid welding is governed in part by the limiting flaw size analysis. The welding method described herein controls the depth of weld deposit for the intermediate passes before the required PT examination is performed. However, the root pass thickness is not addressed by this guidance, as a single layer root pass was assumed. Occasionally, multi-layer root passes are employed to smooth the weld surface to avoid false positives from the PT.

A multi-layer root pass is acceptable provided the same method of limiting the weld deposit depth is followed as for the intermediate weld passes. Stress analysts should note that the intermediate layer critical flaw size calculation assumes a buried flaw, not a surface connected flaw. For the root pass calculation, a surface connected flaw must be assumed. This will result in a smaller critical flaw size, and, consequently a smaller permissible weld deposit thickness before a PT exam is considered necessary.

The staff should verify that if the licensee desires to use a thicker root pass, they must limit the amount of weld deposit to the ratio of the fracture toughness K values (or, J values) for the different flaw types (buried K divided by surface K) multiplied by the maximum depth. This will limit the depth of the root pass to the critical flaw size for a surface connected flaw. Thus, if a licensee desires to use a thicker weld deposit for the root pass, then a limiting flaw size analysis establishes a structural basis.

The staff recognizes that for stainless steel, K, or even J, is not entirely correct for evaluating failure in austenitic stainless steel due to the large capacity for plastic deformation. Generally the result is failure due to net section stress, not fracture. However, the stress intensity ratio suggested above is acceptable for this purpose.

The regulatory requirements governing this review are: 10 CFR 72 122(a), 72.122(h)(5), 72.104(a), 72.106(b), 72.236(d), 72.236(e), 72.236(j), and 72.236(l).

Please refer to the additional information in Section 8.9 of this SRP to supplement the review criteria.

8.4.21   Periodic Inspections (LOW Priority)

Review the SAR operations and acceptance testing chapters for appropriate periodic inspection programs which may be included for the purpose of monitoring materials conditions or performance. Some cask vendors are now anticipating future license renewal for the designs and are incorporating into the SAR the currently specified limited inspections that are required as part of a license renewal application.

- A one-time inspection of normally inaccessible portions of the canister exterior for unanticipated corrosion or other degradation. A single canister (or several) may be selected based upon engineering criteria such as longest time in service, hottest operating temperature, etc. and used to "bound" other canisters of that type of material of construction.

- The periodic (usually monthly) ISFSI radiation survey results should be reviewed to determine if any significant degradation of any neutron shielding material (if used) has occurred.

8.5   Evaluation Findings

The evaluation findings are prepared by the reviewer on satisfaction of the regulatory requirements of Section 8.3. The reviewer should examine these requirements and provide a summary statement for each. These statements should be similar to the following examples:

F8.1   Section(s) _____ of the SAR adequately describe(s) the materials used for SSCs important to safety and the suitability of those materials for their intended functions in sufficient detail to evaluate their effectiveness.

F8.2   The applicant has met the requirements of 10 CFR 72.122(a). The material properties of SSCs important to safety conform to quality standards commensurate with their safety function.

F8.3   The applicant has met the requirements of 10 CFR 72.104(a), 72.106(b), and 72.124. Materials used for criticality control and shielding are adequately designed and specified to perform their intended function.

F8.4   The applicant has met the requirements of 10 CFR 72.122(h)(1) and 72.236(h). The design of the DSS and the selection of materials adequately protect the SNF cladding against degradation that might otherwise lead to damaged fuel.

F8.5   The applicant has met the requirements of 10 CFR 72.236(h) and 72.236(m). The material properties of SSCs important to safety will be maintained during normal, off-normal, and accident conditions of operation so the SNF can be readily retrieved without posing operational safety problems.

F8.6   The applicant has met the requirements of 10 CFR 72.236(g). The material properties of SSCs important to safety will be maintained during all conditions of operation so the SNF can be safely stored for the minimum required years and maintenance can be conducted as required.

F8.7   The applicant has met the requirements of 10 CFR 72.236(h). The [cask designation] employs materials that are compatible with wet and dry SNF loading and unloading operations and facilities. These materials should not degrade over time or react with one another during any conditions of storage.

The reviewer should provide a summary statement similar to the following:

"The staff concludes the material properties of the structures, systems, and components of the [cask designation] are in compliance with 10 CFR Part 72, and that the applicable design and acceptance criteria have been satisfied. The evaluation of the material properties provides reasonable assurance the [cask designation] will allow safe storage of SNF for a licensed (certified) life of _____ years. This finding is reached on the basis of a review that considered the regulation itself, appropriate regulatory guides, applicable codes and standards, and accepted engineering practices."

8.6 Supplemental Information for Methods for Classifying Fuel (HIGH Priority)

A. Grossly Breached SNF Cladding

The regulations in 10 CFR 72.122(h) state "The spent fuel cladding must be protected during storage against degradation that leads to gross ruptures or the fuel must be otherwise confined such that degradation of the fuel during storage will not pose operational safety problems with respect to its removal from storage."

In dry cask storage and transportation systems, a gross cladding breach should be considered as any cladding breach that could lead to the release of fuel particulate greater than the average size fuel fragment. A pellet is ~1.1 centimeters in diameter in 15 x 15 Pressurized-Water Reactor (PWR) assemblies. Pellets from a Boiling-Water Reactor (BWR) are somewhat larger, and those from 17 x 17 PWR assemblies are somewhat smaller. The pellet's length is slightly longer than its diameter. During the first cycle of irradiation in-reactor, the pellet fragments into 25-35 smaller interlocked pieces, plus a small amount of finer powder, due to, pellet-to-pellet abrasion. When the rod breaches, about 0.1 gram of this fine powder may be carried out of the fuel rod at the breach site. Modeling the fragments as either spherical- or pie-shaped pieces indicates that a cladding-crack width of at least 2-3 millimeters would be required to release a fragment. Hence, gross breaches should be considered to be any cladding breach greater than 1 millimeter.

A review of reactor operating records, ultrasonic testing, and sipping (if done in a timely fashion) can be used to classify rods as unbreached or, breached. Evidence of only gaseous or volatile decay products (no heavy metals) in the reactor coolant system is accepted as evidence that a cladding breach is no larger than a pinhole leak or hairline crack. Records that show the presence of heavy metal isotopes that are characteristic of fuel release in the reactor coolant system indicate gross breaches in the cladding. Likewise, visual examination may also be used to determine if a cladding breach is gross, if the breached rod can be positively identified. Because cladding openings larger than 1 millimeter should expose the fuel pellet to visual sighting, visual examination of the breached rod can be used to determine if a breach is gross. However, visual examination is not an acceptable method of confirming intact (undamaged) fuel for assemblies that have indicated leakage.

It should be noted, however, that undamaged spent-fuel rods with pinhole leaks and/or hairline cracks will expose the fuel pellets to the canister or cask atmosphere. If that atmosphere is oxidizing, then the fuel pellet may oxidize and expand, placing stress on the cladding. The expansion may eventually cause a large split in the cladding, resulting in spent fuel that must be classified as damaged (for storage and possibly also for transportation) due to gross breaches in the cladding. Since fuel oxidation and cladding splitting follow Arrhenius time-at-temperature behavior, fuel rods with pinholes or hairline cracks that are exposed to an oxidizing atmosphere may experience this type of additional cladding damage. Section 8.7 of this SRP, "Supplemental Information for Potential Rod Splitting Due to Exposure to an Oxidizing Atmosphere during Short-Term Cask Loading Operations in LWR or other Uranium Oxide Based Fuel," provides information regarding prevention of this phenomenon. Before handling undamaged rods with pinhole leaks and/or hairline cracks in an oxidizing atmosphere, the potential fuel and cladding degradation at the temperature of interest for the duration of the process should be assessed.

B. Fuel Assembly with Defects

Damage under this guidance refers to alterations of the fuel assembly that prevent it from fulfilling its fuel-specific or system-related functions. Defects such as dents in rods, bent or missing structural members, small cracks in structural members, missing rods, etc., need not be considered damaged if the applicant can show that the fuel assembly with these defects still fulfils its fuel-specific and system-related functions. This may be done using calculations based on approved codes, situation-specific data, or reasoned engineering arguments.

C. Canning Damaged Fuel

Spent fuel that has been classified as damaged for storage must be placed in a can designed for damaged fuel, or in an acceptable alternative. The purpose of a can designed for damaged fuel is to (1) confine gross fuel particles, debris, or damaged assemblies to a known volume within the cask; (2) to demonstrate that compliance with the criticality, shielding, thermal, and structural requirements are met; and (3) permit normal handling and retrieval from the cask. The can designed for damaged fuel may need to contain neutron-absorbing materials, if results of the criticality safety analysis depend on the neutron absorber to meet the requirements of 10 CFR 72.124(a).

D. Relationship of Spent Fuel Populations

The applicant will designate the population of spent fuel for which the cask system was designed (e.g., type of fuel, minimum cooling time, burnup limitations, arrays, manufacturers, cladding types, etc.). This population may contain breached rods. Some of these breached rods may be grossly breached. It may also contain assemblies with defects, such as missing rods, missing grid spacers, or damaged spacers. The populations of breached rods, grossly breached rods, and assemblies with defects are determined by in-reactor behavior and ex-reactor handling.

Each of these populations must be classified as damaged or undamaged after the storage or transportation system has been designated. For example, an applicant might propose the use of air as a cover gas in its design of a storage cask. The applicant might also propose this cask for use in storing spent fuel with cladding breaches that are hairline cracks or pinhole leaks. However, if the spent fuel in the cask will operate at a sufficiently high temperature for a long enough time, then oxidation of fuel pellets in breached rods could occur resulting in gross breaches. If this is the case, the breached spent fuel should be considered damaged because grossly breached rods do not meet the requirements of 10 CFR 72.122(h)(1). If an inert atmosphere was used instead of air, only grossly breached rods would be considered damaged for storage. This concept is illustrated in Figure 8-2, "Relationship of Spent Fuel Populations."

Example of Methodology

The following example is given to illustrate the general methodology. This is only an example of the methodology and should not be construed as approved characterization of damaged fuel.

Figure 8-2 Relationship of Spent Fuel Populations

Example of Methodology:

Situation - The vendor of a dual-purpose cask wants to store and transport low-burnup PWR fuel in an inert atmosphere and within the temperature limits recommended in Section 8.4.17.1. The vendor wants to store assemblies having rods with breaches containing only pinholes or hairline cracks, and assemblies having one or more outer grid straps with defects at three or more grid locations without canning them. The vendor is only applying for a storage license at this time but wants to be reasonably certain that the fuel will also be transportable.

Activity - Storage of Spent Fuel

Fuel-specific or system-related functions imposed on rods and assemblies - 10 CFR 72.122(h)(1), regarding gross ruptures, and 10 CFR 72.122(l), concerning retrievability, must be met for storage. 10 CFR 71.55(d), requiring the system to remain subcritical and unchanged during normal transport, must be met. The vendor believes that all the remaining system requirements, except for the subcriticality requirement, can be met, without imposing any limitations on the fuel, if the fuel is within the bounds stated in the situation.

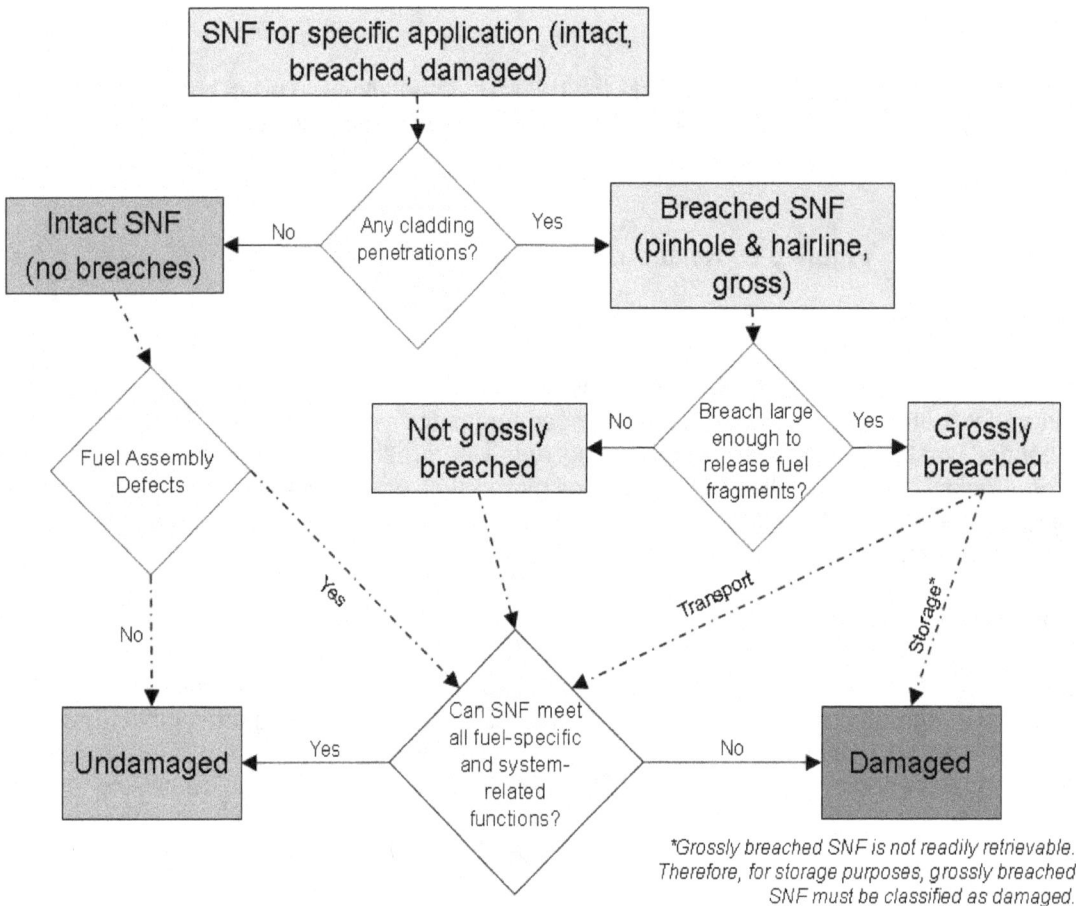

Mechanisms - There are no mechanisms for the pinhole leaks and hairline cracks to evolve into gross breaches since the atmosphere is inert and the temperature is controlled. To be retrievable, the assemblies with missing grid straps must be able to withstand design basis events in a storage cask. Since the applicant also wants these assemblies to be considered undamaged for transportation, the behavior of the assemblies under both normal and hypothetical accident transportation conditions in 10 CFR Part 71 must be evaluated. For example, for normal transportation conditions, the applicant must show that the assemblies with the most missing grid straps in the worst locations can withstand both normal vibration and a one-foot drop and remain in their original physical configuration. Additionally, for hypothetical accident conditions, the analysis must indicate, among other things, that the system will meet shielding and subcriticality requirements when placed under the mechanical and thermal loads specified in 10 CFR Part 71.

Analysis - The applicant conducts an analysis to satisfactorily demonstrate that the assembly with three missing grid straps in the worst configuration remains intact for 1) normal transportation conditions; 2) cask tip-over; and 3) regulatory accident conditions. Further acceptable analysis indicates that all the system-related regulations are met, if the fuel with the characteristic limitations (as noted in Characteristics section below), stays structurally intact.

Characteristics - Assemblies containing breached rods with up to three grid straps missing will be considered undamaged for the purposes of storage. Analysis shows that these assemblies could probably also be considered undamaged for transportation, but fuel with these characteristics will be evaluated and approved as part of a later application for the transportation cask certification.

8.7 Supplemental Information for Potential Rod Splitting Due to Exposure to an Oxidizing Atmosphere During Short-Term Cask Loading Operations in LWR or Other Uranium Oxide Based Fuel (MEDIUM Priority)

The definition of undamaged fuel includes fuel rods containing no cladding defects greater than pinhole leaks or hairline cracks. During the cask water removal process parts of, or all of, the fuel rods will be exposed to a gaseous atmosphere. If the gaseous atmosphere is oxidizing, oxidation of fuel pellets or fuel fragments can occur if a cladding breach exists (such as a pinhole). Oxidation may occur rapidly and cause significant swelling of fuel pellets and fragments, which could result in gross fuel cladding breaches if the time-at-elevated-temperature after water removal is excessive.

8.7.1 Fuel Oxidation and Cladding Splitting

Irradiated uranium dioxide exposed to an oxidizing atmosphere will eventually oxidize to U_3O_8. The time it takes to oxidize is a function of temperature that follows an Arrhenius function and burnup. However, at temperatures that may be expected for some spent fuel, this reaction can occur within a matter of hours.

The grain boundaries of irradiated fuel are highly populated with voids and gas bubbles. Initially the grain boundaries are oxidized to U_4O_9 resulting in a slight matrix shrinkage and further opening of the pellet structure. Oxidation then proceeds into the grain until there is complete transformation of the grains to U_4O_9 [Einziger, 1992]. The grains remain in this phase for a temperature dependent duration until the fuel resumes oxidizing to the U_3O_8 state. The transformation to U_3O_8 occurs with ~33 percent lattice expansion that breaks the ceramic fragment structure into grain sized particles. At higher temperatures, the two transformations occur so rapidly that they are difficult to distinguish. The mechanism of oxidation in irradiated fuel appears to be different than in unirradiated fuel where U_3O_7 is formed and oxidation proceeds from the fragment surface and not down the grain boundaries. This mechanistic change occurs between ~10 and 30 Gwd/MTU.

When the UO_2 is in the form of a fuel rod, the expansion of the fuel, when it transforms to U_3O_8, induces a circumferential stress in the cladding. Due to the swelling of the fuel, the process is usually initially localized to the original cladding crack site. The cladding strains due to this stress range from 2-6 percent before the initial crack starts to propagate along the rod. The incubation time to initiate the propagation and the rate of propagation have an Arrhenius temperature dependence. Axial propagation, spiral propagation and a combination of the modes that result in splitting have been observed in PWR rods [Einziger, 1986].

8.7.2 Data Base

The data base for oxidation was developed mostly in the 1980s in the US, Canada, England, and Germany. The data can usually appear in four forms: 1) O/M ratio (ratio of oxygen to metal content of the oxide) vs. time, 2) time to the $UO_{2.4}$ plateau vs. time, 3) cladding splitting incubation vs. time, and 4) cladding splitting rate vs. time. Some later work was done by the Japanese on the effects of oxygen depletion [Nakamura, 1995], and most recently work is on-going by the French primarily on MOX fuel. Much of the work was done on unirradiated fuel. All the work on cladding splitting was done in the early 1980s by the US [Einziger, 1984, 1986; Johnson, 1984] and Canadians [Novak, 1984; Boase, 1977] and is limited. Recently DOE [Bechtel, 2005] has issued an analysis of the oxidation issue in relationship to handling of potentially breached fuel in their proposed handling facility at the repository. This analysis depends on variables such as the gap between the fuel and the cladding, and burnup in a manner that is currently under technical review. In total, this research has shown that there are a number of variables that can affect the rates at which the fuel oxidizes and the cladding splits: burnup, moisture content of the air, cladding material, and type of initial defect.

The DOE developed a model for fuel oxidation and cladding splitting [Bechtel, 2005] for use during long durations at the Yucca Mountain facility that tries to account for the fuel-to-cladding gap and burnup of the fuel. The gap is the as-measured cold gap and does not account for the closing of the gap due to differential thermal expansion of the cladding and fuel material, which could be calculated. There are inadequate data to verify correctness of the DOE model. Plots in the Einziger document [Einziger, 1986] present actual data and comparisons with the data taken by other researchers at 30 GWd/MTU. The gap closure is implicitly accounted for in the measurements of splitting. However, no burnup effects can be inferred from this data.

No oxidation or cladding splitting studies have been conducted on fuel with burnup greater than 45 GWd/MTU. Data between 30 and 45 GWd/MTU, shows a decrease in the oxidation rate due to the presence of certain actinides and fission products that are burned into the fuel. There is no reason that this should not continue at higher burnups, but the strength of the effect may change with burnup. Higher burnup fuel (>55 GWd/MTU) forms an external rim on the pellets that consists of very fine grains (1 micron). As indicated earlier, the oxidation process is a grain boundary effect. The fuel pellet must be divided into two regions for the purpose of oxidation analysis; the center of the pellet where the grains have grown slightly, and the rim. While the rate of the oxidation may decrease with burnup, the total amount of fuel that is oxidized may increase due to a much greater intergranular surface area in the rim region. The DOE model [Bechtel, 2005] uses a linear decrease in oxidation with burnup but this has, as yet, not been substantiated. A burnup effect is supported by Hanson's analysis [Hanson, 1998] of Einziger and Cook's data from the NRC whole-rod tests in which defect propagation was observed to occur earlier at the defects at the lower end of the rod where the burnup was lower.

Studies using a low partial pressure of water vapor in air have not shown any dependence of the oxidation rate on the moisture content of the air [Ferry, 2005]. On the other hand, there are some studies that have shown a large increase in the oxidation rate when the moisture content is above 50 percent of the dew point. Oxidation in a 100 percent steam atmosphere is a different process. There are also studies that indicate that the oxidation rate will decrease if the oxygen content in the atmosphere drops into the range of a few torr or less [Nakamura, 1995]. It does not appear that there is an effect of oxygen content at higher oxygen levels but the data is sparse.

Oxidation studies on fuel, with few exceptions, have been conducted on LWR fuel [Einziger, 1986; Johnson, 1984]. However, the UO_2 matrix is essentially the same in both PWR and BWR fuel. At the higher burnups, oxidation behavior may vary slightly as the actinide and fission product burn-in varies. The effect of the process on the splitting of the cladding may vary considerably due to the difference in gap size between the cladding types, and the thicker cladding in BWR rods.

The limited cladding splitting studies have been conducted on Zircaloy clad PWR [Einziger, 1984, 1986; Johnson, 1984] and CANDU fuel. Defects were put in the fuel either by an SCC (stress corrosion cracking) process producing small sharp holes more typical of those found in reactor initiated SCC and by drilling that produced a larger duller hole. Most of the defects used in the studies were of the latter type. No measurements were made in cladding above 30 GWd/MTU. Very few data points were measured to determine the splitting rate; therefore, the time to start splitting has to be determined by interpolation. As a result, there is large uncertainty in both measurements. No measurements have been made on other alloy types (e.g., M5 and Zirlo) or at higher burnups where the cladding may be more brittle. Fuel oxidation would introduce uncertainties for fuel performance and fuel retrievability.

8.8 Supplemental Information for Background justification for Cladding Temperature Considerations for the Storage of Spent Fuel (MEDIUM Priority)

8.8.1 Basis for Guidance

Creep is the dominant mechanism for cladding deformation under normal conditions of storage. The relatively high temperatures, differential pressures, and corresponding hoop stress on the cladding will result in permanent creep deformation of the cladding over time. Several laboratory programs have demonstrated that spent fuel has significant creep capacity even after 15 years of dry cask storage. Einziger, et al., [2003] reported that irradiated Surry-2 PWR fuel rods (35.7 GWd/MTU) that were stored for 15 years at an initial temperature of 350°C (with temperatures reaching as high as 415°C for up to 72 hours) experienced thermal creep, which was estimated to be less than 0.1 percent. Post-storage creep tests were conducted to assess the residual creep capacity of the Surry-2 fuel rods. One-rod segment experienced a creep strain of 0.92 percent without rupture at 380°C and 220 MPa in 1820 hours (75.8 days). A different rod segment was tested at 400°C and 190 MPa for 1873 hours (78 days) followed by 693 hours (28.9 days) at 400°C and 250 MPa, and experienced a creep strain of more than 5 percent without failure [Tsai, 2002]. Profilometry measurements on that fuel rod indicated that the creep deformation was uniform around the circumference of the cladding with no signs of localized bulging, which can be a precursor for rupture. A report of the literature [Beyer, 2001] also indicates that some spent fuel cladding can accommodate creep strains of 2.87.5 percent at temperatures between 390 and 420°C and hoop stresses between 225 and 390 MPa. Other significant contributions to the understanding of the effects of creep on spent fuel cladding can be found in several references [Einziger, et al., 1982; Rashid, et al., 2000; Hendricks, 2001; Rashid and Dunham, 2001; Machiels, 2002]. In general, these data and analyses support the conclusions that (1) deformation caused by creep will proceed slowly over time and will decrease the rod pressure, (2) the decreasing cladding temperature also decreases the hoop stress, and this too will slow the creep rate so that during later stages of dry storage, further creep deformation will become exceedingly small, and (3) in the unlikely event that a breach of the cladding due to creep occurs, it is believed that this will not result in gross rupture.

Based on these conclusions, the staff has reasonable assurance that creep under normal conditions of storage will not cause gross rupture of the cladding and that the geometric configuration of the spent fuel will be preserved provided that the maximum cladding temperature does not exceed 400°C (752°F). As discussed below, this temperature will also limit the amount of radially oriented hydrides that may form under normal conditions of storage.

The effects of normal conditions of storage (i.e., the decaying temperature and hoop stress on the cladding with time) can affect the metallurgical condition of spent fuel cladding containing significant amounts of hydrogen (e.g., spent fuel with high burnup levels). As the burnup level of the fuel increases beyond 45 GWd/MTU during reactor operation, the thickness of the oxide layer on the cladding increases. With increasing oxidation during reactor operation, the cladding absorbs more hydrogen. As discussed in Garde, et al., [1996], Chung and Kassner [1997], and Newman [1986], high burnup fuels tend to have relatively higher concentrations of hydrogen in the cladding. The hydrogen is present in the cladding predominantly as zirconium hydride precipitates, or particles. After the fuel is removed from the reactor, the zirconium hydrides are generally elongated and oriented circumferentially and are predominantly present in the outer rim of the cladding. At elevated temperatures, a percentage of the zirconium hydrides will dissolve, and under decreasing temperatures, zirconium hydrides will precipitate, or re-form.

The materials phenomenon of hydride reorientation in zirconium-based alloys usually involves the dissolution of hydrides and the formation of zirconium-hydrides oriented perpendicular to the hoop stress (also referred to as radially oriented or radial hydrides) [Chung, 2000]. This occurs under sufficiently high hoop stresses along with the decrease in solubility of hydrogen that accompanies decreasing temperatures. The extent of the formation of radially oriented hydrides is a function of many parameters including the solubility of hydrogen in irradiated cladding material, cladding temperature, hoop stress, cooling rate, hydrogen concentration, thermal cycling, and materials characteristics. Among these parameters, the formation of radial hydrides is highly dependent on the hoop stress in the cladding. Data obtained from irradiated cladding [Einziger and Kohli, 1984; Cappelaere, et al., 2001; and, Goll, et al., 2001] indicate that stresses greater than 120 MPa (17.4 ksi) are required to initiate the formation of radial hydrides. Other data obtained from unirradiated zirconium-based cladding materials [Kese, 1998] indicate that radial hydrides can form at stresses as low as 90 MPa. Therefore, until the effects of reorientation are better understood, the hoop stress on the cladding should be controlled to preclude the formation of radially oriented hydrides.

In general, a temperature limit of 400°C that is specified for normal conditions of storage and for short-term fuel loading and Part 72 storage operations (which includes drying, backfilling with inert gas, and transfer of the cask to the storage pad) will limit cladding hoop stresses and limit the amount of soluble hydrogen available to form radial hydrides. The use of a 400°C temperature limit for normal conditions of storage and for short-term fuel loading and storage operations will simplify the calculations in SARs while assuring that hydride reorientation will be minimized.

For low burnup fuel, a higher temperature limit may be used for short-term fuel loading and storage operations only, as long as the applicant can demonstrate that the best estimate cladding hoop stresses are equal to or less than 90 MPa for the temperature limit that is justified. For example, if the calculated best estimate hoop stress is equal to 90 MPa at 540°C, then 540°C is the maximum allowable temperature for loading operations. In this example, 570°C is not the maximum allowable temperature limit. If the applicant can show that the best estimate hoop stress is less than or equal to 90 MPa at 570°C, then 570°C is the maximum allowable temperature. For some fuel types, short-term fuel loading and storage operation temperature limits as high as 570°C (1058°F) should be justified by the applicant. The materials reviewer should coordinate with the thermal reviewer to assure that either of the following criteria are used: (1) for low and high burnup fuel, the maximum calculated temperatures for normal conditions of storage and for fuel loading operations do not exceed 400°C, or (2) for low burnup fuel, a higher temperature limit may be used for loading and transfer operations, if the best estimate cladding hoop stress is less than 90 MPa for the temperature specified by the applicant. If the applicants use the latter approach, the materials reviewer should verify that the cladding hoop stresses are less than 90 MPa for each fuel assembly type (e.g., 14x14, 17x17, 9x9, etc.) proposed for storage. Since the hoop stress is dependent on the rod internal pressure, cladding geometry, and the temperature of the gases inside the rod, the materials reviewer should coordinate with the thermal reviewer to verify that the applicant has calculated the best estimate hoop stress corresponding to the rod internal pressure of the highest burnup fuel assemblies of the specific type of assembly. It should be noted that during normal conditions of storage there will be a range of cladding temperatures that are less than the maximum allowable cladding temperature of 400°C, and this leads to a range of the internal rod pressures and the cladding hoop stresses, in any one storage cask. In general, the maximum allowable temperature will be 400°C or the maximum allowable temperature specified and supported (as discussed above) by the applicant. The maximum allowable temperature should be based upon the peak rod temperature, not the average rod temperature. By employing the peak rod temperature, only a small fraction of the rods will experience the temperature and stress conditions that could lead to the formation of radial hydrides during normal conditions of storage.

It also has been observed and reported that thermal cycling (repeated heatup/cooldown cycles) can enhance the amount of hydrogen that eventually re-precipitates in the form of radial hydrides [Kammenzind, et al., 2000]. The extent of the formation of radial hydrides is dependent on many factors including the maximum temperature, change in temperature, number of thermal cycles, applied stress, hydrogen concentration, and solubility of hydrogen in the material. Kammenzind, et al., [2000] indicates that the formation of radial hydrides in spent fuel cladding can be minimized by restricting the change in cladding temperatures to less than 65°C and minimizing the number of cycles to less than 10. The 65°C temperature limit is based upon the temperature drop required to obtain the degree of supersaturation required for the precipitation of hydrides in a short thermal cycle.

For short-term accidents and short-term off-normal conditions that lead to an increase in temperature of the cladding, the dominant cladding failure mechanism is expected to be creep (stress rupture) of the cladding. To limit the amount of spent fuel that could be released from the cladding under off-normal conditions or accidents, the materials reviewer should coordinate with the thermal reviewer to verify that the maximum calculated cladding temperatures are maintained below 570°C (1058°F). The basis for using 570°C is established by the creep tests conducted on irradiated Zircaloy-4 rods [Einziger, et al., 1982]. The results from these experiments indicated that no cladding ruptures were observed for test times of 30 and 73 days.

8.8.2 Review Guidance

Prior to this guidance the short-term cladding temperature limit applicable to fuel loading operations was 570°C. All storage casks were certified using this limit. The current guidance to maintain cladding temperatures less than 400°C during fuel loading operations put into question whether the licensees who use certified storage casks (certified for fuel having average assembly burnups less than 45 GWd/MTU) would have to change their loading procedures and Technical Specifications to comply with this new temperature limit. Based on staff's evaluation, it is expected that fuel assemblies with burnups less than 45 GWd/MTU are not likely to have a significant amount of hydride reorientation due to limited hydride content. Further, most of the low burnup fuel has hoop stresses below 90 MPa. Even if hydride reorientation occurred during storage, the network of reoriented hydrides is not expected to be extensive enough in low burnup fuel to cause fuel rod failures.

Given the conservatism used in calculating peak clad temperatures for low burnup fuel, the staff has reasonable assurance that storage cask systems which use the 570°C temperature limit for low burnup fuel loading operations will continue to perform as expected when the casks were originally certified. Therefore, there is no need to require the licensees of storage-only or dual-purpose cask systems to repackage spent fuel that was loaded using the 570°C temperature limit. Nevertheless, the 400°C limit is intended, with exceptions as stated above, to be generally applicable to all future loadings. Therefore, licensees are not required to modify their Technical Specifications or fuel loading procedures (i.e., vacuum drying) to meet the new 400°C limit for loading low burnup fuel into storage casks previously certified with the 570°C limit. Note that for future amendments to certified designs, the applicants may be required to comply with the 400°C temperature limit as discussed above.

8.9 Supplemental Information for the Design and Testing of Lid Welds on Austenitic Stainless Steel Canisters as Confinement Boundary for Spent Fuel Storage (MEDIUM Priority)

8.9.1 Basis for the Review

10 CFR 72.236(e) states: "The spent fuel storage cask [note: also called "canister"] must be designed to provide redundant sealing of confinement systems." For a bolted lid canister design, the staff has accepted a dual seal arrangement as meeting the intent of this regulation. For a welded canister design, the staff has accepted closure designs employing redundant lids or covers, each with independent field welds. Thus, for either closure type, bolted or welded, a potential leak path must breach two independent seals or welds, sequentially, before the confinement system would be compromised.

The construction codes specify the types of non-destructive examinations (NDE) required for the confinement boundary during canister fabrication and loading operations. In addition to the code required NDE, a helium leakage test of the confinement boundary is considered necessary to satisfy regulatory requirements. Whereas bolted lid canister designs incorporate a helium monitoring system during storage, the welded closure designs must rely on weld integrity to assure continued confinement effectiveness. Consequently, at least one of the redundant welded closures must be helium leak tested per the method of ANSI N 14.5, with one exception permitted.

When the large, multi-pass weld joining the canister shell to the structural lid of an austenitic stainless steel spent fuel canister is executed and examined consistent with the guidance provided herein, the staff has reasonable assurance that no flaws of significant size will exist such that they could impair the structural strength or confinement capability of this weld. For a spent nuclear fuel canister, such a flaw would be the result of improper fabrication or welding technique, as service-induced flaws under normal and off-normal conditions of storage are not credible. Any such fabrication flaw would be reasonably detectable during the in-process and post-weld examination techniques described herein.

Based on evaluation, these described techniques should detect any such flaw which could lead to a failure or credible leakage of radioactive material. Therefore, the staff believes that there is reasonable assurance that no credible leakage of radioactive material would occur through the structural lid to canister shell weld of an austenitic stainless steel canister, and that helium leakage testing of this specific weld is unnecessary provided the weld is executed and examined in accordance with the methods described herein.

Conversely, it is the staff position that other welds associated with the lid assemblies of spent fuel canisters must be subject to the helium leak test of ANSI N 14.5, in addition to the ASME

Code required pressure test and surface NDE in order to demonstrate compliance with 10 CFR 72.236.

Note the criteria outlined above do not supersede the flaw acceptance criteria of any construction code. Instead, this criteria is used to establish the maximum allowable weld deposit depth before an in-process penetrant test (PT) examination is required.

8.9.2   Helium Leak Test

The helium leak test was established to provide assurance that:

- No leakage occurred after the closure welds of the cask system were executed. This was viewed as necessary since no active or passive methods are employed to confirm or monitor the presence of helium within an all-welded spent fuel canister over its licensed lifetime. "No leakage" in this case means measured leak rate performed per ANSI N14.5, at a predetermined sensitivity that shows hypothetical doses would not exceed 10 CFR Part 72 limits.

- If the weld(s) meets the criteria of ANSI N14.5, the staff has assurance that radio nuclide leakage would not exceed the regulatory dose limits in 10 CFR Parts 72.104 and 72.106.

- No oxygen in-leakage could occur, thereby assuring the presence of the inert helium atmosphere which prevents oxidation and corrosion induced degradation of the spent fuel assemblies and enhances cooling of the spent fuel.

Helium Leak-Testing of the Confinement Boundary

The redundant weld requirement for the confinement system closure creates two closure boundaries. The staff should verify that at least one of the redundant boundaries is helium leak tested, or, some closure welds leak tested and the remaining closure welds of the same boundary designed so that the "large weld" exemption criteria of this guidance are met. Only a boundary which is testable or excluded from testing, per this guidance, should be considered the confinement boundary of the redundant closures. Refer to Figures 8-3 and 8-9 and the following narrative for application of this criteria to two currently approved designs:

Leak Testing a Single Lid With Cover Plate Design – Figure 8-3

In Figure 8-3, the dotted line marked (1) defines one closure boundary. Starting on the left side of the sketch, the closure boundary can be traced from the canister wall, up through the large, multi-pass weld joining the canister wall to the heavy section, combined shield and structural lid. The boundary continues through the lid to the small weld joining the heavy lid to the vent-and-drain port closure plate, and back to the heavy lid again. The remainder of the boundary (and sketch) is assumed to be symmetrical with or similar to the half-sketch portion that is shown, for all cases.

This boundary demonstrates confinement integrity by means of the large weld exemption criteria for one weld and by helium leak testing the small cover plate weld.

The large, canister-shell-to-lid weld is exempted from the helium leak test. This is because the canister shell to lid weld is a large, multi-pass weld meeting the flaw tolerance and other appropriate portions of this guidance. Note that this weld is executed prior to filling the canister with helium (excluding purging/welding gas).

Before the remaining welds of this first closure boundary are executed, the canister is drained, dried, purged, and filled with helium to the design operating pressure. The helium line connection is closed off and the cover plate fitted and welded into place. Since the cover plate weld may have potentially been pressurized from underneath due to assumed leakage from the closure valve, it must be helium leak tested in accordance with the methods described in ANSI N14.5-1997. If there are other cover plates and welds, they would also be helium leak tested.

This completes the first closure boundary. Note again that one weld was exempted from the helium leak test by the design criteria. The other weld was leak tested. Thus, this closure boundary demonstrates compliance with regulatory requirements and is consistent with the staff guidance by ensuring at least one of the two redundant closure boundaries is leak tested or exempted from leak testing by conformance with the large-weld exemption guidance. This boundary thus also qualifies as the confinement boundary.

The second boundary, delineated by line 2, can be traced from the canister wall on the left side of the sketch up through the cover plate fillet weld joining the canister wall to the structural lid cover plate. The boundary continues through the cover plate to the fillet weld joining the cover plate to the canister lid. The weld joining the cover plate to the canister wall and lid cannot be helium leak tested since there is no feasible means to do so. However, since the first closure boundary, delineated by line 1, was tested (or exempted thru design), the need to helium leak test at least one of the closure boundaries has been satisfied. Since this second boundary does not meet all the criteria for a confinement boundary, it may not be designated as the confinement boundary. The first closure is thereby the confinement boundary in this design, as it meets all the applicable criteria for a confinement boundary.

Leak Testing a Dual Lid Design – Figure 8-4

In Figure 8-4 of this SRP, the dotted line marked (1) defines one of the redundant closure boundaries. It may be traced from the canister wall on the left side of the sketch. The boundary proceeds through the partial penetration weld joining the canister wall to the shield lid and into the shield lid. It continues through the small fillet weld joining the vent/drain port cover plate, the cover plate, and back through the same fillet weld to the shield lid.

This closure boundary may satisfy the leak test guidance by several methods, depending on details of the weld design. The canister shell to shield lid weld may be designed several ways. The weld may be a small seal weld which would necessitate subsequent helium leak testing. Conversely, it could be a large, multi-pass weld consistent with the guidance described herein. In that case, the weld would qualify for the leak test exemption. Either way, note that this weld (canister to shield lid weld) is executed prior to filling and pressurizing the canister with helium (use of purge or backing gas for welding operations is not considered filling or pressurizing).

Next, the canister is drained, dried, purged, and filled with helium to the design operating pressure. The helium line connection is closed off. The cover plate is fitted and welded into place. Since this weld may potentially be pressurized from underneath due to assumed leakage through the closure valve, it must be helium leak tested regardless of weld size (thickness).
This completes the first closure boundary. Note that one weld was either tested, or, exempted from the helium leak test by the design criteria. The other weld was leak tested. Thus, this closure boundary demonstrates compliance with regulatory requirements and is consistent with

staff guidance by ensuring at least one of the two redundant closures is leak tested or exempted by conformance to this guidance. This closure may therefore be designated as the confinement boundary.

The secondary boundary, delineated by line 2 in sketch B, can be traced from the canister wall on the left side of the sketch up through the canister wall-to-structural lid weld and into the structural lid.

The weld joining the canister wall and structural lid cannot be helium leak tested because helium is not present. Note, however, that this weld complies by design with the criteria described herein due to its size, structural requirements and weld examination requirements of the governing construction code.

In this case, the second closure also qualifies for designation as the confinement boundary because the single large weld involved may be exempted from the helium leak test. In this design, the designer therefore has the freedom to designate either of the redundant closures as the confinement boundary. Only one of the two closures is designated as the confinement boundary.

Figure 8-3 Single Lid with Cover Plate Design

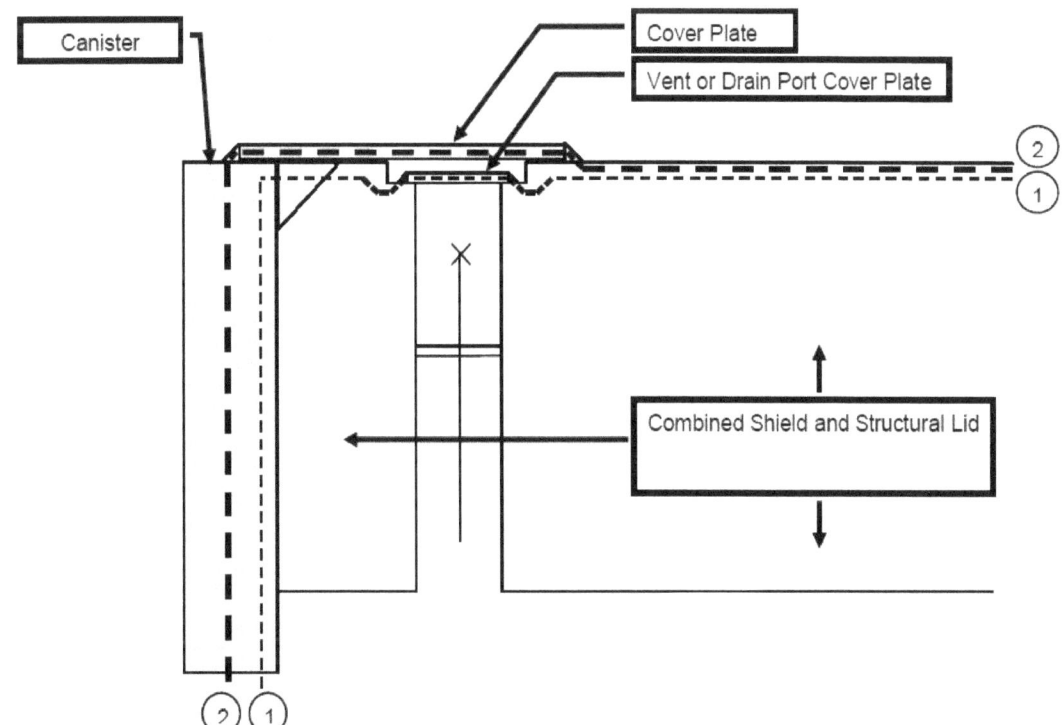

Figure 8-4 Dual Lid Design

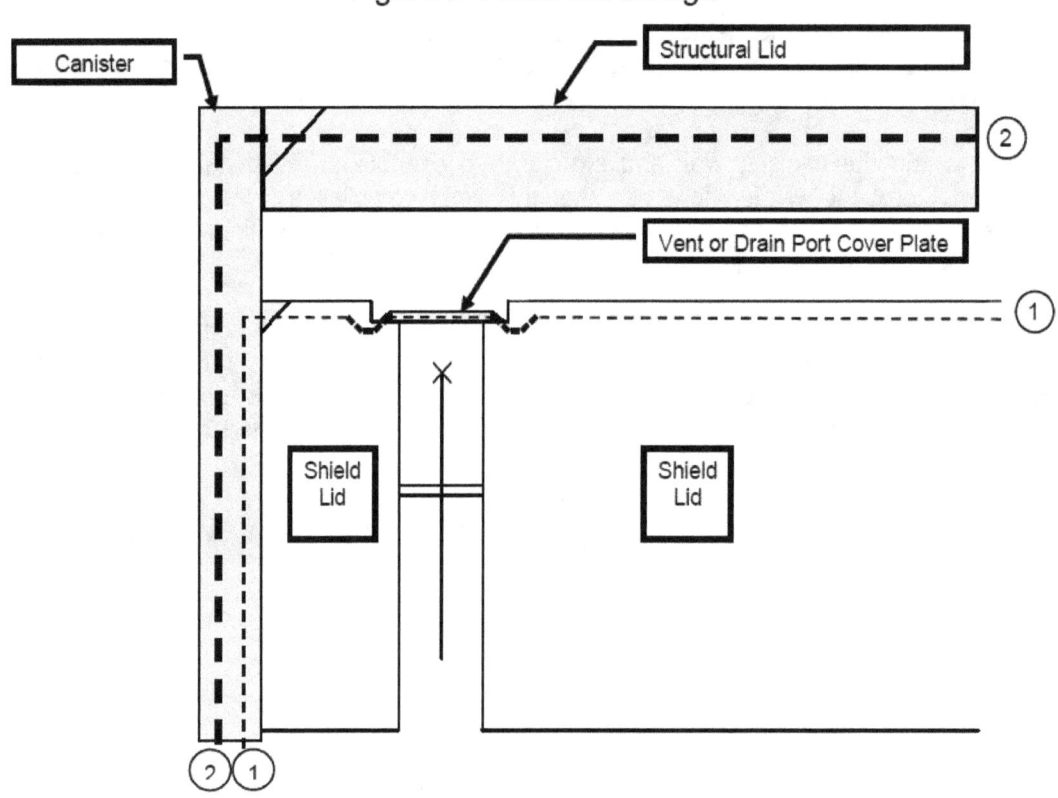

9 OPERATING PROCEDURES EVALUATION

9.1 Review Objective

The operating procedures review ensures that the applicant's safety analysis report (SAR) presents acceptable operating sequences, guidance, and generic procedures for the key operations shown in Section 9.2, "Areas of Review." The review also ensures that the SAR incorporates and is compatible with the applicable operating control limits in the technical specifications.

The operating sequences described in the SAR should provide an effective basis for the development of the more detailed operating and test procedures by the cask user when preparing and implementing detailed site-specific procedures. The NRC normally inspects selected site-specific procedures. Such procedures are important aspects of the site's radiation protection program and allow the cask user to safely store spent nuclear fuel (SNF).

This chapter applies to all discipline reviews. Figure 1-1 presents an overview of the evaluation process and can be used as a guide to assist in coordinating with other review disciplines.

9.2 Areas of Review

This chapter of the dry storage system (DSS) Standard Review Plan (SRP) provides guidance in evaluating the applicant's general operating sequences and generic procedures related to cask operations. Within each area of cask operations, the NRC staff assesses the effectiveness of the applicant's generic procedures on a technical and safety basis for the subsequent development of detailed operating procedures. As required by U.S. Code of Federal Regulations (CFR) Part 72, "Licensing Requirements for the Independent Storage of Spent Nuclear Fuel and High-Level Radioactive Waste," Title 10, "Energy" (10 CFR Part 72) 72.234(f), these procedures are to be provided to each cask user for the subsequent preparation and implementation of detailed site-specific procedures by the cask system user acting under a general license. Areas of review addressed in this chapter include the following:

> ### Loading Operations
> Fuel Specifications
> Damaged Fuel
> Subcriticality Features
> ALARA
> Offsite Release
> Draining and Drying
> Filling and Pressurization
> Welding and Sealing
> Administrative Programs
>
> ### Cask Handling and Storage Operations
>
> ### Cask Unloading
> Damaged Fuel
> Cooling, Venting, and Reflooding
> Fuel Crud
> ALARA
> Offsite Release

9.3 Regulatory Requirements

This section presents a summary matrix of the portions of 10 CFR Part 72 that are relevant to the review areas addressed by this chapter. The NRC staff reviewer should read the exact referenced regulatory language. Table 9-1 matches the relevant regulatory requirements associated with this chapter to the areas of review.

Areas of Review	10 CFR Part 72 Regulations					
	72.104 (b), (c)	72.122(f), (h)(1), (l)	72.212 (b) (9)	72.234 (f)	72.236 (c)	72.236 (h), (i)
Cask Loading Operations	•		•	•	•	•
Cask Handling and Storage Operations	•	•	•	•		•
Cask Unloading	•	•	•	•		•

Table 9-1 Relationship of Regulations and Areas of Review

9.4 Acceptance Criteria

Chapter 9, "Operating Procedures Evaluation," of the SAR should identify and describe the sequence of significant operations and actions that are important to safety for cask loading, cask handling, storage operations, and cask unloading. A sufficient level of detail is needed in Chapter 9 of the SAR for the reviewer to conclude that operating procedures will adequately protect health and minimize danger to life or property, protect the fuel from significant damage or degradation, and provide for the safe performance of tasks and DSS operations.

This portion of the DSS review seeks to ensure that the generic procedure descriptions and operational sequences described in the SAR include the following information:

- Major operating procedures should apply to the principal activities expected to occur during dry storage. The expected scope of activities for the SAR operating procedure descriptions is previously described in Section 9.2 as well as Chapter 8 of Regulatory Guide (RG) 3.61, "Standard Format and Content for a Topical Safety Analysis Report for a Spent Fuel Dry Storage Cask." Operating procedure descriptions should be submitted to address the cask design features and planned operations.

- Operating procedure descriptions should identify measures to control processes and mitigate potential hazards that may be present during planned normal operations. Section 9.5, "Review Procedures," in this chapter discusses previously identified processes and potential hazards.

- Operating procedure descriptions should ensure conformance with the applicable operating controls and limits described in the cask system's Technical Specifications provided in Chapter 13, "Technical Specifications and Operating Controls and Limits Evaluation," of the SAR.

- Operating procedure descriptions should reflect planning to ensure that operations will fulfill the following acceptance criteria:

  - Occupational radiation exposures will remain as low as is reasonably achievable (ALARA).

  - Effective measures will be taken to preclude potential unplanned and uncontrolled releases of radioactive materials.

  - Offsite dose rates will be maintained within the limits of 10 CFR Part 20 and 10 CFR 72.104 for normal operations, and 10 CFR 72.106 for accident-level conditions.

  In addition, the operating procedure descriptions should support and be consistent with the bases used to estimate radiation exposures and total doses as defined in Chapter 11, "Radiation Protection Evaluation," of this SRP.

- Operating procedure descriptions should include provisions for the following activities:

  - Testing, surveillance, and monitoring of the stored material and casks during storage and loading and unloading operations.

  - Contingency actions triggered by inspections, checks, observations, instrument readings, and so forth. Some of these may involve off-normal conditions addressed in Chapter 12, "Accident Analyses Evaluation," of the SAR.

9.4.1 Cask Loading

In addition to the acceptance criteria above, additional acceptance criteria for cask loading are as follows:

- The operating procedure descriptions should facilitate reducing the amount of water vapor and oxidizing material within the confinement cask to an acceptable level to protect the SNF cladding against degradation that might otherwise lead to gross ruptures.

- Operating procedures should specify methods for placing damaged fuel in a damaged-fuel can prior to loading into a cask, if applicable.

9.4.2 Cask Handling and Storage Operations

In addition to the acceptance criteria stated above, operating procedure descriptions should include provisions for maintenance of casks and cask functions during storage.

9.4.3 Cask Unloading

In addition to the acceptance criteria stated above, operating procedures should facilitate ready retrieval of SNF stored in a storage cask.

9.5 Review Procedures

Introduction (MEDIUM Priority)

The interrelationship of the operating procedures evaluation with other disciplines is shown in Figure 9-1.

The review procedures described in this section are presented in a format intended to facilitate an independent review. Even though several individuals may actually be tasked with preparing the chapter of the safety evaluation report (SER) related to operating procedures, all review team members should examine the operating procedure descriptions presented in the SAR. If the descriptions included in the SAR are not sufficiently detailed to allow a complete evaluation concerning fulfillment of the acceptance criteria, reviewers should request additional information from the applicant.

The operating procedure sequences are described in Chapter 9 of the SAR, and the direct dose rate information in Chapter 6, "Shielding Evaluation," of the SAR is used to assess compliance with radiation protection requirements in Chapter 11 of the SAR. The reviewer should verify that the evaluation of Chapter 9 of the SAR is coordinated with the shielding and radiation protection evaluations covered in Chapters 6, "Shielding Evaluation" and 11, "Radiation Protection Evaluation," of this SRP.

In addition, the following review procedures are based on the assumption that the ISFSI operations are at a reactor facility licensed under 10 CFR Part 50, "Domestic Licensing of Production and Utilization Facilities," and that loading and unloading activities will be performed in the facility's SNF pool. Review procedures for dry fuel transfers and/or ISFSI operations at sites away from a reactor will be developed at a later date, if necessary.

Reviewers should be familiar with ANSI/ANS 57.9, "Design Criteria for an Independent Spent Fuel Storage Installation (Dry Type)," which applies to DSS operating procedures. Background information is available in NUREG/CR-4775, "Guide for Preparing Operating Procedures for Shipping Packages," which provides guidance on preparing operating procedures for shipping packages. Although NUREG/CR-4775 specifically addresses 10 CFR Part 71, most of the guidance can be adapted for storage casks that are governed by 10 CFR Part 72. Consequently, reviewers should be familiar with this information before initiating the DSS operating procedures review.

Since many of the detailed procedures may be developed by facilities licensed under 10 CFR Part 50 or 72, further background information on site-specific procedure requirements may be found in RG 1.33, "Quality Assurance Program Requirements (Operation)," and its associated standard ANSI/ANS 3.2. Reviewers of Chapter 9, "Operating procedures Evaluation" of the SAR should also be familiar with Chapter 11, "Conduct of Operations Evaluation," of NUREG-1567, "Standard Review Plan for Spent Fuel Dry Storage Facilities." Specifically, Section 11.4.3, "Normal Operations," in NUREG-1567 provides NRC review acceptance criteria for facility-developed procedures.

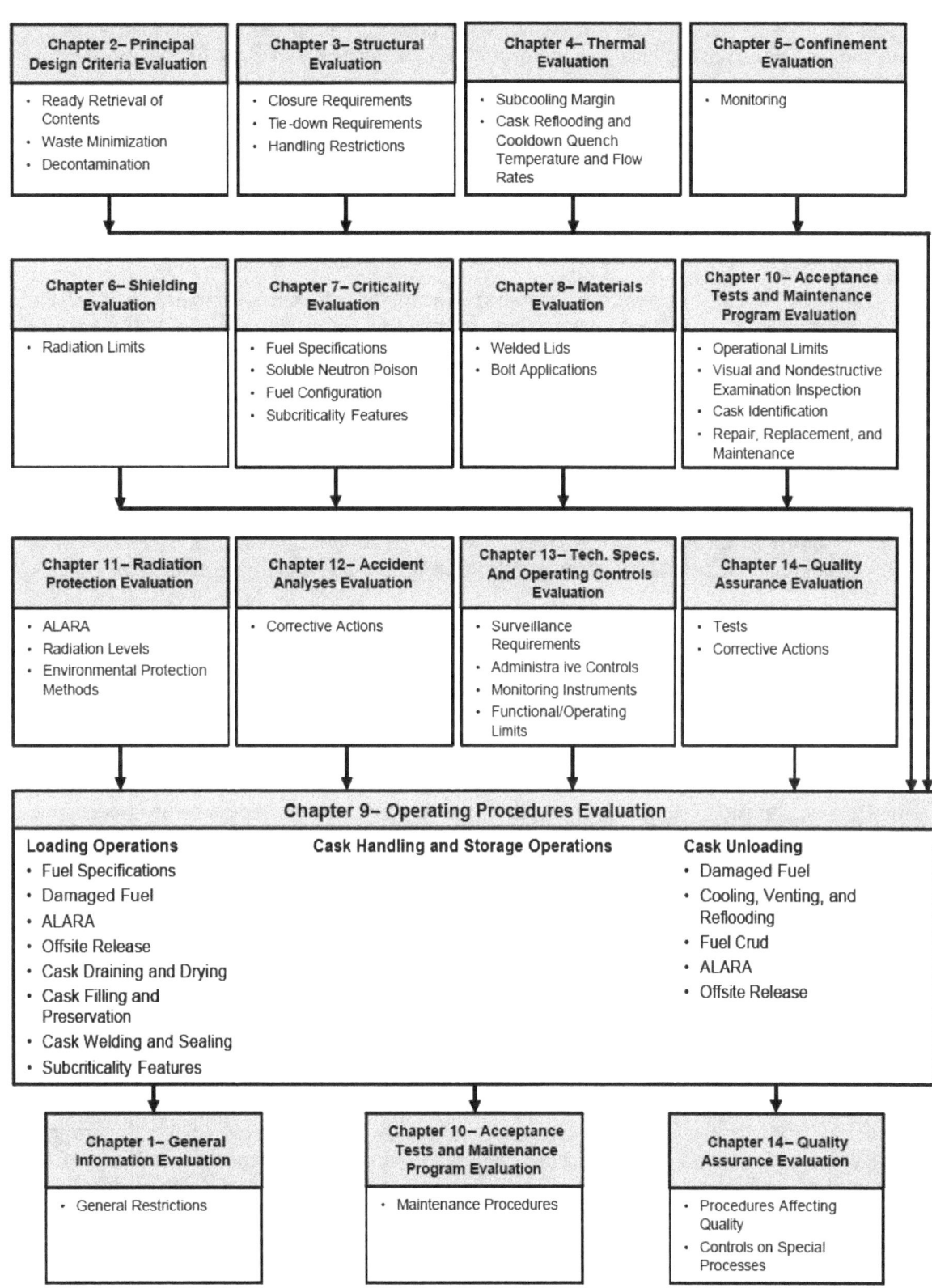

Figure 9-1 Overview of Operating Procedures Evaluation

In general, reviewers should perform the following steps in the process of evaluating all of the operating procedure descriptions and operational sequences provided in the SAR.

- Verify that the proposed operating procedure descriptions incorporate and are compatible with the applicable operating limits and controls in Chapter 13, "Technical Specifications and Operational Controls and Limits Evaluation" of the SAR. Coordinate with the review of operating controls and limits, as described in Chapter 13, "Technical Specifications and Operating Controls and Limits Evaluation," of this SRP.

- Ensure that the proposed operating procedure descriptions properly consider the prevention of hydrogen gas generation from any cause (including the reaction of zinc primer coating with acidic pool water, radiolysis, or other causes). Prevention of hydrogen generation or adequate purging of hydrogen is essential during loading and unloading operations that involve seal welding, seal cutting, grinding, or other forms of hot work.

- Determine whether the descriptions include appropriate precautions to minimize occupational radiation exposures in accordance with ALARA principles and the limits given in 10 CFR Part 20, as mandated by 72.126(a)(5). Provisions may include use of remotely controlled equipment, monitoring, and use of portable shielding.

- Verify that the operating procedure descriptions include a general listing of the major tools and equipment needed to support ISFSI loading, storage, and unloading operations (including those at the pool facility). The descriptions should also address installation, use, and removal of the cask and fuel, tools, and equipment. In addition, the descriptions should describe any specialized tools and equipment in sufficient detail to enable users to understand their function. Examples include lifting yokes, transporter equipment, welding and cutting equipment, and vacuum drying equipment. The use of any such equipment that is classified as being important to safety is subject to approval as part of the application review. Such equipment should be identified and described in detail, its performance characteristics should be defined, and the design should be evaluated.

In addition to these generic review procedures, all disciplines should evaluate each of the specific areas of operating procedure review as described in the following subsections.

9.5.1   Cask Loading (Priority - as indicated)

(MEDIUM Priority) The operating procedure descriptions in the SAR should present the activities sequentially in the anticipated order of performance. The generic procedures in Chapter 9, "Operating Procedures Evaluation" of the SAR should be reviewed to ensure that they include appropriate key prerequisite, preparation, and receipt inspection activities to be accomplished before cask loading. The reviewer should verify that tests, inspections, verifications, and cleaning procedures required in preparation for cask loading are specified. In addition, where applicable, the reviewer should verify that the procedure descriptions include actions needed to ensure that any fluids such as shield water and primary coolants fill their respective cavities according to design specifications.

Fuel Specifications (MEDIUM Priority)

The reviewer should verify that the loading procedure description appropriately addresses the SNF specifications (e.g., burnup, cooling period, source terms, heat generation, cladding damage, associated non-fuel hardware, etc.) in Chapter 2, "Principal Design Criteria," and Chapter 13, "Technical Specifications and Operation Controls and Limits Evaluation" of the SAR. For cask systems relying upon burnup credit, the loading procedure description should include verification that assemblies selected for loading meet the specifications for assembly operational history and the loading curve as well as include performance of measurements to confirm assembly burnup values. Depending on the types and specifications of fuel assemblies stored in the reactor SNF pool, detailed site-specific procedures may be necessary to ensure that all fuel loaded in the cask meets the fuel specifications for the cask design. These procedures can be evaluated only on a site-specific basis and will generally be evaluated through inspections rather than during the licensing review. The SAR should indicate, however, that such procedures may be necessary.

Damaged Fuel (MEDIUM Priority)

The reviewer should verify that the SAR includes appropriate measures for the loading of damaged fuel, if damaged fuel is included in the proposed cask contents. Chapter 2, "Principal Design Criteria Evaluation," and Chapter 8, "Materials Evaluation," of this SRP provide criteria for the storage of damaged fuel. Information in Section 8.6, "Supplemental Information for Methods for Classifying Fuel," of this SRP should be used to identify the conditions that determine when SNF is to be classified as damaged fuel. Section 8.4.17.2 of this SRP should be reviewed to determine the classification, documentation, and special handling requirements for damaged fuel and determine if operating procedures address these requirements.

Subcriticality Features (MEDIUM Priority)

Where applicable, the reviewer should verify that the procedure descriptions include the use of features important to criticality safety that may require installation by the DSS user. Such items include fuel spacers and items (e.g., blocks) used to prevent loading of contents in selected basket locations. The procedure descriptions should include installation, or verification of the installation, of these items prior to cask loading for casks that rely upon these features in the criticality analysis. Additionally, the procedure descriptions should include verification, in accordance with Technical Specification requirements, of the minimum soluble boron level necessary for cask loading for casks requiring soluble boron to meet subcriticality.

ALARA (LOW Priority)

The reviewer should verify that the procedure descriptions incorporate ALARA principles and practices. These may include provisions to perform radiological surveys as well as exposure and contamination control measures, temporary shielding, and suggested caution statements related to actions that could change radiological conditions. In addition, the reviewer should verify that any recommended surveys incorporate the applicable operating controls and limits described in Chapter 13, "Technical Specifications and Operating Controls and Limits Evaluation" of the SAR.

Offsite Release (LOW Priority)

Where applicable, the reviewer should verify that the SAR describes methods to minimize offsite releases such as decontamination, filtered ventilation, temporary containments (tents), and so forth. The procedure descriptions should also provide for minimizing generation of radioactive waste.

Draining and Drying (MEDIUM Priority)

The reviewer should evaluate the descriptions related to methods for use in draining and drying the cask for ISFSI operations at a reactor facility or at sites away from a reactor with a transfer pool. In particular, the descriptions should clearly describe the procedures for removing water vapor and oxidizing material to an acceptable level, and the reviewer should assess whether those procedures are appropriate.

The NRC staff has accepted vacuum drying methods comparable to those recommended in PNL-6365 (Knoll, 1987). This report evaluates the effects of oxidizing impurities on the dry storage of light-water reactor (LWR) fuel and recommends limiting the maximum quantity of oxidizing gasses (such as O_2, CO_2[4], and CO) to a total of 1 gram-mole per cask. This corresponds to a concentration of 0.25 volume percent of the total gases for a $7.0m^3$ (about 247 ft^3) cask gas volume at a pressure of about 0.15 MPa (1.5 atm) at 300°K (80.3°F). This 1 gram-mole limit reduces the amount of oxidants below levels where any cladding degradation is expected. Moisture removal is inherent in the vacuum drying process, and levels at or below those evaluated in PNL-6365 (about 0.43 gram-mole H_2O) are expected if adequate vacuum drying is performed.

If alternative methods other than vacuum drying are used (such as forced helium recirculation), the reviewer should ensure that additional analyses or tests are provided to sufficiently justify that cover gas moisture and impurity levels as specified in Chapter 9, "Operating Procedures Evaluation" of the SAR are met and will not result in unacceptable cladding degradation.

The following examples illustrate the accepted methods for cask draining and drying in accordance with the recommendations of PNL-6365 (Knoll, 1987):

- The cask should be drained of as much water as practicable and evacuated to less than or equal to 4.0E-04 MPa (4 millibar, 3.0 mm Hg or Torr). After evacuation, adequate moisture removal should be verified by maintaining a constant pressure over a period of about 30 minutes without vacuum pump operation (or the vacuum pump is running but it is isolated from the cask with its suction vented to atmosphere). The cask is then backfilled with an inert gas (e.g., helium) for applicable pressure and leak testing. Care should be taken to preserve the purity of the cover gas and, after backfilling, cover gas purity should be verified by sampling.

- The procedures should reflect the potential for blockage of the evacuation system or masking of defects in the cladding of non-intact rods, as a result of icing during evacuation. Icing can occur from the cooling effects of water vaporization and system depressurization during evacuation. Icing is more likely to occur in the evacuation system lines than in the cask because of decay heat from the fuel. A staged draw down or other means of preventing ice blockage of

---

[4] Can be broken down by radiolysis.

the cask evacuation path may be used (e.g., measurement of cask pressure not involving the line through which the cask is evacuated).

- The procedures should specify a suitable inert cover gas (such as helium) with a quality specification that ensures a known maximum percentage of impurities to minimize the source of potentially oxidizing impurity gases and vapors and adequately remove contaminants from the cask.

- The process should provide for repetition of the evacuation and repressurization cycles if the cask interior is opened to an oxidizing atmosphere following the evacuation and repressurization cycles (as may occur in conjunction with remedial welding, seal repairs, etc.).

Reviewers should ensure that the drying specifications are consistent with the proposed operating controls and limits described in the technical specifications provided in Chapter 13 of the SAR. In addition, reviewers should assess the need for any additional technical specifications.

Welding and Sealing (HIGH Priority)

Structural and materials disciplines should coordinate their review of welded lids as described in Section 8.4.7, "Weld Design/Inspection," of this SRP for application of the proper weld joint, welding procedures, and non-destructive examination methods (NDE) to ensure the appropriate operating procedures are in place and acceptable. Reviewers should verify that procedures are acceptable for NDE and welding of the closure welds. While the NRC accepts progressive surface examinations utilizing dye penetrant testing (PT) or magnetic particle (MT) examination, it is only permitted if unusual design or loading conditions exist. In addition, if a PT or MT examination is used, a stress-reduction-factor of 0.8 is imposed on the weld strength for the reasons presented in Section 8.4.7.3. The SAR should also ensure ALARA principles are followed and include acceptable provisions for correcting weld defects and any additional drying and purging that may be necessary.

The reviewer should verify that provisions for placing and tightening any closure bolts, such as those associated with concrete casks, are consistent with information presented in Chapters 2, 3, and 10 of the SAR that address applicable design criteria, structural evaluation, and the acceptance tests and maintenance program, respectively. The materials discipline should ensure that the closure bolts satisfy the conditions given in Section 8.4.10, "Bolt Applications," of this SRP. The SAR should specify the torque required to properly seal the closure lid. The inner seal should be tested using a helium leak test with the interior of the cask pressurized as previously described. The outer seal should also be tested using a helium leak test with the between-seal volume pressurized as required by the respective subsection of the ASME B&PV Code, Section III.

Filling and Pressurization (LOW Priority)

The reviewer should verify that the procedure recommendations address steps to fill and pressurize the cask with inert gas such as helium with a known maximum percentage of impurities. The operating procedures should state that the filling and pressurization (or evacuation and backfill) process be repeated if the cask cavity is exposed to the atmosphere. Also, the reviewer should ensure that the procedure recommendations include the requirements in Chapter 13, "Technical Specifications and Operation Controls and Limits Evaluation" of the SAR.

The SAR should specify the leak rate criteria (e.g., total leakage, leakage per closure, sensitivities of tests, etc.), and the reviewer should verify that these criteria are consistent with those presented in Chapters 2, 9, and 13 of the SAR. In addition, the reviewer should assess the general methods of leak testing (e.g., pressure rise, mass spectrometry) as they apply to the leak rate being tested. Particular attention should be paid to the possible use of quick-disconnect fittings for draining and filling operations. Although no credit is usually taken for these devices as part of the confinement boundary, their presence can negate the results of the leak test, and the SAR should provide guidance regarding their use. In addition, the guidelines presented in the SAR should note that leak testing is in accordance with ANSI N14.5, "Radioactive Materials – Leakage Tests on Packages for Shipment."

The reviewer should ensure that the SAR presents applicable pressure testing criteria (e.g., test pressure, hold periods, inspections) and that these criteria are consistent with those presented in Chapter 9 of the SAR.

Administrative Programs (HIGH Priority)

The applicant may request that one or more administrative programs be approved by the NRC in lieu of the requirements set forth in Section 9.5.1 above for offsite releases, draining and drying, filling and pressurization, and welding and sealing. Requirements for such administrative programs are provided in NUREG-1745, "Standard Format and Content for Technical Specifications for 10 CFR Part 72 Cask Certificates of Compliance," and are summarized in this section.

The applicant may request the NRC approve an administrative program for offsite releases. In this case, the reviewer should verify that the SAR describes a Radioactive Effluent Control Program and related operating procedures that shall be established, implemented, and maintained to:

- Implement the requirements of 10 CFR 72.126.

- Limit the surface contamination and verification of meeting those limits prior to removal of the cask from the Part 50 structure.

- Limit the leakage rate and verification of meeting those limits prior to removal of the cask from the Part 50 structure.

- Show compliance with the requirements of 10 CFR 72.104 and 72.106.

In addition, the applicant may request the NRC approve an administrative program for cask loading. In this case, the reviewer should verify that the SAR requirements are implemented for loading fuel and components into the cask and preparing the cask for storage. The requirements of the program for loading and preparing the cask should be completed prior to removing the cask from the 10 CFR Part 50 structure. (Items 1, 5, and 6 below are associated with requirements that will remain in the technical specifications; however, the process for establishing the specified action limit may be moved to this administrative program if a method of evaluation acceptable to the NRC is presented in the SAR. Items 2, 3, and 4 have been relocated from the Limiting Conditions of Operations [LCO] section to this administrative program because it is felt that NRC-approved methods of evaluation will be relatively easy to develop. If appropriate methods are not presented in the SAR, these items will retain LCOs.)

At a minimum, the cask-loading program shall establish criteria that need to be verified to address SAR commitments and regulatory requirements for:

1. Vacuum drying times and pressures, or forced helium drying criteria,, to assure that the short-term fuel temperature limits are not violated and the cask is adequately dry.

2. Inerting pressure and purity to assure adequate heat transfer and corrosion control.

3. Leak testing to assure adequate cask integrity and consistency with the offsite dose analysis.

4. Surface dose rates to identify significant problems with shielding fabrication, gross misloads, and verify consistency with the offsite dose analysis.

5. Ambient and pool water temperature to assure adequate subcriticality and material ductility.

6. SNF pool boron concentration to verify the acceptable subcriticality margin.

7. Clad oxidation thickness for high-burnup fuel in accordance with SRP Chapter 8, :Materials Evaluation" or other NRC-approved methodology if high-burnup fuel is included in the contents.

The program shall include compensatory measures and appropriate completion times if the program requirements are not met.

9.5.2 Cask Handling and Storage Operations (LOW Priority)

The reviewer should examine the recommendations associated with procedures necessary to transfer the cask to the storage location. The reviewer should pay particular attention to ensuring that all accident events applicable to such transfer are bounded by the design events analyzed in Chapters 2, "Principal design Criteria", 3, "Structural Evaluation" and 12, "Accident Analyses Evaluation" of the SAR. This includes procedures to be specified in the SAR for use after a design-basis accident for testing the effectiveness of the shielding. The structural and thermal disciplines should coordinate their review to ensure that all conditions for lifting and handling methods are bounded by the evaluations in their respective Chapters 3 and 4 of the SAR. There may be technical specifications associated with cask transfer operations such as restricting lift heights and environmental conditions (e.g., high/low temperatures, etc.) requiring coordination with the review in Chapter 13, "Technical Specifications and Operating Controls and Limits Evaluation," of this SRP.

The reviewer should verify that the procedure recommendations discuss the inspection, surveillance, and maintenance requirements that are applicable during ISFSI storage. Surveillance and monitoring requirements should also be included in Chapter 13 of the SAR, and maintenance should be included in Chapter 10 of the SAR. Reviewers should note that if the confinement vessel closure is bolted, the NRC staff generally requires that the successful operation of the seals be demonstrated with an initial leak test and a monitoring system and/or a surveillance program as discussed in Chapter 10, "Acceptance Tests and Maintenance Program Evaluation," of this SRP.

The shielding and radiation protection reviewers should verify that proposed procedures give due consideration to maintaining doses ALARA during cask handling and storage operations.

The applicant may request that an ISFSI Operations Program be approved by the NRC. Requirements for such an administrative program are provided in NUREG-1745. The reviewer should verify that such a program establishes criteria for:

- Minimum cask center-to-center spacing.

- Pad parameters (i.e., pad thickness, concrete strength, soil modulus, reinforcement, etc.) that are consistent with the SAR analysis.

- Maximum lifting heights for the cask system to ensure that the gravity load limits are met for the design-basis events.

9.5.3   Cask Unloading (Priority – as indicated)

(LOW Priority) The reviewer should verify that the SAR adequately describes the necessary unloading procedure recommendations. The unloading procedure descriptions should present the activities sequentially in the anticipated order of performance, including those key prerequisite and preparation tasks that must be accomplished before cask unloading. Where applicable, the reviewer should verify that the procedure guidance ensures that any fluids, such as shield or borated water, fill their respective cavities according to design specifications. Additionally, for casks that require borated water to maintain subcriticality, the reviewer should ensure that the procedure guidance includes verification that the water to be used for cask reflood meets the minimum soluble boron content required by the Technical Specifications.

Damaged Fuel (LOW Priority)

The SAR should include appropriate additional measures for the potential presence of damaged fuel. Procedures should be designed to maximize worker protection from unanticipated radiation exposures or contaminants due to damaged fuel in accordance with ALARA principles and, to the maximum extent possible, prevent any uncontrolled releases to the environment. The following points outline the relevant safety concerns and an acceptable approach to address damaged fuel contingencies in cask unloading:

- The procedure descriptions should provide for fuel unloading under normal conditions.

- The unloading process should ensure that the fuel can be safely unloaded with regard to structural, criticality, thermal, and radiation protection considerations. This includes the provision for safe maintenance of the fuel and cask while any additional measures needed to address suspected damaged fuel are planned and implemented.

- The unloading process should reflect the potential for damaged fuel and changing radiological conditions.

- The process should include measures to check for and detect damaged fuel conditions (such as atmosphere samples) before opening the cask. (Note that fuel oxidation resulting from exposure to air at temperatures typical for dry storage is a known form of fuel degradation. Therefore, the presence of air in a cask designed to maintain an inert atmosphere indicates that the fuel may be degraded. The detection of fission gases is another indicator that the fuel may be degraded.)

The process may establish sample result thresholds above which damaged fuel is suspected. Other technically sound methods may be used to check for potential air leakage paths. Such methods may include designs that monitor cask internal pressure or seal integrity and alert the licensee to a problem before oxidation could occur. However, this method may not address detection of potential fuel degradation resulting from other mechanisms (such as a cask drop accident).

- If the sample indicates normal conditions, the normal unloading process should be followed.

- If damaged fuel is suspected or found, the procedure description should stipulate that additional measures, appropriate for the specific conditions that include the canning of the damaged fuel, are to be planned, reviewed, and approved by the designated approval authority and implemented to minimize exposures to workers and radiological releases to the environment. These additional measures may include provision of filters, respiratory protection, and other methods to control releases and exposures in accordance with ALARA.

Cooling, Venting, and Reflooding (LOW Priority)

The reviewer should verify that the SAR describes applicable operational measures to control cask cooling, venting, and reflooding (when appropriate). Also, the reviewer should verify that these measures are consistent with the results of the structural, materials, and thermal evaluations in the SAR, respectively. Cask cooling, venting, and reflooding should not result in damage to the fuel. Operational measures may include external cooling of the confinement cask for initial temperature reduction, restricting reflood flow rates to control and limit internal cask pressure from steam formation, and limiting cooldown rates.

Special attention should be devoted to reviews in this area since analysis of existing designs have predicted fuel temperatures during storage and transfer in excess of 533.15°K (500°F) for design-basis heat loads. Operational controls may be required to address the following potential effects during a cooldown and reflood evolution:

- Cask pressurization may occur as a result of steam formation as reflood water contacts hot surfaces.

- Excessive cooling rates may cause fuel cladding and fuel rod component damage and release of radioactive material as a result of stress (thermal, internal pressure, etc.) beyond material strengths (see SRP Section 8.4.17.1, "Cladding Temperature Limits").

- Excessive cooling rates may induce thermal stress that causes gross deformation of the fuel assembly components and subsequent binding with the basket.

- Cask supply and vent line failures from inadequate design for pressure and temperature could result in radiological exposures and personnel hazards (e.g., steam burns).

Fuel Crud (LOW Priority)

The reviewer should verify that the procedure descriptions include contingencies for protection from fuel crud particulate material. Appendix E of ANSI/ANS 57.9 provides a short discussion of crud with respect to dry transfer systems. The unloading procedures should alert cask users to wait until any loose particles have settled and to slowly move the fuel assemblies to minimize crud dispersion in the SNF pool. Experience with wet unloading of boiling-water reactor (BWR) fuel after transportation has involved handling significant amounts of crud. This fine crud, which includes ^{60}Co and ^{55}Fe, will remain suspended in water or air for extended periods. The dry cask reflood process, during unloading of BWR fuel, has the potential to disperse crud into the fuel transfer pool and the pool area atmosphere, thereby creating airborne exposure and personnel contamination hazards. By contrast, no significant crud dispersal problems have been observed in handling pressurized-water reactor (PWR) fuel due to differences in the characteristics of crud on this type of fuel.

ALARA (LOW Priority)

The reviewer should verify that the procedure descriptions incorporate ALARA principles and practices. These may include provisions to perform radiological surveys, exposure and contamination control measures, temporary shielding, and suggested caution statements related to specific actions that could change radiological conditions. The reviewer should verify that any recommended surveys incorporate the applicable operating controls and limits described in Chapter 13, "Technical Specifications and Operation Controls and Limits Evaluation" of the SAR.

Offsite Release (LOW Priority)

Where applicable, the reviewer should verify that the SAR describes methods such as filtered ventilation, decontamination, or temporary containments to minimize offsite releases. The procedures should also provide for minimizing generation of radioactive waste.

Administrative Programs (HIGH Priority)

The applicant may request that the NRC approve an administrative program for cask unloading. NUREG-1745 provides requirements for such an administrative program. The reviewer should verify the proposed administrative program meets the requirements summarized in Section 9.5.1 of this SRP.

9.6   Evaluation Findings

The reviewer should examine the 10 CFR Part 72 acceptance criteria and provide a summary statement for each. These statements should be similar to the following model, as applicable:

F9.1   The [cask designation] is compatible with [wet/dry] loading and unloading. General procedure descriptions for these operations are summarized in Chapter(s) _____ of the applicant's safety analysis report (SAR). Detailed procedures will need to be developed and evaluated on a site-specific basis.

F9.2   The [welded/bolted lids or other features] of the cask allow ready retrieval of the spent fuel for further processing or disposal as required.

F9.3   The smooth surface [or other feature] of the cask is designed to facilitate decontamination. Only routine decontamination will be necessary after the cask is removed from the spent fuel pool.

F9.4   No significant radioactive waste is generated during operations associated with the independent spent fuel storage installation (ISFSI). Contaminated water from the spent fuel pool will be governed by the 10 CFR Part 50 license conditions.

F9.5   No significant radioactive effluents are produced during storage. Any radioactive effluents generated during the cask loading will be governed by the 10 CFR Part 50 license conditions.

F9.6   The content of the general operating procedures described in the SAR are adequate to protect health and minimize damage to life and property. Detailed procedures will need to be developed and approved on a site-specific basis.

F9.7   The radiation protection chapter of this SER assesses the operational restrictions to meet the limits of 10 CFR Part 20. Additional site-specific restrictions may also be established by the site licensee.

The reviewer should provide a summary statement similar to the following:

"The staff concludes that the generic procedures and guidance for the operation of the [cask designation] are in compliance with 10 CFR Part 72 and that the applicable acceptance criteria have been satisfied. The evaluation of the operating procedure descriptions provided in the SAR offers reasonable assurance that the cask will enable safe storage of spent fuel. This finding is based on a review that considered the regulations, appropriate regulatory guides, applicable codes and standards, and accepted practices."

10 ACCEPTANCE TESTS AND MAINTENANCE PROGRAM EVALUATION

10.1 Review Objective

The acceptance tests and maintenance program review ensures that the applicant's Safety Analysis Report (SAR) includes the appropriate acceptance tests and maintenance programs for the system. A clear, specific listing of these commitments will help avoid ambiguities concerning design, fabrication, and operational testing requirements when the U.S. Nuclear Regulatory Commission (NRC) staff conducts subsequent inspections. Acceptance tests may also be described in the applicable chapter of this Standard Review Plan (SRP).

10.2 Areas of Review

This chapter of the dry storage system (DSS) SRP provides guidance for use in evaluating the acceptance tests and maintenance programs outlined in the SAR. The acceptance tests demonstrate that the cask has been fabricated in accordance with the design criteria and that the initial operation of the cask complies with regulatory requirements. The maintenance program describes actions that the licensee needs to implement during the storage period to ensure that the cask performs its intended functions.

As defined in Section 10.5, "Review Procedures," a comprehensive evaluation *may* encompass the following acceptance tests and maintenance programs:

Acceptance Tests
Structural/Pressure Tests
Leak Tests
Visual and Nondestructive Examination Inspections
Shielding Tests
Neutron Absorber Tests
Thermal Tests
Cask Identification

Maintenance Program
Inspection
Tests
Repair, Replacement, and Maintenance

10.3 Regulatory Requirements

This section presents a summary matrix of the portions of U.S. Code of Federal Regulations (CFR), Part 72, "Licensing Requirements for the Independent Storage of Spent Nuclear Fuel High-Level Radioactive Waste and Reactor-Related Greater Than Class C Waste," Title 10, "Energy" (10 CFR Part 72) that are relevant to the review areas addressed by this chapter. The NRC staff reviewer should read the exact referenced regulatory language. Table 10-1 matches the relevant regulatory requirements associated with this chapter to the areas of review identified in the previous section.

Table 10-1 Relationship of Regulations and Areas of Review

Areas of Review	10 CFR Part 72 Regulations							
	72.82 (d)	72.122 (a), (f)	72.124 (b)	72.162	72.212 (b)(8)	72.232 (b)	72.236 (c)	72.236 (g), (j), (k), (l)
Acceptance Tests	•	•	•	•		•		•
Maintenance Program	•	•						•
Design Verification	•	•			•	•	•	•

10.4 Acceptance Criteria

In general, the acceptance tests and maintenance programs outlined in the SAR should cite appropriate authoritative codes and standards. The staff has previously accepted the following as the regulatory basis for the design, fabrication, inspection, and testing of DSS components:

System/Component	Acceptable Regulatory Basis*
Confinement System	• American Society of Mechanical Engineers (ASME), "Boiler and Pressure Vessel (B&PV) Code," Section III, Division 1, 2007 • "American National Standard for Radioactive Materials – Leakage Tests on Packages for Shipment" (ANSI N14.5)
Confinement Internals (e.g. basket)	• ASME B&PV Code, Section III, Subsection NG
Metal Cask Overpack	• ASME B&PV Code, Section VIII
Concrete Cask Overpack	• American Concrete Institute (ACI), "Code Requirements for Structural Concrete" (ACI-318), "Code Requirements for Nuclear Safety Related Concrete" (ACI-349), as appropriate
Other Metal Structures	• ASME B&PV Code, Section III, Subsection NF • American Institute of Steel Construction (AISC), "Manual of Steel Construction"
* The SAR should clearly identify any exceptions to the listed codes and standards (see SRP Chapter 13, "Technical Specifications and Operating Controls and Limits Evaluation").	

10.5   Review Procedures

Introduction

Figures 10-1 and 10-2 present an overview of the evaluation process and can be used as a guide to assist in coordinating with the review disciplines.

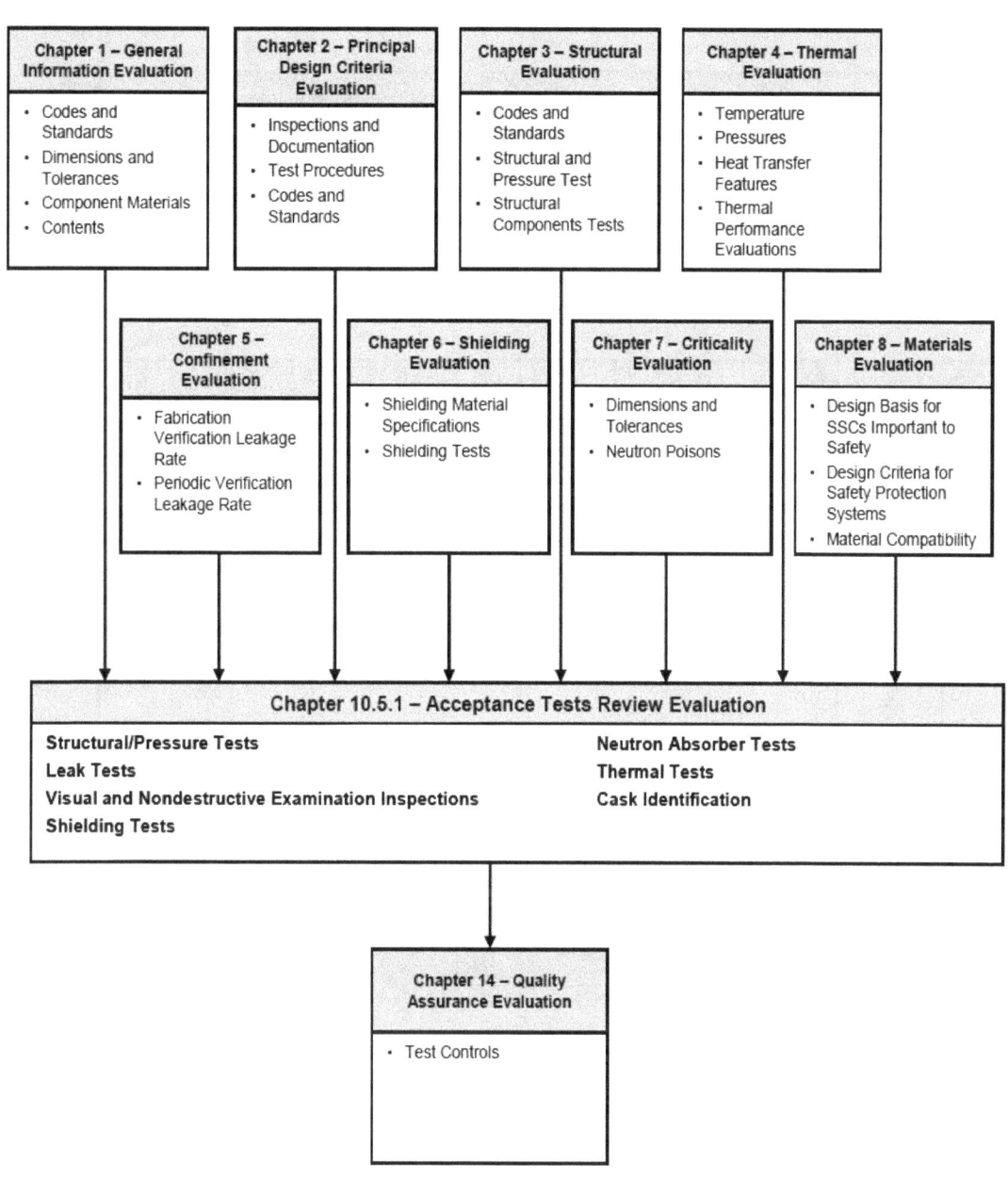

Figure 10-1 Overview of Acceptance Test Review Evaluation

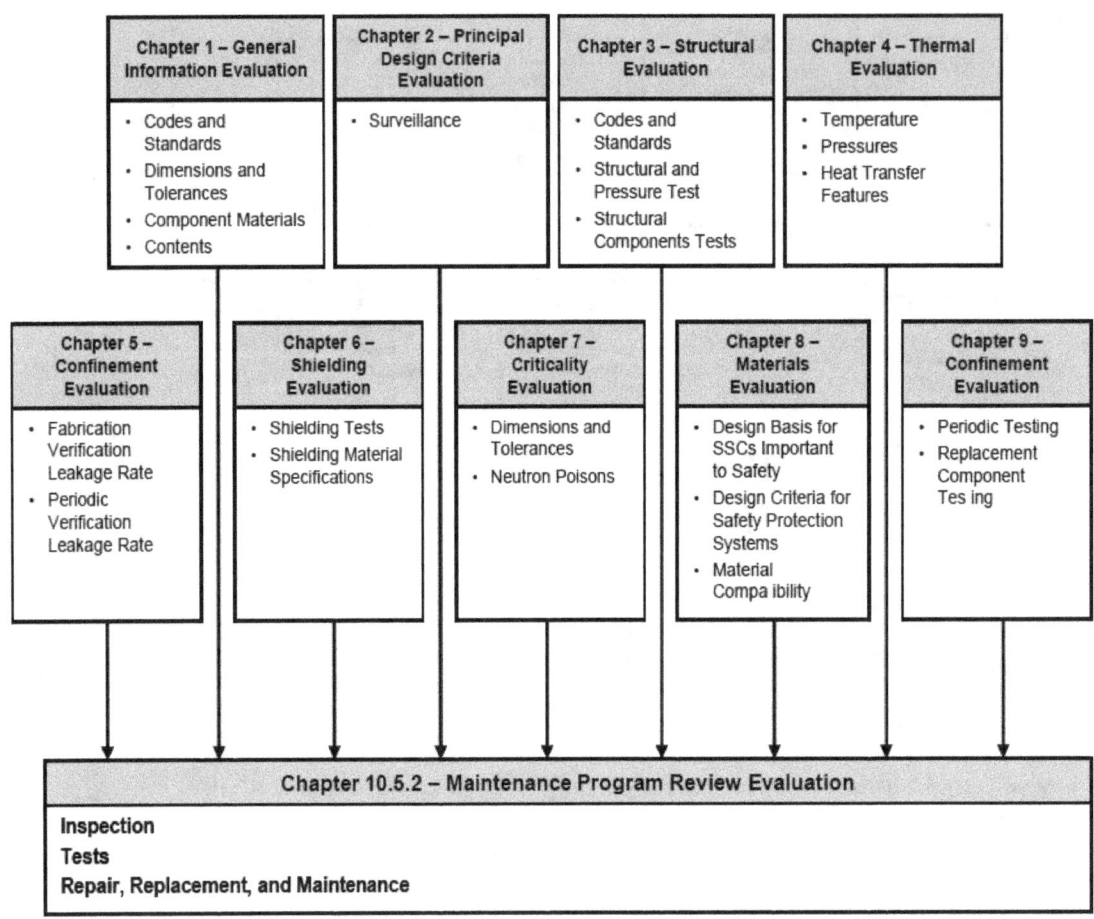

Figure 10-2 Overview of Maintenance Program Review Evaluation

The review procedures described in this section are presented in a format intended to facilitate a single, independent review. Although one or more individual(s) may be tasked with preparing the corresponding section of the safety evaluation report (SER) related to the proposed acceptance tests and maintenance program, all review team members should examine the related information presented in the SAR. Information in the SAR related to the acceptance tests may be located in the chapters related to specific disciplines (e.g. SAR Chapter 4, "Thermal Evaluation") and/or in SAR Chapter 10, "Acceptance Tests and Maintenance Program." Reviewers should devote special attention to those tests (or the lack of tests) that affect their functional area of review. If the descriptions included in the SAR are not sufficiently detailed to allow a complete evaluation concerning fulfillment of the acceptance criteria, reviewers should request additional information from the applicant.

In general, applicants commit to design, construct, and test the system under review to the codes and standards identified in SAR Chapter 2, "Principal Design Criteria." The NRC does not generally review specific test and maintenance procedures as part of the licensing process; however, the applicant is expected to describe (in the SAR) certain elements of the proposed test and maintenance programs. The staff may inspect selected portions of test procedures as part of its onsite activities.

The following subsections provide *representative examples* of test and maintenance program elements that should be subject to licensing review. If included in the SAR, each of these tests and maintenance elements should be reviewed to ensure that the applicant has identified the purpose of the test, explained the proposed test method (including any applicable standard to which the test will be performed), defined the acceptance criteria and bases for the test, and described the actions to be taken if the acceptance criteria are not satisfied.

10.5.1 Acceptance Tests (Priority – as indicated)

The following guidance is presented on the basis of tests deemed acceptable by the staff in previous SAR reviews. The guidance is based on operational experience and the knowledge from past licensing reviews. Alternative tests and criteria may be used if the SAR provides appropriate explanation and adequate justification. Additional tests and criteria may be needed, depending on the operational experience and uniqueness of the design proposal.

10.5.1.1 Structural/Pressure Tests

(MEDIUM Priority) Lifting trunnions should be fabricated and tested in accordance with ANSI N14.6, "American National Standard for Radioactive Materials-Special Lifting Devices for Shipping Containers Weighing 10,000 pounds (4,500 Kilograms) or More." Site-specific details of the pool and lifting procedures may enable the cask to be considered a non-critical load, as defined by this standard. Generally, however, the cask is considered a critical load during its handling in the pool. Consequently, trunnion testing should be performed at a minimum of 150 percent of the maximum service load, if redundant lifting is employed or 300 percent of the service load if non-redundant lifting applies. These load tests should be performed to ensure that the trunnions and cask are conservatively constructed and provide an adequate margin of safety when filled with SNF. Trunnion load testing should also be performed annually for the transfer cask and at least one year before use for the storage cask. Load testing of integral trunnions is not required once the loaded storage cask has been placed on the pad. Restrictions on cask lifting resulting from these tests should be included in Chapter 13, "Technical Specifications and Operating Controls and Limits Evaluation," of the SAR and the related SER prepared by the NRC staff. SAR Chapter 10, "Acceptance Tests and Maintenance

Program Evaluation" should explicitly state the testing values. Periodical NDE, in lieu of annual load tests, is acceptable for the trunnion provided that other conditions as specified in ANSI N14.6 are also met.

(MEDIUM Priority) The entire confinement boundary should be pressure tested hydrostatically or pneumatically to 125 or 110 percent of the design pressure, respectively. The pressure test should be performed in accordance with governing code associated with the confinement boundary, which typically has been ASME B&PV Code, Section III, Division 1, Subsection NB or NC. The test pressure should be maintained for a minimum of 10 minutes, after which a visual inspection should be performed to detect any leakage. SAR Chapter 10, "Acceptance tests and Maintenance Program Evaluation" should clearly specify the hydrostatic and pneumatic test pressures. The helium leakage test, per ANSI 14.5 is not considered as a substitute for the Code required pressure test, and conversely, the Code pressure test is not a substitute for the helium leakage test. If a shop pressure test isn't performed and only a field pressure test is performed after the first closure weld is made, the staff has accepted the shop helium leakage test as meeting the pressure test acceptance criteria of no leakage for the shell welds since they are generally inaccessible in the field.

(LOW Priority) Some casks contain a neutron shielding material that may off-gas at higher temperatures. Such material is usually contained inside a thin steel shell to prevent loss of mass and provide protection from minor accidents and natural phenomenon events. Rupture disks or relief valves are generally provided to prevent catastrophic failure of this shell. The shell should be tested to 125 percent of the rupture disk burst pressure, which is usually equivalent to 125 percent of the shell design pressure. The SAR should clearly specify the burst pressure for the rupture disk, along with its coincident burst temperature and tolerance on burst pressure.

(HIGH Priority) Some cask designs use ferritic steels that are subject to brittle fracture failures at low temperature. ASME B&PV Code, Section II, Part A, contains procedures for testing ferritic steel used in low temperature applications. NUREG/CR-1815, "Recommendations for Protecting Against Failure by Brittle Fracture in Ferritic Steel Shipping Containers Up to Four Inches Thick," provides staff guidance concerning materials and thickness ranges subject to brittle fracture testing. On the basis of guidance in NUREG/CR-1815, Section 5.1.1, the NRC established two methods for identifying suitable materials:

- The nil-ductility transition (NDT) temperature must be determined by either direct measurement, (American Society for Testing and Materials' (ASTM) "Method of Conducting Drop Weight Test to Determine Nil-ductility Transition Temperature for Ferritic Steel" [ASTM E-208]) or indirect measurement ("Dynamic Tear Testing of Metallic Materials" [ASTM E-604]), and the minimum operating temperature of the steel must be specified as 28°C (50°F) higher than the NDT.

- The NRC staff accepts ASME Charpy testing procedures for verification of the material's minimum absorbed energy. Acceptable energy absorption values and test temperatures of Charpy, V-Notch impact tests are listed in the ASME B&PV Code, Section II, SA-20, "Specifications for General Requirements for Steel Plates for Pressure Vessels" Table A1.15. Coordinate with the thermal review (Chapter 4 of this SRP) to ensure that the applicant selected the correct temperatures for the tests and that the SAR specifies the method of testing.

For casks with ferritic steel walls thicker than 102 mm (4 in.), the guidance provided in NUREG/CR-3826, "Recommendations for Protecting Against Failure by Brittle Fracture in Ferritic Steel Shipping Containers Greater than Four Inches Thick," should be followed.

10.5.1.2   Leak Tests (MEDIUM Priority)

The licensee should perform leak tests on all confinement boundaries except as excluded in Chapter 8, "Materials Evaluation" - Section 8.9.2, which only applies to the closure welds typically made in the field. For all-welded cask confinements, the NRC staff has, with adequate justification, considered it acceptable for licensees to omit leak testing of the second cask closure weld and the seal welds for the closure plates of the purge and vent valves (if not potentially pressurized at the time of welding). For such cases, leak testing must show that the inner closure weld meets the leakage limits. A fabrication leak test should be performed on every canister in the shop to ensure that the tested leakage rate is compatible with the regulatory dose limits at the controlled area boundary, 10 CFR 72.236(d), (i), and (j).

Leakage criteria in units of $Pa.m^3/s$ or reference cm^3/s must be at least as restrictive as those specified in the principal design criteria (in SAR Chapter 2). The SAR should also indicate the general testing methods (e.g., pressure increase, mass spectrometer) and required sensitivities. If cask closure depends on more than one seal (e.g., lid, vent port, drain port), the leakage criteria should ensure that the total leakage is within the design requirements. Leak testing should be conducted in accordance with ANSI N14.5.

10.5.1.3   Visual and Nondestructive Examination Inspections

(HIGH Priority) Reviewers should verify the applicant's commitment to fabricate and examine cask components in accordance with an accepted design standard such as ASME B&PV Code, Section III or VIII. These sections define the examination requirements mentioned in Section II, "Materials Specifications and Properties"; Section V, "NDE Specifications and Procedures"; and Section IX, "Qualification Standard for Welding and Brazing Procedures, Welders, Brazers, and Welding and Brazing Operators." The following guidance assumes that the ASME B&PV Code is applicable to the cask being reviewed.

(HIGH Priority) The nondestructive examination (NDE) of weldments must be well-characterized on drawings, using standard NDE symbols and/or notations (see American Welding Society's (AWS) "Standard Symbols for Welding, Brazing, and Nondestructive Examination" [AWS A2.4]). Each fabricator should be required to establish and document a detailed, written weld inspection plan in accordance with an approved quality assurance (QA) program that complies with 10 CFR Part 72, Subpart G. The inspection plan should include visual (VT), dye penetrant (PT), magnetic particle (MT), ultrasonic (UT), and radiographic (RT) examinations, as applicable. The inspection plan should identify welds to be examined, the examination sequence, type of examination, and the appropriate acceptance criteria as defined by either the ASME B&PV Code, or an alternative approach proposed and justified by the applicant. Inspection personnel should be qualified, in accordance with the current revision of the American Society for Nondestructive Testing's (SNT) "Personnel Qualification and Certification in Nondestructive Testing" (SNT-TC-1A), as specified by the ASME B&PV Code. All weld-related NDE should be performed in accordance with written and approved procedures. Fabrication controls and specifications should be in-place and field tested to prevent post-welding operations (such as grinding) from compromising the design requirements (such as wall thickness).

(HIGH Priority) Confinement boundary non-closure welds should meet the requirements of ASME B&PV Code, Section III, Division 1, Subsections NB or NC, Article NB/NC-5200, "Required Examination of Welds for Fabrication and Preservice Baseline." This section requires volumetric examination and either PT or MT for all Category A and most Category B or Category C welded joints in vessels, and longitudinal or full penetration welded joints in other components. The ASME-approved specifications for RT, UT, PT, and MT are detailed in ASME B&PV Code, Section V, Articles 2, 4, 6, and 7, respectively.

(HIGH Priority) Acceptance standards for nondestructive testing should be in accordance with ASME B&PV Code, Section III, Division 1, Subsection NB or NC -5300. Testers should reject unacceptable imperfections (such as a crack, a zone of incomplete fusion or penetration, elongated indications with lengths greater than specified limits, and rounded indications in excess of the limits in ASME B&PV Code, Section III, Division 1, Appendix VI). Repaired welds should be reexamined in accordance with the original examination method and associated acceptance criteria.

(HIGH Priority) For confinement welds that cannot be volumetrically examined using RT, the licensee may use 100 percent UT. The ASME-approved UT specifications are detailed in ASME B&PV Code, Section V, Article 4. Acceptance criteria should be defined in accordance with ASME B&PV Code, Section III, Division 1, Subsection NB or NC-5330, "Ultrasonic Acceptance Standards." Cracks, lack of fusion, or incomplete penetration are unacceptable, regardless of length.

(HIGH Priority) The NRC has accepted multiple surface examinations of welds, combined with helium leak tests for inspecting the final redundant seal welded closures.

(HIGH Priority) For confinement internals, the licensee should perform all NDE testing in accordance with ASME B&PV Code, Section III, Division 1, Subsection NG.

(LOW Priority) Nonconfinement welds (which exclude welds of confinement internals) should meet the requirements of ASME B&PV Code, Section III, Subsection NF, or Section VIII, Division 1, as applicable. The required volumetric examination of welds is either RT or UT, as discussed in ASME B&PV Code, Section III, NF-5200, and Section VIII, UW-11. The appropriate specifications from ASME B&PV Code, Section V, are invoked in Article 2 for RT and in Article 5 for UT. Acceptance standards for RT are detailed in ASME B&PV Code, Section III, Subsection NF, NF-5320, "Radiographic Acceptance Standards," and for UT in NF-5330, "Ultrasonic Acceptance Standards." For Section VIII weldments, RT acceptance criteria should be in accordance with ASME B&PV Code, Section VIII, Division 1, UW-51, and the repair of unacceptable defects should be in accordance with UW-38. Repaired welds should be reexamined in accordance with the original acceptance criteria.

(LOW Priority) Nonconfinement welds that cannot be examined using RT should undergo UT in accordance with ASME B&PV Code, Section V, Article 4. Acceptance criteria should be in accordance with ASME B&PV Code, Section VIII, Division 1, UW-53 and Appendix 12, and the repair of unacceptable defects should be in accordance with UW-38. Repaired welds should be reexamined in accordance with the original examination methods and associated acceptance criteria. If applicable, the SAR should also justify the rationale for not requiring RT examination of these welds.

(LOW Priority) Nonconfinement welds for cask system components that are designed and fabricated in accordance with ASME B&PV Code, Section III, that cannot be examined using RT or UT should undergo PT or MT examination in accordance with ASME B&PV Code, Section V, Articles 6 and 7, respectively. Acceptance criteria should be in accordance with Articles NF-5350 and NF-5340, respectively. Repaired welds should be reexamined in accordance with the original acceptance criteria. If applicable, the SAR should also justify the rationale for not requiring volumetric inspection techniques (RT or UT) for these welds.

(Low Priority) Nonconfinement welds may also be welded, repaired and examined in accordance with AWS D1.1, Structural Welding Code – Steel, D1.3, Structural Welding Code – Sheet Steel and D1.6, Structural Welding Code – Stainless Steel. Use of these standards shall be called out on the licensing drawings.

(LOW Priority) Finished surfaces of the cask should be visually examined in accordance with the ASME B&PV Code Section V, Article 9. For welds examined using VT, the acceptance criteria should be in accordance with ASME B&PV Code, Section VIII, Division 1, UW-35 and UW-36, or NF-5360, "Acceptance Standards for Visual Examination of Welds."

(HIGH for confinement/LOW for non-confinement) The licensee should use PT to detect discontinuities (such as cracks, seams, laps, laminations, and porosity) that open to the surface of nonporous metals. PT should be performed in accordance with ASME B&PV Code, Section V, Article 6. Acceptance criteria for PT examination of confinement welds should be in accordance with ASME B&PV Code, Section III, Subsection NB/NC, Article NB/NC-5350. Repair procedures should be in accordance with NB/NC-4450 of the ASME B&PV Code, Section III. Acceptance criteria for PT examination of nonconfinement welds should be in accordance with ASME B&PV Code, Section VIII, Division 1, Appendix 8, or NF-5350, "Liquid Penetrant Acceptance Standards." Repair procedures should be in accordance with ASME B&PV Code, Section III or NF-2500, "Examination and Repair of Material," and NF-4450, "Repair of Weld Material Defects."

10.5.1.4   Shielding Tests (LOW Priority)

The materials that comprise the DSS should sufficiently maintain their physical and mechanical properties during all conditions of operations. DSS gamma shielding materials (e.g., lead) should not experience slumping or loss of shielding effectiveness to an extent that compromises safety. The shield should perform its intended function throughout the licensed service period.

DSS materials used for neutron shielding should be designed to perform their safety function without degradation, gas release, or physical alteration for the full term of the license. Tests are required to ensure these conditions are met.

Tests of the effectiveness of both the gamma and neutron shielding may be required if, for example, the cask contains a poured lead shield or a special neutron absorbing material. In such instances, the SAR should describe any scanning or probing with an auxiliary source for the purpose of characterizing the shielding. This shield testing should be done for every cask that uses poured shielding material, to demonstrate proper fabrication in accordance with the design drawings. Alternatively, the applicant may propose an alternate testing program for fabricated casks with appropriate justification.

Dose measurements of loaded SNF, in lieu of an auxiliary source, may be used to verify shielding effectiveness with appropriate scanning of the shield and appropriate testing program that considers the actual source strength of the loaded contents..

10.5.1.5   Neutron Absorber Tests (HIGH Priority)

Neutron absorber materials require both qualification and acceptance testing to provide assurance that the control of criticality by absorbing thermal neutrons will be assured in systems designed for nuclear fuel storage, transportation or both. Both qualification and acceptance testing are in general as described in ASTM Designation C1671, "Standard Practice for Qualification and Acceptance of Boron Based Metallic Neutron Absorbers for Nuclear Criticality Control for Dry Storage Systems and Transportation Packaging."

Acceptance tests are used to ensure that material properties for plates and other shapes produced in a given production run are in compliance with the materials requirements of the application. In one sense, acceptance tests verify that the material of a given production run has yielded products that have been shown to be like the products that were used in the qualification testing. Acceptance tests are used to ensure that the production process is operating in a satisfactory manner, and they use statistical data for selected measurable parameters. For all boron-containing absorber materials, acceptance tests should (a) verify ^{10}B content and uniformity, (b) require visual examinations to establish only acceptable levels of defects are present from cracks, porosity, blisters, or foreign inclusions, and (c) make dimensional (e.g., plate thickness which is important to the areal density).

Some materials may obtain 100 percent credit for the amount of ^{10}B that is shown to be present in the absorber materials. This level of credit is sometimes called 90 percent credit because the credit level refers to a manner in which K-effective calculations are conducted and in these calculations, any absorber is given a 10 percent penalty before being used in the calculation. Likewise other materials that are given only 82 percent credit are called materials with 75 percent credit. For purposes of obtaining high levels (100 percent) of credit, the amount of ^{10}B, which is the absorber species, is assessed in boron-containing absorber materials using neutron attenuation testing.

Neutron attenuation tests are calibrated using appropriate standards such as those based on (coated with) zirconium diboride (ZrB_2) plates to ensure the accuracy of the measured values. Approved substitutes may be used for the attenuation tests. These include tests such as chemical analysis, provided that (1) both the neutron attenuation tests and the alternative tests have at least the sensitivity of tests specified in C-1671 and (2) the alternate form of testing is regularly bench marked against calibrated neutron attenuation tests. Chemical analyses should also include spectrochemical analysis for material impurity levels and ^{10}B content. Uniformity is assessed using statistical sampling techniques that ensure that the entire plate of material and all plates in a lot meet a 95/95 criterion, which means that a test result has a 95 percent likelihood of containing the minimum required amount of ^{10}B and that this is known at the 95 percent confidence level.

The reviewer should confirm that the calculation of minimum poison content (e.g., poison areal density) conservatively accounts for tolerance limits on material density, poison concentration, and component dimensions. It is noted that thickness tolerances on rolled plates, sheets or shape are typically on the order of ± 10 percent. The acceptance testing should provide a representative sampling of coupons for plates or sheets from a given lot. Statistical sampling can be used to the extent practical, using test locations on a coupon that will account for local variations or anomalies within the coupon and hence within the plates represented by the coupon. Adequate numbers of samples should be taken to ensure the confidence level required for the application.

<u>Acceptance Testing of Fabricated Materials for 75-Percent Boron Credit</u>

For multi-phase absorber materials analyzed with 75-percent poison credit (or less) the reviewer should confirm that acceptance testing is consistent with the following:

- The effective ^{10}B content should be verified from plate coupons by either (a) neutron attenuation testing, or (b) chemical assay to determine boron content with mass spectrometric analysis for isotopic composition.

- Sufficient coupons should be taken for acceptance testing to justify the level of credit given. Rejection of a coupon should result in rejection of the plate from which it is taken. Sampling may be reduced to lesser percentages of the coupons taken (e.g., to 50 percent of all coupons) after acceptance of all coupons in the first 25 percent of the lot. A rejection during reduced inspection should invoke a 100 percent inspection for coupons from that lot.

- A visual examination of all plates for defects should be conducted.

<u>Acceptance Testing for Greater Than 75 Percent Boron Credit</u>

For acceptance testing of borated absorbers at levels of poison credit beyond 75 percent, the extent of the acceptance testing and inspection is enhanced. Some of the data helpful in meeting the guidance in C-1671 Sec 5.3.4 are as follows:

- The effective ^{10}B content is verified by neutron attenuation testing of coupons. An adequate number of coupons should be acceptance tested for each lot of materials to statistically demonstrate that the 95/95 criterion is satisfied for the minimum required ^{10}B content. The minimum areal density is specified in the SAR. Note that if the coupon from a plate fails the single neutron attenuation measurement, the associated plate is rejected unless acceptable alternative testing is done with acceptable results.

- Sufficient coupons should be taken to satisfy the 95/95 criterion. For example, coupons are taken from at least every other plate unless justification for fewer is given. Measurements are made on samples taken from 100 percent of all coupons. Rejection of a coupon should result in rejection of the plate. Sampling may be reduced to 50 percent of all coupons after acceptance of all coupons in the first 25 percent of the lot. A rejection during reduced inspection should invoke a return to 100 percent inspection for that lot.

- A statistical analysis of the neutron attenuation results should be performed by the applicant for all plates in a lot. This analysis shall show that the lot meets the 95/95 criterion. That is, using a one-sided tolerance limit factor for a normal distribution with at least 95 percent probability, the areal density is greater than or equal to the specified minimum value with 95 percent confidence level. Failure to meet this acceptance criterion of this statistical analysis shall result in rejection of the entire lot for use at the 100 percent (90 percent credit in K-effective calculations). Applicants may choose to convert all areal densities determined by neutron attenuation to a volume density by dividing by the thickness of the coupon. The one side tolerance limit of volume density with 95 percent probability and 95 percent confidence may then be determined. The minimum specified value of the areal density may be divided by the 95/95 lower tolerance limit of ^{10}B volume density to arrive at the minimum plate thickness. Hence, all plates which have any locations thinner than this minimum shall be rejected, and those equal to or thicker may be accepted.

- A visual examination of all plates for defects should be verified.

The reviewer should refer to Section 8.4.13.2 of this SRP regarding how to compute per level of credit.

10.5.1.6   Thermal Tests (LOW Priority)

Depending on the details of the DSS design and the ability to determine its heat removal capability through thermal analysis, testing may be required to verify performance. The applicant should establish acceptance criteria on the basis of the conditions of the test (e.g., test heat loading, ambient conditions). SAR Chapter 4, "Thermal Evaluation," should discuss the correlation between test performance and actual loading conditions to avoid ambiguous or unreviewed analysis after the test data are obtained.

10.5.1.7   DSS Identification (LOW Priority)

The vendor/licensee must mark the DSS with a model number, unique identification number, and empty weight. Generally this information will appear on a data plate, which should be detailed in one of the drawings included in SAR Chapter 1, "General Description." In addition, the vendor/licensee should mark the exterior of shielding casks or other structures that may hold the confinement while it is in storage. This marking should provide a unique, permanent, and visible number to permit identification of the cask stored therein.

10.5.2   Maintenance Program (LOW Priority)

DSSs are typically designed as passive units requiring minimal maintenance. The SAR should address the following areas, as applicable:

10.5.2.1   Inspection

Usually, the DSS has at least one monitoring system (e.g., pressure, temperature, dosimetry). The SAR should discuss how such systems will be used to provide information regarding possible off-normal events and what surveillance actions may be necessary to ensure that these systems function properly. Detailed procedures will be developed and implemented by the licensee at the site.

The SAR should describe routine periodic visual surface and weld inspections, which should be limited to the readily accessible surfaces (i.e., the exterior surface of the DSS and all surfaces of empty transfer casks). In addition, the SAR should discuss inspection of lifting and rotating trunnion load-bearing surfaces.

10.5.2.2   Tests

The SAR should describe any periodic tests of DSS components or calibration of monitoring instrumentation, as well as periodic tests to verify shielding, thermal, and confinement capabilities. The applicant should otherwise justify that aging and degradation of materials related to the shielding, confinement, and thermal designs are not credible during the licensed period of the DSS. The SAR should also describe procedures for any applicable periodic testing of neutron poison effectiveness. As an alternative to the licensee's periodic testing of neutron poison effectiveness, the applicant may show continued poison effectiveness in the manner described in Section 7.5.3.2 of this SRP. The qualification tests of the poison material, discussed in Section 8.4.13.3 of this SRP, may also be useful in showing the material's continued effectiveness.

In addition, the SAR should discuss any routine testing of support systems (e.g., vacuum drying, helium backfill, and leak testing equipment).

10.5.2.3   Repair, Replacement, and Maintenance

The SAR should discuss the repair and replacement of DSS components, as may be required during the lifetime of the storage and transfer casks. This discussion should include methods of repair or replacement, testing procedures, and acceptance criteria. The SAR should also describe procedures for routine maintenance (such as lubrication and re-application of corrosion inhibiting materials in the event of scratches) through the expiration of the service life of the equipment. Such information is also often included in SAR Chapter 12, "Accident Analyses," which describes actions to be taken following an off-normal event or accident-level condition.

10.6   Evaluation Findings

The 10 CFR Part 72 acceptance criteria should be reviewed with a summary statement provided for each. These statements should be similar to the following model, as applicable:

F10.1   Chapter(s) _____ of the SAR describe(s) the applicant's proposed program for preoperational testing and initial operations of the (DSS designation). Chapter(s) _____ discuss the proposed maintenance program.

F10.2   Structures, systems, and components (SSCs) important to safety will be designed, fabricated, erected, tested, and maintained to quality standards commensurate with the importance to safety of the function they are intended to perform. Chapter _____ of the SAR identifies the safety importance of SSCs, and Chapter(s) _____ present(s) the applicable standards for their design, fabrication, and testing.

F10.3   The applicant/licensee will examine and/or test the (DSS designation) to ensure that it does not exhibit any defects that could significantly reduce its confinement effectiveness. Chapter(s) _____ of the SAR describe(s) this inspection and testing.

F10.4   The applicant/licensee will mark the DSS with a data plate indicating its model number, unique identification number, and empty weight. Drawing _____ in SAR Chapter ____ illustrates and/or describes this data plate.

The reviewer should provide a summary statement similar to the following:

"The staff concludes that the acceptance tests and maintenance program for the (DSS designation) are in compliance with 10 CFR Part 72 and that the applicable acceptance criteria have been satisfied. The evaluation of the acceptance tests and maintenance program provides reasonable assurance that the cask will allow safe storage of spent fuel throughout its licensed or certified term. This finding is reached on the basis of a review that considered the regulation itself, appropriate regulatory guides, applicable codes and standards, and accepted practices."

11   RADIATION PROTECTION EVALUATION

11.1   Review Objective

This chapter describes the radiation protection evaluation requirements and considerations of the proposed dry storage system (DSS).   As used here, radiation protection refers to organizational, design, and operational elements that are primarily intended to limit radiation exposures from normal operations and anticipated occurrences.   The evaluation of the radiological consequences for accidents is addressed in Chapter 12, "Accident Analyses Evaluation" of this SRP.

The primary objectives of the radiation protection evaluation are to determine whether the design features and proposed operations meet the following criteria:

- the proposed DSS radiation protection features meet the U.S. Nuclear Regulatory Commission (NRC) design criteria for direct radiation;

- the applicant has proposed engineering features and operating procedures for the DSS that will ensure occupational exposures will remain ALARA; and

- the radiation doses to the general public will meet regulatory standards during both normal conditions and anticipated occurrences.

In independent spent fuel storage installation (ISFSI) operations, the major mode of radiation exposure associated with spent nuclear fuel (SNF) storage cask handling is from direct radiation.   Because of the cask design requirements, radionuclides are not expected to be released from the cask during either normal operations or design-basis accidents (DBAs).

11.2   Areas of Review

This chapter of the DSS Standard Review Plan (SRP) provides guidance for use in evaluating the radiation protection capabilities of the proposed cask system.   The following outline shows the areas of review addressed in Section 11.4, "Acceptance Criteria," and Section 11.5, "Review Procedures," that may be encompassed in a comprehensive radiation protection review:

Radiation Protection Design Criteria and Features

Occupational Exposures

Exposures at or Beyond the Controlled Area Boundary
   Normal Conditions
   Accident Conditions and Natural Phenomenon Events

ALARA
   Design Considerations
   Engineering Controls and Procedures

11.3   Regulatory Requirements

This section presents a summary matrix of the portions of U.S. Code of Federal Regulations (CFR) Parts 20 and 72 that are relevant to the review areas addressed by this chapter. The NRC staff reviewer should read the exact referenced regulatory language. Virtually the entire contents of 10 CFR 20 "Standards for Protection Against Radiation" are also applicable to this review. Tables 11-1 and 11-2 match the relevant regulatory requirements associated with this chapter to the areas of review identified in the previous section.

Table 11-1 Relationship of 10 CFR Part 20 Regulations and Areas of Review

Areas of Review	10 CFR Part 20 Regulations									
	20.1101	20.1201 (a)	20.1207	20.1208	20.1301 (a), (b), (d)	20.1302 (a)	20.1406	20.1501 (a)(1)	20.1701	20.1702
Radiation Protection Design Criteria and Features	●						●	●	●	●
Occupational Exposures	●	●	●	●				●		●
Exposures at or Beyond the Controlled Area Boundary	●				●	●		●		
ALARA	●						●	●		●

Table 11-2 Relationship of 10 CFR Part 72 Regulations and Areas of Review

Areas of Review	10 CFR Part 72 Regulations			
	72.104	72.106(b)	72.126 (a), (d)	72.236(d)
Radiation Protection Design Criteria and Features	●	●	●	●
Occupational Exposures	●		●	
Exposures at or Beyond the Controlled Area Boundary	●	●	●	●
ALARA	●		●	●

11.4   Acceptance Criteria

This section describes the acceptance criteria used for review of radiation protection features of and programs proposed for use with a DSS.   These criteria are organized according to the areas of review specified in Section 11.2 of this chapter.   The reviewer should note that some overlap exists between acceptance criteria for radiation protection and those related to Chapter 5, "Confinement Evaluation," and Chapter 6, "Shielding Evaluation," of this SRP; therefore, the reviews of these chapters should be coordinated.

11.4.1   Radiation Protection Design Criteria and Features

Limitations on dose rates associated with direct radiation from the cask are established on the basis of the shielding and confinement evaluations to satisfy the regulatory requirements for dose limits to individuals located beyond the controlled area boundary (10 CFR 72.104).

11.4.2   Occupational Exposures

Estimated dose rates should be provided in Chapter 6, "Shielding Evaluation," of the Safety Analysis Report (SAR) for representative points within the restricted area(s) (e.g. on and near the transfer cask and storage overpack surfaces) as well as at or beyond the perimeter of the controlled area.   The radiation protection review includes a dose assessment that incorporates findings of the shielding review, as applicable.   Individual and collective doses should be calculated.

All individual doses to workers should be well below the dose limits specified in 10 CFR 20.1201.   Collective doses should be consistent with the objectives contained in a well-structured ALARA program.   The information provided by the applicant should allow for the determination of compliance with these criteria.   To assess the applicant's occupational dose calculations, the reviewer should check such things as the number of workers specified for a task and the time specified for performing the task being reasonable.

11.4.3   Exposures at or Beyond the Controlled Area Boundary

a.   Normal Conditions:

For normal operations and anticipated occurrences, the estimated dose to any real individual located at or beyond the controlled area boundary may not exceed the following values specified in 10 CFR 72.104(a):

Whole body	0.25 mSv/yr (25 mrem/yr)
Thyroid	0.75 mSv/yr (75 mrem/yr)
Other organ	0.25 mSv/yr (25 mrem/yr)

For purposes of the DSS review, the calculated doses must include both direct radiation and any planned discharges of radioactive material.

b.   Accident and Natural Phenomenon Events:

Radiation shielding and confinement features should be provided sufficient to meet the requirements of 10 CFR 72.106(b). Any individual located on or beyond the nearest boundary of the controlled area may not receive the following dose from any DBA:

The more limiting of	
TEDE or Sum of the DDE and the CDE to any individual organ or tissue (other than the lens of the eye)	0.05 Sv (5 rem) 0.5 Sv (50 rem)
Lens of the eye	0.15 Sv (15 rem)
Shallow Dose Equivalent (SDE) to skin or any extremity	0.5 Sv (50 rem)

11.4.4   ALARA

For any new design or design change, the ALARA discussion should demonstrate how the design or design change

- accounted for radiation protection, technological, and economic considerations; and

- to the extent practicable, employed engineering controls and procedures that were founded upon sound radiation protection principles.

11.5   Review Procedures

The interrelationship of the radiation protection review with other disciplines is shown in Figure 11-1.

11.5.1   Radiation Protection Design Criteria and Features for the Transfer Cask and Storage Cask (MEDIUM Priority)

The reviewer should read the general description and functional features of the cask presented in Chapter 1, "General Description," of the SAR. In addition, Chapter 2, "Principal Design Criteria," of the applicant's SAR should be reviewed as well as any additional detail regarding radiation protection provided in the Shielding and Confinement chapters of the SAR. If not previously discussed, the following additional criteria should be presented in Chapter 11, Radiation Protection, of the SAR.

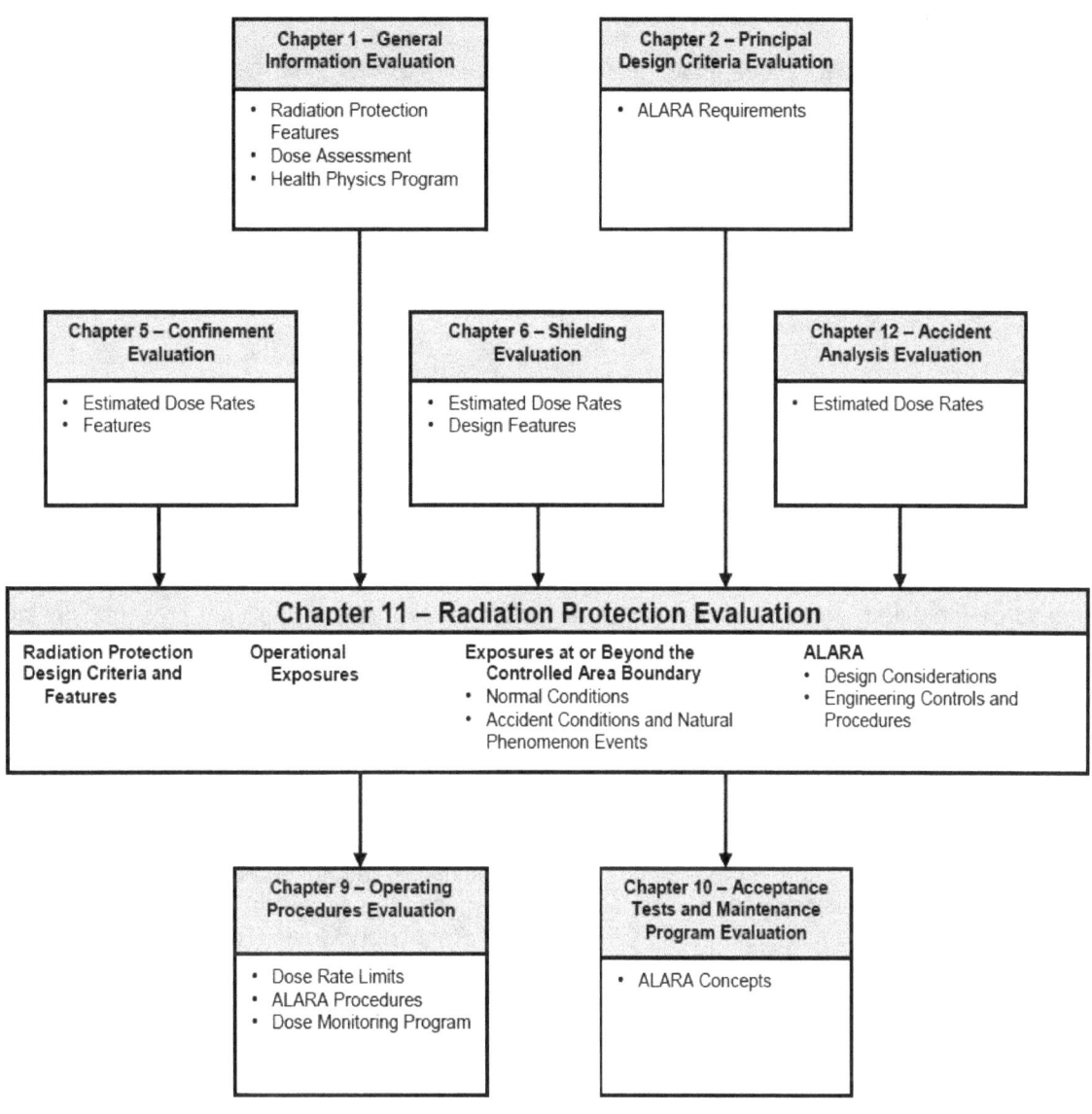

Figure 11-1 Overview of the Radiation Protection Evaluation

- The cask system design should satisfy ALARA and other occupational exposure requirements of 10 CFR Part 20, and

- The sum of the doses from direct radiation and from release of radioactive materials to the atmosphere should satisfy the requirements of 10 CFR 72.104(a) and 72.106(b). Because of the stringent design requirements for SNF cask systems, the release of radionuclides into the atmosphere is expected to be insignificant under both normal and accident conditions. Direct radiation is the major mode of exposure.

11.5.2   Occupational Exposures (MEDIUM Priority)

The reviewer should analyze Chapter 9, "Operating Procedures," of the SAR and direct radiation dose calculations in Chapter 6, "Shielding Evaluation" of the SAR. These data should be used in Chapter 11, "Radiation Protection" of the SAR to estimate the doses received by occupational personnel, including minors, during cask loading and transfer to the ISFSI. Any significant differences from these doses that may occur during cask retrieval and unloading should be identified. In addition, the reviewer should verify that the applicant presents similar dose estimates for periodic or routine maintenance as well as surveillance activities. These estimates may require additional assumptions concerning adjacent casks for a typical storage configuration.

The reviewer should verify that the applicant presents the rationale used to justify the bases for various exposure times, personnel locations relative to the casks (including hot spots), number of personnel required, and appropriate gamma and neutron dose rates. In addition, the reviewer should verify that the calculated doses are consistent with these estimates. The actual operations will be performed under an active dose-monitoring program that further ensures compliance with the requirements of 10 CFR Part 20. Regulatory Guide (RG) 8.34, "Monitoring Criteria and Methods to Calculate Occupational Radiation Doses," which was developed to implement revisions to 10 CFR Part 20, can be used to determine the acceptability of the applicant's occupational exposure evaluation and monitoring recommendations.

11.5.3   Exposures at or Beyond the Controlled Area Boundary (MEDIUM Priority)

As required by 10 CFR 72.236(d), the application must demonstrate that the shielding and confinement features of the cask are sufficient to meet the requirements for real individuals in 10 CFR 72.104, and for DBA conditions in 10 CFR 72.106. These demonstrations in the application facilitate future site-specific evaluations for each general ISFSI licensee. The real individual is an individual at or beyond the controlled area. Dose to any real individual must not exceed the limits specified in 10 CFR 72.104 from both the storage facility and other surrounding fuel cycle activities. For example, a real individual may be anyone living, working, or recreating close to the facility for a significant portion of the year.

However, for approval of a cask design, the reviewer should ensure that the applicant evaluates the shielding and confinement features of a single cask and a theoretical array of casks, assuming design-basis source terms and full-time occupancy. Supplemental shielding that may be used at an ISFSI to meet the exposure requirements to a real individual should also be appropriately evaluated. The reviewer should coordinate the review of supplemental shielding with the Chapter 13, "Technical Specifications and Operating Controls and Limits Evaluation," of this SRP review.

11.5.3.1   Normal Conditions

The single-cask analysis should identify the minimum distance that is required to meet the doses in 10 CFR 72.104. Past applications have shown this distance to be typically within 200m of the cask. A dose or dose rate versus distance curve for a single cask should be included to facilitate site-specific evaluations for general ISFSI licensees. To satisfy section 10 CFR 72.106(b), dose evaluations should be determined at a minimum of 100m (328 ft) distance to the closest boundary of the controlled area. However, the applicant may use a longer distance provided that the longer distance is made a condition of use. In addition, the SAR should determine the degree to which the normal condition dose rates could change for the identified off-normal conditions.

The reviewer should verify that the applicant includes a dose or dose rate versus distance curve in its evaluation of offsite dose for a hypothetical cask array. The theoretical cask array should consist of at least 20 storage casks (2x10 array), and the analysis may include the effect of shielding among casks in the array. The reviewer should examine predicted dose rates and compare them to the dose rates from previously approved casks, and any associated annual doses that have been observed for the casks at existing ISFSIs.

It is important to note that the general ISFSI licensee is permitted to use either distance between the ISFSI and the controlled area boundary or engineered features (supplemental shielding) such as berms to mitigate doses to real individuals near the site. The SAR needs to provide sufficient information to support informed choices on the part of the general licensee. If the SAR analyses were performed for the minimum 100-meter distance and did not use any additional shielding, and the projected dose at 100 meters exceeded the regulatory limits, the reviewer should verify that the application contains a justification for how a general licensee could reasonably meet the requirements of Section 72.104. If the dose versus distance curves for the single cask and hypothetical array in the SAR were only evaluated at distances greater than 100 m, or assumed some engineered feature, then the CoC should contain a condition of use to that effect.

An example of such a condition may be similar to the following: "The use of this system may require more than the minimum 100-meter distance between the ISFSI and the controlled area boundary, or engineered features (i.e., berms or shield walls), or both to ensure the dose limits in 10 CFR 72.104 can be met. In cases where engineered features are used to ensure that the requirements of 10 CFR 72.104(a) are met, such features are to be considered important to safety [ITS] and must be evaluated to determine the applicable [QA] category."

If an engineered feature is used in the SAR evaluations, then that feature is to be considered to be part of the system. As such, it should be described in the CoC.

As required by 72.212(b)(2)(i)(C), a general licensee must perform a written evaluation to demonstrate that the requirements of 72.104 are met. An evaluation similar to that for a site-specific ISFSI should be performed. The licensee may use information provided in the cask SAR as well as site-specific information to perform the evaluation. Evaluations performed by the general ISFSI licensee are not submitted to NRC for approval; however, they are subject to NRC inspection and should be recorded and maintained by the general licensee.

The general licensee should establish measures in the radiological protection program, environmental monitoring program, and/or operating procedures to identify and re-evaluate potential increases in exposure to the real individuals. Compliance with the dose limits in 10 CFR 72.104 will be verified by the environmental monitoring program with direct radiation measurements and/or effluent measurements, as appropriate.

11.5.3.2   Accident Conditions and Natural Phenomenon Events

The direct dose rate associated with accident conditions at the boundary of the controlled area should be reviewed as discussed in Chapter 6, "Shielding Evaluation," of this SRP. Also, the dose rate resulting from accidental release of radionuclides, as presented in Chapter 5, "Confinement Evaluation," of this SRP, should be reviewed. The accident-related radionuclide release dose should account for both air and liquid pathways as appropriate. In addition, the reviewer should verify that the applicant has evaluated the source terms for both SNF fission product and cask surface contamination. The sum of these should satisfy the requirements of 10 CFR 72.106(b). For purposes of demonstrating compliance with 10 CFR 72.106(b) and evaluation against the Environmental Protection Agency Protective Action Guides in the *Manual of Protective Action Guides and Protective Actions for Nuclear Incidents* (EPA 410R-92-001), the skin, extremities, and the lens of the eye may be considered separately from other organs.

As noted in Chapter 6, "Shielding Evaluation," of this SRP, the time-integrated dose at the boundary of the controlled area may be small. Consequently, the reviewer should verify that the applicant estimates the doses at 100m (328 ft.) from the storage location to the nearest boundary of the controlled area unless the SAR specifies a greater minimum distance that is also made a condition of use for the proposed DSS. Alternatively, applicants may depict dose estimation using a curve showing dose versus distance from an assumed array of casks. For those systems for which these curves indicate the need for greater distance and/or engineered features, such as berms, a condition similar to that described in the preceding section of this SRP may be needed, with the requirement being 72.106(b) in this instance.

11.5.4   ALARA (MEDIUM Priority)

Further information on ALARA can be found in RG 8.8, "Information Relevant to Ensuring that Occupational Radiation Exposures at Nuclear Power Stations Will Be As Low As is Reasonably Achievable," and RG 8.10, "Operating Philosophy for Maintaining Occupational Radiation Exposures As Low As is Reasonably Achievable."

11.5.4.1   Design Considerations

The cask design features should be reviewed to ensure that the features for which credit is taken in radiation protection analyses are clearly identified on the drawings. Also, the reviewer should ensure the application includes commitments to implement those features that have been credited in analyses to show compliance with regulatory requirements or ALARA goals. The reviewer should coordinate with the reviewers of SRP Chapters 5, "Confinement Evaluation" and 6, "Shielding Evaluation."

11.5.4.2   Procedures and Engineering Controls

The reviewer should determine that the descriptions of proposed DSS operations adequately demonstrate that ALARA principles have been incorporated into operational procedures and engineering controls. The reviewer should ensure that plans and procedures have been developed in accordance with applicable requirements and guidance.

11.6 Evaluation Findings

Evaluation findings are prepared by the reviewer upon determination that the regulatory requirements related to radiation protection as identified in Section 11.3 of this chapter have been satisfied. Some of these determinations can be made only after evaluating the results of reviews performed under other chapters of this SRP. If the documentation submitted with the application fully supports positive findings for each of the regulatory requirements, the statements of findings should be similar to the following:

F11.1 The [cask designation] provides radiation shielding and confinement features that are sufficient to meet the requirements of 10 CFR 72.104 and 72.106.

F11.2 The design and operating procedures of the [cask designation] provide acceptable means for controlling and limiting occupational radiation exposures within the limits given in 10 CFR 20 and for meeting the objective of maintaining exposures ALARA.

A summary statement similar to the following should be made:

"The staff concludes that the design of the radiation protection system of the [cask designation] is in compliance with 10 CFR Part 72 and that the applicable design and acceptance criteria have been satisfied. The evaluation of the radiation protection system design provides reasonable assurance that the [cask designation] will allow safe storage of SNF. This finding is reached on the basis of a review that considered the regulation itself, appropriate regulatory guides, applicable codes and standards, accepted health physics practices, and the statements and representations in the application."

12 ACCIDENT ANALYSES EVALUATION

12.1 Review Objective

In this portion of the dry storage system (DSS) review, the U.S. Nuclear Regulatory Commission (NRC) evaluates the applicant's identification and analysis of hazards as well as the summary analysis of system responses to both off-normal and accident or design-basis events.

Normal conditions are the intended operations, planned events, and environmental conditions, that are known or reasonably expected to occur with high frequency during storage operations.

Off-normal events are those man-made events or natural phenomena expected to occur with moderate frequency or once per calendar year. ANSI/ANS 57.9 refers to these events as Design Event II.

Design-basis accident events are considered to occur infrequently, if ever, during the lifetime of the facility. ANSI/ANS 57.9-92 subdivides this class of accidents into two categories – Design Events III and IV. Design Event III is a set of infrequent events that could be expected to occur during the lifetime of a DSS, and Design Event IV is a set of events that establishes a conservative design basis for structures, systems, and components (SSC) important to safety. The effects of natural phenomena such as earthquakes, tornadoes, hurricanes, floods, tsunami, and seiches, with severity frequencies consistent with Design Event III and IV, are considered to be design-basis accident events, in addition to design-basis man-made events.

This review ensures that the applicant has conducted thorough accident analyses as reflected by the following factors:

- Identified all credible accidents.
- Provided complete information in the safety analysis report (SAR).
- Analyzed the safety performance of the cask system in each review area.
- Fulfilled all applicable regulatory requirements.

12.2 Areas of Review

This portion of the DSS review evaluates the applicant's identification and analysis of hazards with particular emphasis on the safety performance of the cask system under off-normal events and conditions, and accident or design-basis events. Consequently, this chapter of the DSS Standard Review Plan (SRP) provides guidance for use in reviewing the applicant's identification and analysis of hazards as well as the summary analysis of system responses. A comprehensive accident analysis evaluation may encompass the following areas of review:

Cause of the Event
Detection of the Event
Summary of Event Consequences and Regulatory Compliance
Corrective Course of Action

12.3 Regulatory Requirements

This section presents a summary matrix of the portions of U.S. Code of Federal Regulations (CFR), Part 72, "Licensing Requirements for the Independent Storage of Spent Nuclear Fuel and High-Level Radioactive Waste," Title 10, "Energy" (10 CFR Part 72) that are relevant to the review areas addressed by this chapter. The NRC staff reviewer should read the exact referenced regulatory language. Table 12-1 matches the relevant regulatory requirements associated with this chapter to the areas of review identified in the previous section.

Areas of Review	10 CFR Part 72 Regulations				
	72.104 (a)	72.106 (b)	72.122(b)(1),(3), (d), (g), (h)(4), (i), (l)	72.124(a)	72.236(c), (d), (l)
Cause of the Event			●		
Detection of the Event			●	●	
Summary of Event Consequences and Regulatory Compliance	●	●	●	●	●
Corrective Course of Action			●		

Table 12-1 Relationship of Regulations and Areas of Review

12.4 Acceptance Criteria

Accidents and natural phenomena events may share common regulatory and design limits. Consequently, the following sections sometimes refer to these scenarios collectively as accident conditions.

By contrast, off-normal conditions (anticipated occurrences) are distinguished, in part, from accidents or natural phenomena by the appropriate regulatory guidance and design criteria. For example, the radiation dose from an off-normal event must not exceed the limits specified in 10 CFR Part 20, "Standards for Protection Against Radiation," and 10 CFR 72.104(a), whereas the radiation dose from an accident or natural phenomenon must not exceed the specifications of 10 CFR 72.106(b). Accident conditions may also have different allowable structural criteria.

In general, this portion of the DSS review seeks to ensure that the DSS design and the applicant's hazard identification and analyses of related system responses fulfill the following acceptance criteria:

12.4.1 Dose Limits for Off-Normal Events

During normal operations and off-normal conditions, the requirements specified in 10 CFR Part 20 must be met. In addition, the annual dose equivalent to any individual located beyond the controlled area must not exceed 0.25 mSv (25 mrem) to the whole body, 0.75 mSv (75 mrem) to the thyroid, and 0.25 mSv (25 mrem) to any other organ as a result of exposure to the following sources (10 CFR 72.104):

- Planned discharges to the general environment of radioactive materials (with the exception of radon and its decay products).

- Direct radiation from operations of the ISFSI.

- Any other cumulative radiation from uranium fuel cycle operations (i.e., nuclear power plant) in the affected area.

12.4.2   Dose Limit for Design-Basis Accidents

The dose from any credible design basis accident to any individual located on or beyond the nearest boundary of the controlled area may not exceed the limits specified in 10 CFR 72.106. Specifically, these are: the more limiting of a total effective dose equivalent of 0.05 Sv (5 rem), or the sum of the deep dose equivalent to and the committed dose equivalent to any individual organ or tissue (other than the lens of the eye) of 0.5 Sv (50 rem); a lens dose equivalent of 0.15 SV (15 rem); and a shallow dose equivalent to skin or any extremity of 0.5 Sv (50 rem).

12.4.3   Criticality

The spent nuclear fuel (SNF) must be maintained in a subcritical condition under credible conditions (i.e., k_{eff}, including all biases and uncertainties, equal to or less than 0.95). At least two unlikely, independent, and concurrent or sequential changes in the conditions essential to nuclear criticality safety should occur before a nuclear criticality accident is deemed to be possible (double contingency).

12.4.4   Confinement

The cask and its systems important to safety must be evaluated using appropriate tests or by other means acceptable to the NRC to demonstrate that they will reasonably maintain confinement of radioactive material under credible accident conditions.

12.4.5   Recovery and Retrievability

Recovery is the capability to return the stored radioactive material to a safe condition after an accident event without endangering public health and safety. This generally means ensuring that any potential release of radioactive materials to the environment or radiation exposures is not in excess of the limits in 10 CFR Part 20 during post accident recovery operations.

Retrievability is specified in 10 CFR 72.122(I) and requires that storage systems must be designed to allow ready retrieval of spent fuel, high-level radioactive waste, and reactor-related GTCC waste for further processing or disposal. Ready retrieval is the ability to move a canister containing spent fuel to either a transportation package or to a location where the spent fuel can be removed. Ready retrieval also means maintaining the ability to handle individual or canned spent fuel assemblies by the use of normal means. Retrievability applies to normal conditions and off-normal events, and not to design-basis accident events.

12.4.6   Instrumentation

The SAR must identify all instruments and control systems that must remain operational under accident conditions.

12.5 Review Procedures

<u>Introduction</u>

Figure 12-1 presents an overview of the evaluation process and can be used as a guide to assist in coordinating between the review disciplines.

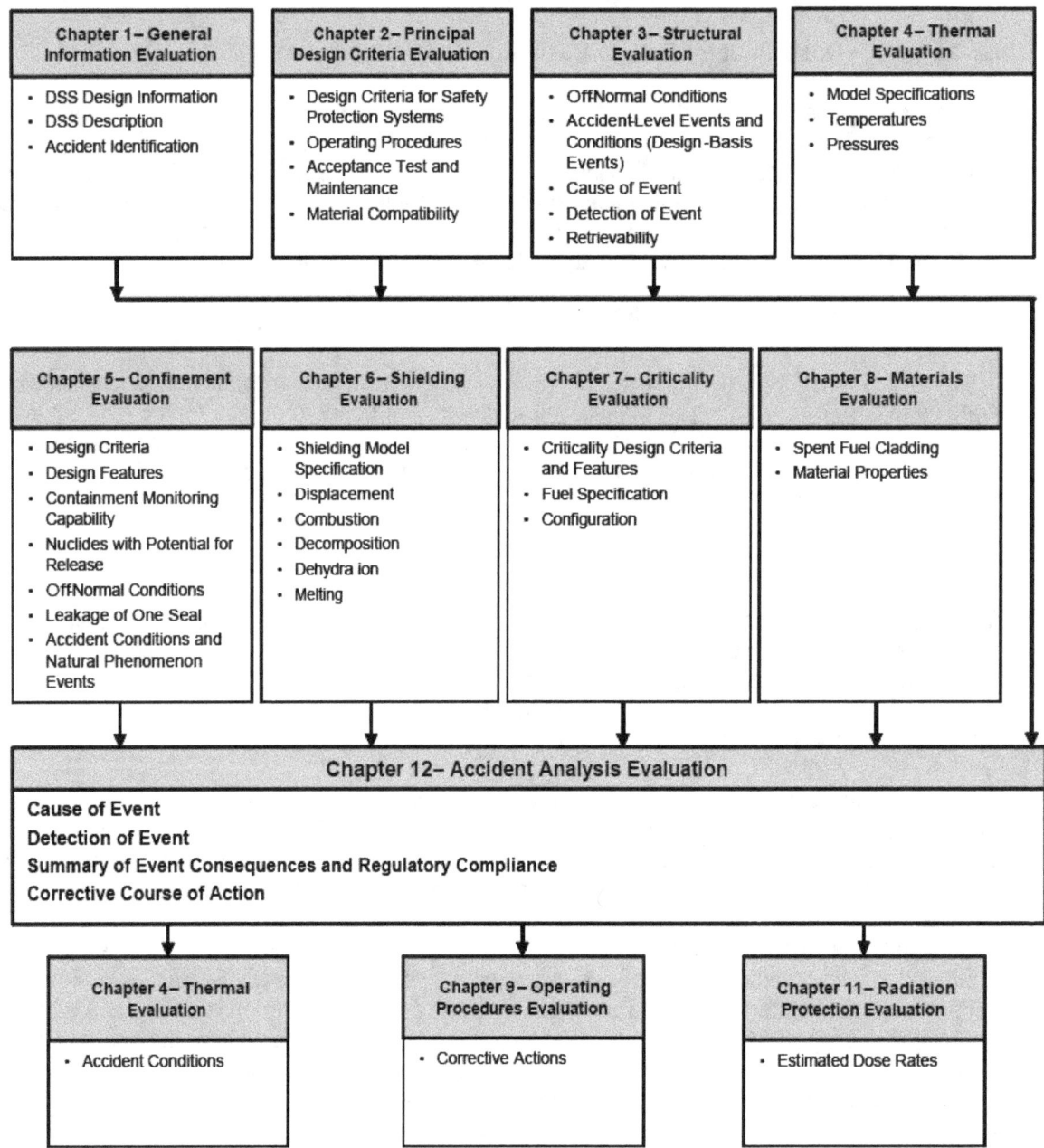

Figure 12-1 Overview of Accident Analysis Evaluation

The review procedures presented here describe general procedures for reviewing a DSS submittal. The review procedures in Chapter 15 of NUREG-1567, "Standard Review Plan for Spent Fuel Dry Storage Facilities," provide more detailed procedures and, where applicable, may be used as a guide to supplement the review procedures presented herein.

The off-normal conditions, accidents, and natural phenomena events identified in SAR Chapter 2, "Principal Design Criteria" should be reviewed by all disciplines, especially those accidents with potential consequences resulting in the failure of the confinement boundary. Off-normal conditions should be evaluated against the requirements of 10 CFR 72.104. Accidents and natural phenomena events should be evaluated against the requirements of 10 CFR 72.106 and 72.122(b). Recovery methods or the need for overpacks or dry transfer systems to maintain safe storage conditions would then not be considered and evaluated as part of the NRC approval process. For each type of event, this discussion should include the applicant's evaluation of the following areas, as applicable.

12.5.1   Cause of the Event (MEDIUM Priority)

The cause of the accident should be described. The description should include the chain of events that leads to the credible accident condition and any bounding conditions.

12.5.2   Detection of the Event (MEDIUM Priority)

The licensee may detect an event through surveillance programs or monitoring instrumentation and alarms. Surveillance programs and monitoring instrumentation and alarms should have reasonable flexibility to allow for the identification of an accident condition or noncompliance situation that has not been previously considered in the SAR. The method of detection will be intuitively obvious for some events, whereas other events (e.g., fuel rod rupture) may remain undetected for a significant period of time.

DSS monitoring equipment (such as a pressure monitoring system) are classified as not important to safety, but are classified as Category B under the guideline of NUREG/CR-6407, "Classification of Transportation Packaging and Dry Spent Fuel Storage Components According to Importance to Safety." Reviewers should refer to Chapter 5, "Confinement Evaluation," of this SRP.

12.5.3   Summary of Event Consequences and Regulatory Compliance (MEDIUM PRIORITY)

The applicant should address event consequences in each functional area corresponding to earlier chapters of the SAR (i.e., structural, thermal, shielding, criticality, confinement, materials, and radiation protection). This discussion should refer back to each SAR chapter in which the individual consequences are evaluated in detail. The applicant should provide a summary of the accident dose calculations and show that the consequences comply with the applicable regulatory criteria. For off-normal conditions, the applicant should demonstrate compliance with Part 20 as well as Part 72.

12.5.4   Corrective Course of Action (MEDIUM Priority)

The applicant should identify what action(s), if any, would be necessary to recover from the event. If various courses of action are possible, the applicant should present a discussion concerning the selection of the most appropriate action. Because the fuel must be readily retrievable, returning the cask to the fuel handling building and reloading the SNF into a new cask is a viable option. If corrective courses of action are to be included in operating procedures or administrative programs, then the applicable sections of SAR Chapter 9, "Operating Procedures," should be referenced.

12.6 Evaluation Findings

Review the 10 CFR Part 72 acceptance criteria and provide a summary statement for each. These statements should be similar to the following model:

F12.1 Structures, systems, and components of the [cask designation] are adequate to prevent accidents and to mitigate the consequences of accidents and natural phenomena events that do occur.

F12.2 The spacing of casks, discussed in Chapter _____ of the safety evaluation report (SER) and included as an operating limit in Chapter 13, "Technical Specifications and Operation Controls and Limits Evaluation" of the SAR will ensure accessibility of the equipment and services required for emergency response.

F12.3 Table _____ of the SER lists the Technical Specifications for the [cask system designation]. These Technical Specifications are further discussed in Chapter _____ of the SER.

F12.4 The applicant has evaluated the [cask designation] to demonstrate that it will reasonably maintain confinement of radioactive material under credible accident conditions.

F12.5 An accident or natural phenomena event will not preclude the ready retrieval of SNF for further processing or disposal.

F12.6 The SNF will be maintained in a subcritical condition under accident conditions.

F12.7 Neither off-normal nor accident conditions will result in a dose to an individual outside the controlled area that exceeds the limits of 10 CFR 72.104(a) or 72.106(b), respectively.

F12.8 No instruments or control systems are required to remain operational under accident conditions [as applicable].

The reviewer should provide a summary statement similar to the following:

"The staff concludes that the accident design criteria for the [DSS designation] are in compliance with 10 CFR Part 72, and the accident design and acceptance criteria have been satisfied. The applicant's accident evaluation of the cask adequately demonstrates that it will provide for safe storage of SNF during credible accident situations. This finding is reached on the basis of a review that considered independent confirmatory calculations, the regulation itself, appropriate regulatory guides, applicable codes and standards, and accepted engineering practices."

13 TECHNICAL SPECIFICATIONS AND OPERATING CONTROLS AND LIMITS EVALUATION

13.1 Review Objective

The technical specifications and operating controls and limits review ensures that the operating controls and limits or the technical specifications, including their bases and justification, meet the requirements of the U.S. Code of Federal Regulations (CFR), Part 72, "Licensing Requirements for the Independent Storage of Spent Nuclear Fuel, High-Level Radioactive Waste and Reactor-Related Greater Than Class C Waste," Title 10, "Energy" (10 CFR Part 72). This evaluation is based on information that the applicant presents in Safety Analysis Report (SAR) Chapter 13, "Technical Specifications and Operation Controls and Limits Evaluation" as well as accepted practices and the applicant's commitments discussed in other chapters of the SAR or in correspondence subsequent to submission of the application. The NRC staff should also describe in the Safety Evaluation Report (SER) any additional operating controls and limits that the staff deems necessary and has added them, as appropriate, to the cask system's Technical Specifications.

For simplicity in defining the acceptance criteria and review procedures, the term "technical specifications" may be considered synonymous with "operating controls and limits." The technical specifications define the conditions that are deemed necessary for safe dry storage system (DSS) use. Specifically, they define operating limits and controls, monitoring instruments and control settings, surveillance requirements, design features, and administrative controls that ensure safe operation of the DSS. As such, these technical specifications are included in a DSS Certificate of Compliance (CoC). Each specification should be clearly documented and justified in the technical evaluation sections of the SAR and the associated SER as necessary for safe DSS operation.

If a reviewer determines that a design feature, content specification, analytical assumption, operating assumption, limiting condition of operation, element of reactor programmatic controls, or other SAR item is important and should not be changed without NRC staff approval, then it should be further evaluated and considered as a potential CoC condition or technical specification. The reviewer should consider, in part, risk-insights, safety margins, operational experience, defense-in-depth considerations, design novelty, and other issues that are unique to each proposed design. The reviewer should also implement the guidance in this chapter for establishing such conditions and technical specifications in the CoC.

13.2 Areas of Review

This chapter of the DSS Standard Review Plan (SRP) provides guidance for use in evaluating the technical specifications that the applicant deems necessary for safe use of the proposed DSS system. As defined in Section 13.5, "Review Procedures," a comprehensive review of the proposed technical specifications would assess the applicant's compliance with the regulations to provide a level of control commensurate with that specified by 10 CFR 72.234 and 72.236. These requirements represent the following areas of review:

> ### Functional/Operating Limits, Monitoring Instruments, and Limiting Control Settings

> ### Limiting Conditions

Surveillance Requirements

Design Features

Administrative Controls

13.3   Regulatory Requirements

This section presents a summary matrix of the portions of 10 CFR Part 72 that are relevant to the review areas addressed by this chapter. The U.S. Nuclear Regulatory Commission (NRC) staff reviewer should read the exact referenced regulatory language. Table 13-1 matches the relevant regulatory requirements associated with this chapter to the areas of review identified in the previous section.

Table 13-1 Relationship of Regulations and Areas of Review											
Areas of Review	**10 CFR Part 72 Requirements**										
	72.234 (a)	72.236									
		(a)	(b)	(c)	(d)	(e), (f), (h)	(g)	(i)	(j)	(l)	
Functional/Operating Limits, Monitoring Instruments, and Limiting Control Settings	•	•		•	•					•	
Limiting Conditions	•	•		•	•					•	
Surveillance Requirements	•				•		•		•		
Design Features	•		•		•	•	•	•		•	
Administrative Controls	•	•			•			•		•	

13.4   Acceptance Criteria

The reviewer should verify that the applicant identifies proposed technical specifications necessary to maintain subcriticality, confinement, shielding, heat removal, and structural integrity under normal, off-normal, and accident-level conditions. In addition, the reviewer should ensure that the applicant identifies the basis for each of the proposed technical specifications by reference to the analysis in the SAR. The NRC staff can use NUREG-1745, "Standard Format and Content for Technical Specifications for 10 CFR Part 72 Cask Certificates of Compliance," as an appropriate template in the review of the technical specifications. However, the staff may impose alternative technical specifications to NUREG-1745 guidance, based on operational experience, and the Office of General Counsel legal interpretations that have been made since issuance of NUREG-1745.

13.4.1 Functional/Operating Limits, Monitoring Instruments, and Limiting Control Settings

Acceptance criteria for functional and operating limits, monitoring instruments, and limiting control settings include limits placed on fuel, waste handling, and storage conditions to protect the integrity of the fuel and container, to protect the employees against occupational exposures, and to guard against the uncontrolled release of radioactive materials.

13.4.2 Limiting Conditions

Acceptance criteria for functional and operating limits, monitoring instruments, and limiting control settings include limits placed on fuel, waste handling, and storage conditions to protect the integrity of the fuel and container, to protect the employees against occupational exposures, and to guard against the uncontrolled release of radioactive materials. Acceptance criteria for limiting conditions are the lowest levels required for safe operation.

13.4.3 Surveillance Requirements

Acceptance criteria for establishing surveillance requirements include the frequency and scope of surveillance requirements to verify performance and availability of structures, systems, and components (SSCs) important to safety, and the verification of the bases for the proposed limiting conditions.

13.4.4 Design Features

Acceptance criteria for design features include commitments to specified codes. The condition or technical specification should also describe a process to address deviations from the applicable codes that may be necessary. In such cases, the licensee should request an alternative to the requirements of the applicable code from the NRC. If the staff finds that the deviation does not adversely impact safety, it may authorize the requested alternative in writing.

Currently, there is an existing code for the design and construction of metallic nuclear fuel storage casks and the document is identified as Subsection WC of Division 3 of Section III of the ASME Boiler and Pressure Vessel Code. This was first issued as the 2005 addenda to the 2004 Code. The current Code edition is 2007. As of February 2008, NRC staff had not taken a position regarding the acceptability of this document. In the past, Division 1 of the ASME B&PV Code had been used by NRC staff allowing alternatives to some provisions of that document which were judged to not be applicable to spent nuclear fuel storage casks. Early SNF dry storage licenses and certificates of compliance were issued without documenting which specific alternatives to ASME B&PV Code, Section III, were approved. Poor quality assurance practices during design and fabrication sometimes led to significant deviations from the Code without appropriate certificate holder design review or NRC review and approval. Therefore, the applicant should document commitments to ASME B&PV Code, Section III, with proposed alternatives in the application.

Likewise the NRC should document these commitments in the 10 CFR Part 72 licenses, certificates of compliance, or technical specifications and its approval of the proposed alternatives in the SER. Also, the NRC should include a statement (in the CoC or technical specifications) that refers the reader to the SAR and applicable SERs for any alternatives to the codes. In addition, to ensure that similar problems do not exist in other areas, all other codes and standards applied to components important to safety should be identified in the SAR and should be included in the CoC or technical specifications. Figure 13-1 presents an example of a provision for allowing alternatives to applicable codes.

#.#.#   Codes and Standards
The American Society of Mechanical Engineers (ASME) Boiler and Pressure Vessel (B&PV) Code, Section III, 1992 Edition with Addenda through 1994 is the governing Code for the storage system.

#.#.#.# Design Alternatives to Codes, Standards, and Criteria
Table #-# lists all approved alternatives for the design of the DSS.

#.#.#.# Construction/Fabrication Alternatives to Codes, Standards, and Criteria
Proposed alternatives to ASME B&PV Code Section III, 1992 Edition with Addenda through 1994, including alternatives referenced in Section 4.3.1, may be used when authorized by the Director of the Office of Nuclear Material Safety and Safeguards or designee.

The proposal to the NRC must demonstrate that the alternatives would provide an acceptable level of quality and safety, or that compliance with the specified requirements of ASME B&PV Code, Section III, 1992 Edition with Addenda through 1994 would result in hardship or unusual difficulty without a compensating increase in the level of quality and safety.

Figure 13-1 Provision Example

In addition, acceptance criteria for design features include specifications important to criticality safety. Where criticality analyses rely upon the condition that the assemblies' active fuel length remains within the cask region containing the solid neutron absorbers, the applicant should commit to ensuring the cask features fulfill this analysis assumption. One common method is the installation of fuel spacers, upper and/or lower spacers as needed, to maintain the assemblies' position under all credible conditions. The minimum Boron-10 content of the solid neutron absorbers is another important design feature specification together with the qualification and acceptance testing method for ensuring the neutron absorbers meet the required minimum Boron-10 content throughout the absorber material. The proximity of fuel assemblies to each other also affects the cask's reactivity, generally with reactivity increasing as the assemblies are brought closer together; therefore, a minimum dimension(s) between adjacent assembly locations is specified. This dimension may be a minimum flux trap width or a minimum fuel cell pitch. These design parameters and commitments should also be included in the license, certificate of compliance, or technical specifications.

13.4.5 Administrative Control

Acceptance criteria for administrative controls include organizational and management procedures, recordkeeping, review and audit systems, and reporting necessary to ensure that the DSS is managed in a safe and reliable manner. Administrative action that must be taken in the event of noncompliance with a limit or condition should be specified.

13.5 Review Procedures (HIGH Priority)

Figure 13-2 presents an overview of the evaluation process and can be used as a guide to assist in coordinating between review disciplines.

Reviewers should evaluate each chapter of the SAR with the goal of establishing the technical specifications. The variability of designs and operations makes it impossible to define each instance for which a technical specification is necessary. For this reason, it is important that the NRC staff conduct a coordinated, detailed, and thorough evaluation of each technical section of the SAR. Reviewers should note all instances in which the SAR either makes an assumption or imposes a condition that should be identified as a technical specification. Reviewers should also note any instances in which the SAR requests alternatives or exemptions from regulatory requirements, or other conditions that the reviewer identifies as an operational limit or condition. Such limits and exemptions should be clearly identified and documented in SAR Chapter 13. "Technical Specifications and Operation Controls and Limits Evaluation".

The various technical disciplines should review the results of their specific evaluations and compare their list of technical specifications to those identified by the applicant. The NRC staff should ensure that the conditions for use, as evaluated and approved by the technical reviewers, complement one another and are not contradictory. In addition, the staff will coordinate the resolution of any disputed condition, limit, or specification. The staff is responsible for identifying any unique specifications (e.g., administrative) that may not be covered in the technical sections, although input may be solicited from the technical reviewers regarding any topic.

All reviewers should be familiar with the technical specifications of similar cask designs previously approved by the NRC staff. For example, the staff has previously approved cask designs and issued technical specifications regarding a variety of items including, but not limited to, the following examples:

- General requirements and conditions regarding site-specific parameters, operating procedures, quality assurance, heavy loads, training, etc.

- A preoperational training exercise and demonstration of most cask operations including loading, sealing, and drying (using mockups as appropriate); placement in storage; and return of fuel to the SNF pool.

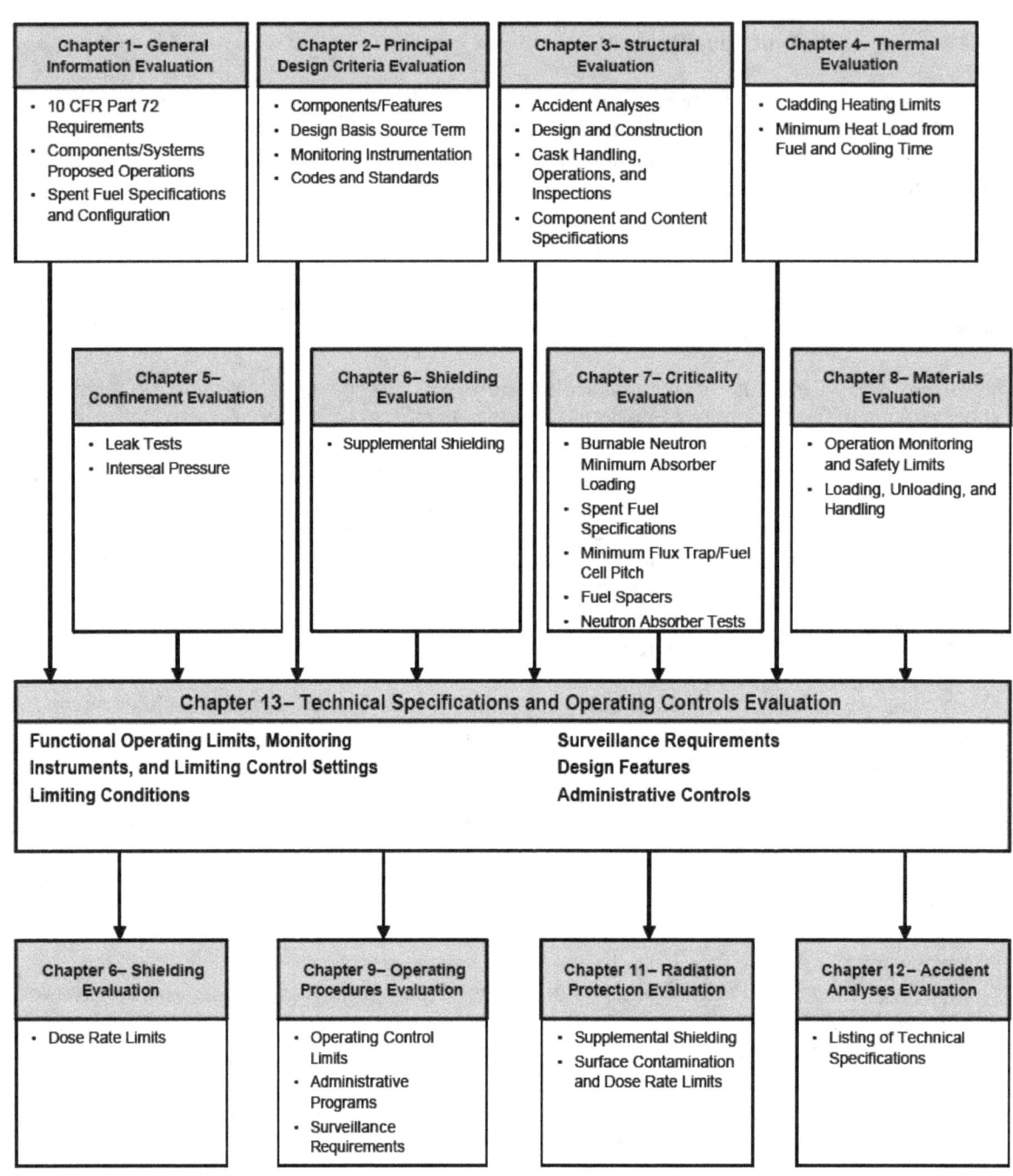

Figure 13-2 Overview of Technical Specifications and Operating Controls and Limits Evaluation

- Specifications for the SNF to be stored in the cask, including, but not limited to, the type of SNF (i.e., boiling water reactor [BWR], pressurized water reactor [PWR], or both), the minimum and maximum allowable enrichments of the fuel before irradiation, burnup (i.e., megawatt-days/MTU), the minimum acceptable cooling time of the SNF before storage in the cask, the maximum heat designed to be dissipated, the maximum SNF loading limit, condition of the SNF (i.e., intact assembly or consolidated fuel rods, allowable cladding condition), associated non-fuel hardware, and physical parameters (e.g., length, width, depth, weight, etc.). The reviewer should be aware that additional SNF specifications regarding operational history parameters (e.g., average moderator temperature, average in-core soluble boron concentrations, and operations under control rod banks or with control rod insertion) will need to be included in the technical specifications for cask systems relying on burnup credit

- Criticality controls such as cask water boron concentrations, minimum flux trap/fuel cell pitch, use of fuel spacers, minimum neutron absorber loading, and neutron absorber tests.

- The inerting atmosphere requirements during vacuum drying and helium backfill parameters.

- Cask handling restrictions such as lift height limits and ambient temperature (high/low) conditions.

- Confinement barrier requirements such as helium leak rate limits.

- Thermal performance parameters such as maximum temperatures or delta-temperatures.

- Radiological controls such as radiation dose rates and contamination limits.

- Cask array and/or spacing limits for thermal performance and radiological considerations.

- Definition of damaged fuel

- Code of record and alternatives to specific Code requirements

- Specification/requirements for alternative materials for ITS components

- Manufacture and testing of neutron poison material(s) for criticality control

- Hydrogen monitoring/mitigation during wet loading/unloading

- Maintaining inert atmosphere during canister draining/flooding to prevent Oxidation

- Use of copper bearing or weathering steel for structural steel components at coastal marine ISFSI sites (or other corrosion mitigation measures)

- Operational controls to maintain cladding temperature limits

- Low Temperature Ductility of Ferritic Steels

All disciplines should coordinate their review of the proposed technical specifications to assure the operational limitations are measurable and inspectible. Other topics may include:

- Frequency and scope proposed for the surveillance requirements.

- Administrative controls that include organization and administrative systems and procedures, record-keeping, review, and audit systems required to ensure that the DSS is managed in a safe and reliable manner.

- Administrative action that must be taken in the event of noncompliance with a limit or condition.

The reviewer should verify that the applicant includes a written description in a condition to the CoC or technical specification that documents the codes to which the applicant has committed. In addition, the condition or technical specification should describe a process to address any deviations from the ASME B&PV Code or other codes that may be needed. Likewise, the reviewer should verify that these commitments are documented in the 10 CFR Part 72 CoC or technical specifications. A list of proposed alternatives to code requirements should also be provided in the SAR. This list should be revised as necessary to reflect all NRC-authorized alternatives.

NUREG-1745 provides a recommended format for use by applicants in presenting technical specifications. However, this format may not be applicable to all controls. Since the basis for the control may be extensively discussed in earlier chapters of the SAR, the applicant may use an abbreviated format in SAR Chapter 13.

Reviewers should ensure that all necessary technical specifications are explicitly delineated in SER Chapter 13, "Technical Specifications and Operating Controls and Limits Evaluation," and in the CoC. These delineations typically restate the technical specifications defined in the SAR but may be modified or supplemented as the staff deems appropriate. Reviewers should also ensure that limits and exemptions requested by the applicant are clearly identified and documented in the SER. The staff may prepare a separate table or appendix for SER Chapter 13 to explicitly designate the technical specifications that are applicable to the cask. Applicable drawings from the SAR should be identified by number and revision.

13.6   Evaluation Findings

NRC staff reviewers prepare evaluation findings regarding satisfaction of the regulatory requirements related to technical specifications. Evaluation findings developed or included in all SER sections relating to technical specifications are also listed in this section. These statements should be similar to the following model:

F13.1   The staff concludes that the conditions for use for [DSS name] identify necessary technical specifications to satisfy 10 CFR Part 72 and that the applicable acceptance criteria have been satisfied. The proposed technical specifications provide reasonable assurance that the DSS will allow safe storage of SNF. This finding is based on the regulation itself, appropriate regulatory guides, applicable codes and standards, and accepted practices. The technical specifications identified by the applicant include the following: [Reviewer to specify].

The reviewer should provide a summary statement similar to the following:

"The proposed technical specifications provide reasonable assurance that the cask will allow safe storage of spent fuel. This finding is reached on the basis of a review that considered the regulation itself, appropriate regulatory guides, applicable codes and standards, accepted practices and the statements and representations in the application."

14 QUALITY ASSURANCE EVALUATION

14.1 Review Objective

The objective of the review is to determine whether the applicant for a dry storage system (DSS) certificate has submitted a quality assurance (QA) program description (QAPD) that demonstrates that the applicant's QA program complies with the requirements of 10 CFR Part 72, Subpart G (Part 72), "Quality Assurance."

The basis for that determination is developed from an evaluation of the applicant's high level QAPD against the criteria provided in Section 14.4, Review Procedures below, Part 72, and any associated information found in the Federal Register since the last rulemaking has been completed, as applicable. (Note: The scope of review does not include actual procedures and instructions that implement the QA program, but may be described in the QAPD).

Determination of compliance for the applicant's QA program occurs during NRC inspection activities where implementation of the QA plan is evaluated. (Note: The scope of an inspection does include actual procedures and instructions that implement the QA program).

14.2 Areas of Review

This SRP provides guidance for use by a reviewer to perform an evaluation of a QAPD in terms of the 18 criteria defined in 10 CFR Part 72, Subpart G and Section 14.4, "Review Procedures" below, and the Federal Register, as applicable.

14.3 Regulatory Requirements

This section identifies the reviewer's need to review the exact regulatory language found in Part 72 relevant to quality assurance as applied to a DSS. Refer to Subpart G -Quality Assurance of 10 CFR Part 72.

14.4 Acceptance Criteria

The acceptance criteria below reflect the 18 quality criteria of Part 72, Subpart G. These criteria are presented in the form of descriptions of information to be included in the applicant's QAPD. For each criterion shown in Sections 14.5.1 through 14.5.18 of this SRP, examples of measures have been provided which may assist the reviewer in determining if the QAPD indicates that it meets the applicable criterion. For each of the activities and items identified as important to safety, the applicant should identify the applicable QA programmatic elements and include, as applicable, provisions for meeting each of the following quality criteria itemized in Section 14.5.

14.5 Review Procedures (All items in this section are HIGH Priority)

The purpose of the review is to obtain reasonable assurance that the applicant has developed and described a QA program for design, fabrication, construction, testing, operations, modification, and decommissioning activities associated with important-to-safety DSS systems, structures and components (SSCs).

It is important that the applicant's QAPD and associated portions of the safety analysis report (SAR) provide sufficient detail to enable the reviewer to assess that the applicant has committed to comply with the program and the QA program complies with the applicable requirements of 10 CFR Part 72, Subpart G. If the reviewer determines that sufficient detail does not exist in the QAPD, the reviewer should refer to Section 14.6, Evaluation Findings for further direction. If the QAPD indicates commitment to follow certain standards, codes, etc., then the reviewer should consider the commitments as an integral part of the QA program.

The reviewer should recognize that application for QA program approval may either be separate from the SAR or may exist as a section in the applicant's SAR. Since it is possible that some aspects of the QA program are described in various portions of the application (the SAR or a submittal separate from the SAR) the reviewer should consider these aspects when evaluating the program against the acceptance criteria of Section 14.4. Therefore, if possible, the QAPD evaluation should be coordinated with other aspects of the DSS review. Such coordination will allow reviewers to derive a more accurate and complete assessment of the applicant's level of commitment to the overall QA program, the selection of quality criteria and quality levels, and the proposed implementation methods.

The applicant's QA program may be structured to apply QA measures and controls to all activities and items in proportion to their importance to safety, commonly referred to as a graded approach. A graded approach for the application of QA should be described in the QAPD by adequately assigning appropriate grading classifications and providing an associated justification. However, an applicant may choose to apply the highest level of QA and control to all activities and items. The QA program should identify the activities and items that are important to safety and the degree of their importance. For application of a graded approach, the highly important-to-safety activities and items must have a high level of control, while those less important may have a lower level of control. If the QA program is graded, the staff should be able to conclude that the structure of the graded program is acceptable and that the highest levels of QA are applied to those SSCs that are most important to safety. In making determinations about the application of QA to those SSCs that are listed in the description as important to safety, the reviewer of the QA program description should coordinate with the appropriate NRC project manager and associated technical staff to compare those SSCs described in other portions of the applicant's submittal.

If after review, the reviewer finds the QAPD acceptable, the acceptance of the evaluation should be documented in the Safety Evaluation Report (SER) for QAPDs submitted as part of a SAR. If the applicant's QAPD was submitted prior to the applicant's SAR submittal, the acceptance of the evaluation should be documented in a letter to the applicant and if possible included in the SER at a later time. In either case, the documentation of the review should include the basis for acceptance as noted in the example in Section 14.6 Evaluation Findings. Any recommendations for modifications in the application that are required before the application can be accepted should be addressed by referring to Section 14.6 for initiation of a request for additional information (RAI).

Figure 14-1 presents an overview of the evaluation process and can be used as a guide to assist in coordinating with other review disciplines.

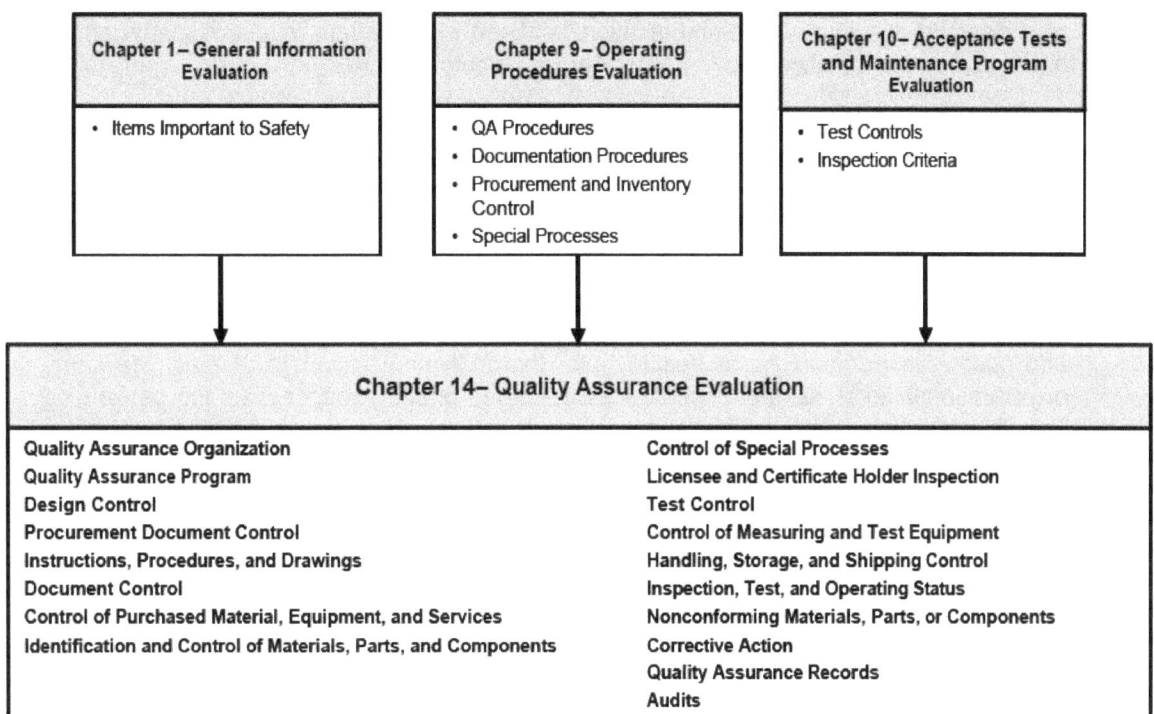

Figure 14-1   Quality Assurance Evaluation

14.5.1   Quality Assurance Organization

The QAPD should describe the structure, interrelationships, and areas of functional responsibility and authority for all organizational elements that will perform activities related to quality and safety. The following are examples of areas/items that may be addressed to support implementation of the quality criteria:

a.   Measures to retain and exercise responsibility for the QA program. The assignment of responsibility for the overall QA program in no degree relieves line management of their responsibility for the achievement of quality.

b.   Measures to identify and describe the QA functions performed by the applicant's QA organization or delegated to other organizations that will provide controls to ensure implementation of the applicable elements of the QA criteria.

c.   Measures to provide clear management controls and effective lines of communication should exist between the applicant's QA organizations and suppliers to ensure proper direction of the QA program and resolution of QA problems.

d.   Measures to identify onsite and offsite organizational elements that will function under the purview of the QA program and the lines of responsibility.

e.   Measures to ensure that high-level management is responsible for documenting and promulgating the applicant's QA policies, goals, and objectives, and this management level should maintain a continuing involvement in QA matters. The application should also describe the lines of communication between intermediate levels of management and between this position and the Manager (or Director) of QA.

f.   Measures to designate a position that retains overall authority and responsibility for the QA program.

g.   Measures to provide authority and independence of the individual responsible for managing the QA program should be such that he or she can direct and control the organization's QA program, effectively ensure conformance to quality requirements, and remain sufficiently independent of undue influences and responsibilities of schedules and costs. In addition, measures to have this individual report to at least the same organizational level as the highest line manager directly responsible for performing activities affecting quality.

h.   Measures for individuals or groups responsible for defining and controlling the content of the QA program and related manuals to have appropriate organizational position and authority, as should the management level responsible for final review and approval.

I.   Measures describing the qualification requirements for the principal QA management positions so as to demonstrate management and technical competence commensurate with the responsibilities of these positions.

j.   Measures to ensure conformance to established requirements be verified by individuals or groups who do not have direct responsibility for performing the work being verified. The quality control function may be part of the line organization provided that the QA organization performs periodic surveillance to confirm sufficient independence from the individuals who performed the activities.

k.   Persons and organizations performing QA functions should have direct access to management levels that will ensure accomplishment of quality-affecting activities. These individuals should have sufficient authority and organizational freedom to perform their QA functions effectively and without reservation. In addition, they should be able to identify quality problems; initiate, recommend, or provide solutions through designated channels; and verify implementation of solutions.

I.   Designated QA individuals or organizations should have the responsibility and authority, delineated in writing, to stop unsatisfactory work and control further processing, delivery, or installation of nonconforming material. In addition, the application should describe how stop-work requests will be initiated and completed.

m.   Measures to determine the extent of QA controls to be determined by the QA staff in combination with the line staff and to depend upon the specific activity or item complexity and level of importance to safety.

14.5.2   Quality Assurance Program

The QAPD should provide acceptable evidence that the applicant's proposed QA program will be well-documented, planned, implemented, and maintained to provide the appropriate level of control over activities and SSCs consistent with their relative importance to safety. The following are examples of areas/items that may be addressed to support implementation of the quality criteria:

a. Measures used to ensure that the QA program meets applicable acceptance criteria.

b. Measures for management to regularly assess the effectiveness of the QA program. In addition, measures for management (above and beyond the QA organization) to regularly assess the scope, status, adequacy, and compliance of the QA program to the requirements of 10 CFR Part 72. Measures to provide for management's frequent contact with program status through reports, meetings, and audits as well as performance of a periodic assessment that is planned and documented with corrective action identified and tracked.

c. Measures used to ensure that trained, qualified personnel within the organization will be assigned to determine that functions delegated to contractors are properly accomplished.

d. Summarizations of the corporate QA policies, goals, and objectives and establishment of a meaningful channel for transmittal of these policies, goals, and objectives down through the levels of management.

e. Measures to designate responsibilities for implementing the major activities addressed in the QA manuals.

f. Measures to control the distribution of the QA manuals and revisions.

g. Measures for communicating to all responsible organizations and individuals that policies, QA manuals, and procedures are mandatory requirements.

h. Measures to provide a comprehensive listing of QA procedures, plus a matrix of these procedures cross-referenced to each of the QA criteria, to demonstrate that the QA program will be fully implemented by documented procedures.

I. Identification of the structures, systems, and components (SSCs) that are important to safety and how they will be controlled by the QA program.

j. Measures for review and documents to show agreement with the QA program provisions of its suppliers to ensure implementation of a program meeting the QA criteria.

k. Measures for the resolution of disputes involving quality arising from a difference of opinion between QA/Quality Control (QC) personnel and personnel from other departments (engineering, procurement, manufacturing, etc.).

l. Measures for indoctrination, training, and qualification programs that fulfill the following criteria:

   • Personnel responsible for performing activities affecting quality should be instructed as to the purpose, scope, and implementation of the quality-related manuals, instructions, and procedures.

   • Personnel performing activities affecting quality should be trained and qualified in the principles and techniques of the activities being performed.

- Maintenance of the proficiency of personnel performing quality-affecting activities by retraining, reexamining, and re-certifying.

- Preparation and maintenance of documentation of completed training and qualification.

- Qualification of personnel in accordance with accepted codes and standards.

14.5.3   Design Control

The QAPD should describe the approach that the applicant will use to define, control, and verify the design and development of the DSS. The following are examples of areas/items that may be addressed to support implementation of the quality criteria:

a.   Measures to carry out design activities in a planned, controlled, and orderly manner.

b.   Measures to correctly translate the applicable regulatory requirements and design bases into specifications, drawings, written procedures, and instructions.

c.   Measures to describe how the applicant will specify quality standards in the design documents and control deviations and changes from these quality standards.

d.   Measures to describe how the applicant will review designs to ensure that design characteristics can be controlled, inspected, and tested and that inspection and test criteria are identified.

e.   Measures to describe how the applicant will establish both internal and external design interface controls. These controls should include review, approval, release, distribution, and revision of documents involving design interfaces with participating design organizations.

f.   Measures to describe how they will properly select and perform design verification processes such as design reviews, alternative calculations, or qualification testing. When a test program is to be used to verify the adequacy of a design, the measures should be developed to describe how they will use a qualification test of a prototype unit under adverse design conditions.

g.   Design verification constitutes confirmation that the design of the SSC is suitable for its intended purpose. Measures to ensure design verifications are completed by an individual with a level of skill at least equal to that of the original designer, recognizing design checking can be performed by a less experienced person. (As an example, design checking, which should also be performed, includes confirmation of the numerical accuracy of computations and the accuracy of data input to computer codes. Confirmation that the correct computer code has been used is part of design verification.) Measures to describe how design verification will be performed by persons other than those performing design checking. In addition, measures to include how individuals or groups responsible for design verification will not include the original designer and normally not include the designer's immediate supervisor.

h.   Measures to ensure design and specification changes are subject to the same design controls and the same or equivalent approvals that were applicable to the original design.

I.   Measures to ensure the documentation of all errors and deficiencies in the design or the design process that could adversely affect SSCs important to safety. In addition, the applicant should provide measures for adequate corrective action, including root cause evaluation of significant errors and deficiencies, to preclude repetition.

j.   Before selecting materials, parts, and equipment that are standard, commercial (off-the-shelf), or have been previously approved for a different application, measures should be provided to review the suitability of any materials, parts, and equipment for the intended application.

k.   Measures to provide written procedures to identify and control the authority and responsibilities of all individuals or groups responsible for design reviews and other design verification activities.

I.   Measures that include the use of valid industry standards and specifications for the selection of suitable materials, parts, equipment, and processes for SSCs that are important to safety.

14.5.4   Procurement Document Control

Documents used to procure SSCs or services should include or reference applicable design bases and other requirements necessary to ensure adequate quality. The following are examples of areas/items that may be addressed to support implementation of the quality criteria:

a.   Measures to establish procedures that clearly delineate the sequence of actions to be accomplished in the preparation, review, approval, and control of procurement documents.

b.   Measures to ensure that qualified personnel review and concur with the adequacy of quality requirements stated in procurement documents. These measures should also ensure that the quality requirements are correctly stated, inspectible, and controllable; there are adequate acceptance and rejection criteria; and the procurement document has been prepared, reviewed, and approved in accordance with QA program requirements.

c.   Measures to document the review and approval of procurement documents before they are released, and the documentation should be available for verification.

d.   Procurement documents should identify the applicable QA requirements that should be compiled and described in the supplier's QA program. In addition, the applicant should review and concur with the supplier's QA program.

e.   Measures to ensure procurement documents contain or reference the regulatory requirements, design bases, and other technical requirements.

f.   Measures to ensure procurement documents identify the documentation (e.g., drawings, specifications, procedures, inspection and fabrication plans, inspection and test records, personnel and procedure qualifications, and chemical and physical test results of material) to be prepared, maintained, and submitted to the purchaser for review and approval.

g.   Measures to ensure procurement documents identify records to be retained, controlled, and maintained by the supplier and those records to be delivered to the purchaser before use or installation of the hardware.

h.   Measures to ensure procurement documents specify the procuring agency's right of access to the supplier's facilities and records for source inspection and audit.

I.   Measures to ensure that changes and revisions to procurement documents are subject to the same or equivalent review and approval as the original documents.

14.5.5   Instructions, Procedures, and Drawings

The QAPD should define the applicant's proposed procedures for ensuring that activities affecting quality will be prescribed by, and performed in accordance with, documented instructions, procedures, or drawings of a type appropriate for the circumstances. The following are examples of areas/items that may be addressed to support implementation of the quality criteria:

a.   Measures to ensure activities affecting quality are prescribed and accomplished in accordance with documented instructions, procedures, or drawings.

b.   Measures to establish provisions that clearly delineate the sequence of actions to be accomplished in the preparation, review, approval, and control of instructions, procedures, and drawings.

c.   Measures to ensure instructions, procedures, and drawings specify the methods for complying with each of the applicable QA criteria.

d.   Measures to ensure instructions, procedures, and drawings include quantitative acceptance criteria (such as dimensions, tolerances, and operating limits) as well as qualitative acceptance criteria (such as workmanship samples) as verification that activities important to safety have been satisfactorily accomplished.

e.   Measures to ensure the QA organization reviews and concurs with the procedures, drawings, and specifications related to inspection plans, tests, calibrations, and special processes as well as any subsequent changes to these documents.

14.5.6   Document Control

The QAPD should define the applicant's proposed procedures for preparing, issuing, and revising documents that specify quality requirements or prescribe activities affecting quality. The following are examples of areas/items that may be addressed to support implementation of the quality criteria:

a. The QAPD should identify all documents to be controlled under this subsection. As a minimum, this should include design specifications; design and fabrication drawings; procurement documents; QA manuals; design criteria documents; fabrication, inspection, and testing instructions; and test procedures.

b. Measures to ensure establishment of procedures to control the review, approval, and issuance of documents and changes thereto before release to ensure that the documents are adequate and applicable quality requirements are stated.

c. Measures to ensure establishment of provisions to identify individuals or groups responsible for reviewing, approving, and issuing documents and revisions thereto.

d. Measures to ensure document revisions receive review and approval by the same organizations that performed the original review and approval or by other qualified responsible organizations designated by the applicant.

e. Measures to ensure that approved changes be included in instructions, procedures, drawings, and other documents before the change is implemented.

f. Measures to ensure the control of obsolete or superseded documents to prevent inadvertent use.

g. Measures to ensure documents are available at the location where the activity is performed.

h. Measures to ensure establishment of a master list (or equivalent) to identify the current revision number of instructions, procedures, specifications, drawings, and procurement documents. In addition, measures to ensure updating of the list and distribution of it to predetermined, responsible personnel to preclude use of superseded documents.

14.5.7   Control of Purchased Material, Equipment, and Services

The QAPD should define the applicant's proposed procedures for controlling purchased material, equipment, and services to ensure conformance with specified requirements. The following are examples of areas/items that may be addressed to support implementation of the quality criteria:

a. Measures to ensure qualified personnel evaluate the supplier's capability to provide services and products of acceptable quality before the award of the procurement order or contract. In addition, measures to ensure QA and engineering groups participate in the evaluation of those suppliers providing critical items and services important to safety, and the applicant should define the responsibilities for each group's participation.

b. Measures to ensure evaluation of suppliers on the basis of one or more of the following criteria:

  • The supplier's capability to comply with the elements of the QA criteria that are applicable to the type of material, equipment, or service being procured.

  • Review of previous records and performance of suppliers who have provided similar articles or services of the type being procured.

- A survey of the supplier's facilities and QA program to assess the capability to supply a product that meets applicable design, manufacturing, and quality requirements.

c. Measures to ensure documentation and filing of the results of supplier evaluations.

d. Measures to ensure planning and performing adequate surveillance of suppliers during fabrication, inspection, testing, and shipment of materials, equipment, and components in accordance with written procedures to ensure conformance to the purchase order requirements. In addition the measures should ensure that the procedures provide the following information:

- Instructions that specify the characteristics or processes to be witnessed, inspected or verified, and accepted; the method of surveillance and the extent of documentation required; and those responsible for implementing these instructions.

- Procedures for audits and surveillance to ensure that the supplier complies with the quality requirements (surveillance should be performed for SSCs for which verification of procurement requirements cannot be determined upon receipt).

e. Measures to ensure the supplier furnish the following records to the purchaser:

- Documentation that identifies the purchased material or equipment and the specific procurement requirements (e.g., codes, standards, and specifications) met by the items.

- Documentation that identifies any procurement requirements that have not been met and a description of any nonconformances designated "accept as is" or "repair."

f. Measures to describe the proposed procedures for reviewing and accepting these documents and, as a minimum, to ensure that this review and acceptance will be undertaken by a responsible QA individual.

g. Measures to ensure the conduct periodic audits, independent inspections, or tests to ensure the validity of the suppliers' certificates of conformance.

h. Measures to ensure the performance of a receiving inspection of the supplier-furnished material, equipment, and services to ensure fulfillment of the following criteria:

- The material, component, or equipment should be properly identified in a manner that corresponds with the identification on the purchasing and receiving documentation.

- Material, components, equipment, and acceptance records should be inspected and judged acceptable in accordance with predetermined inspection instructions before installation or use.

- Inspection records or certificates of conformance attesting to the acceptance of material, components, and equipment should be available before installation or use.

- Items accepted and released should be identified as to their inspection status before they are forwarded to a controlled storage area or released for installation or further work.

i.   Measures to assess the effectiveness of suppliers' quality controls at intervals consistent with the importance to safety, complexity, and quantity of the SSCs procured.

14.5.8   Identification and Control of Materials, Parts, and Components

The QAPD should define the applicant's proposed provisions for identifying and controlling materials, parts, and components to ensure that incorrect or defective SSCs are not used. The following are examples of areas/items that may be addressed to support implementation of the quality criteria:

a.   Measures to establish procedures to identify and control materials, parts, and components (including partially fabricated subassemblies).

b.   Measures to determine identification requirements during generation of specifications and design drawings.

c.   Measures to ensure that identification will be maintained either on the item or on records traceable to the item to preclude use of incorrect or defective items.

d.   Measures to ensure Identification of materials and parts of important-to-safety items are traceable to the appropriate documentation (such as drawings, specifications, purchase orders, manufacturing and inspection documents, deviation reports, and physical and chemical mill test reports).

e.   Measures to ensure the location and method of identification do not affect the fit, function, or quality of the item being identified.

f.   Measures to verify and document the correct identification of all materials, parts, and components before releasing them for fabrication, assembly, shipping, and installation.

14.5.9   Control of Special Processes

The QAPD should describe the controls that the applicant will establish to ensure the acceptability of special processes (such as welding, heat treatment, nondestructive testing, and chemical cleaning) and that the proposed controls are performed by qualified personnel using qualified procedures and equipment. The following are examples of areas/items that may be addressed to support implementation of the quality criteria:

a.   Measures to establish procedures to control special processes (such as welding, heat treating, nondestructive testing, and cleaning) for which direct inspection is generally impossible or disadvantageous. In addition, the applicant should provide a listing of these special processes.

b.   Measures to qualify procedures, equipment, and personnel connected with special processes in accordance with applicable codes, standards, and specifications.

c.   Measures to ensure qualified personnel perform special processes in accordance with written process sheets (or the equivalent) with recorded evidence of verification.

d.   Measures to establish, file, and keep current qualification records of procedures, equipment, and personnel associated with special processes.

14.5.10   Licensee Inspection

The QAPD should define the applicant's proposed provisions for inspection of activities affecting quality to verify conformance with instructions, procedures, and drawings. The following are examples of areas/items that may be addressed to support implementation of the quality criteria:

a.   Measures to establish, document, and conduct an inspection program that effectively verifies conformance of quality-affecting activities with requirements in accordance with written, controlled procedures.

b.   Measures to ensure inspection personnel are sufficiently independent from the individuals performing the activities being inspected.

c.   Measures to ensure inspection procedures, instructions, and check lists provide the following details:

•   Identification of characteristics and activities to be inspected.

•   Identification of the individuals or groups responsible for performing the inspection operation.

•   Acceptance and rejection criteria.

•   A description of the method of inspection.

•   Procedures for recording evidence of completing and verifying a manufacturing, inspection, or test operation.

•   Identification of the recording inspector or data recorder and the results of the inspection operation.

d.   Measures to ensure the use of inspection procedures or instructions with the necessary drawings and specifications when performing inspection operations.

e.   Measures to qualify inspectors in accordance with applicable codes, standards, and company training programs and in addition keeping inspector's qualifications and certifications current.

f.   Measures to inspect modifications, repairs, and replacements in accordance with the original design and inspection requirements or acceptable alternatives.

g.   Measures to establish provisions that identify mandatory inspection hold points for witnessing by a designated inspector.

h.   Measures to identify the individuals or groups who will perform receiving and process verification inspections, and should demonstrate that these individuals or groups have sufficient independence and qualifications.

I.   Measures to establish provisions for indirect control by monitoring processing methods, equipment, and personnel if direct inspection is not possible.

14.5.11   Test Control

The QAPD should define the applicant's proposed provisions for tests to verify that SSCs conform to specified requirements and will perform satisfactorily in service. The following are examples of areas/items that may be addressed to support implementation of the quality criteria:

a.   Measures to establish, document, and conduct a test program to demonstrate that the item will perform satisfactorily in service in accordance with written, controlled procedures.

b.   Measures to ensure written test procedures incorporate or reference the following information:

   •   Requirements and acceptance limits contained in applicable design and procurement documents.

   •   Instructions for performing the test.

   •   Test prerequisites.

   •   Mandatory inspection hold points.

   •   Acceptance and rejection criteria.

   •   Methods of documenting or recording test data results.

c.   Measures to ensure a qualified, responsible individual or group document test results and evaluate their acceptability. When practicable, the measures should ensure testing of the SSC occurs under conditions that will be present during normal and anticipated off-normal operations.

14.5.12   Control of Measuring and Test Equipment

The QAPD should define the applicant's proposed provisions to ensure that tools, gauges, instruments, and other measuring and testing devices are properly identified, controlled, calibrated, and adjusted at specified intervals. The following are examples of areas/items that may be addressed to support implementation of the quality criteria:

a.   Measures to ensure documented procedures describe the calibration technique and frequency, maintenance, and control of all measuring and test equipment (instruments, tools, gauges, fixtures, reference and transfer standards, and nondestructive test equipment) that will be used in the measurement, inspection, and monitoring of SSCs that are important to safety.

b.   Measures to ensure measuring and test equipment are identified and traceable to the calibration test data.

c.   Measures to ensure the use of labels, tags, or documents for measuring and test equipment to indicate the date of the next scheduled calibration and to provide traceability to calibration test data.

d.   Measures to calibrate measuring and test instruments at specified intervals on the basis of the required accuracy, precision, purpose, degree of usage, stability characteristics, and other conditions that could affect the accuracy of the measurements.

e.   Measures to assess the validity of previous inspections when measuring and test equipment is found to be out of calibration. In addition, measures should also be provided to document the assessment and take control of the out of calibration equipment.

f.   Measures to document and maintain the complete status of all items under the calibration system.

g.   Measures to ensure reference and transfer standards are traceable to nationally recognized standards; where national standards do not exist, the applicant should establish provisions to document the basis for calibration.

14.5.13   Handling, Storage, and Shipping Control

The QAPD should define the applicant's proposed provisions to control the handling, storage, shipping, cleaning, and preservation of SSCs in accordance with work and inspection instructions to prevent damage, loss, and deterioration. The following are examples of areas/items that may be addressed to support implementation of the quality criteria:

a.   Measures to establish and accomplish special handling, preservation, storage, cleaning, packaging, and shipping requirements in accordance with predetermined work and inspection instructions.

b.   Measures to control the cleaning, handling, storage, packaging, shipping, and preservation of materials, components, and systems in accordance with design and specification requirements to preclude damage, loss, or deterioration by environmental conditions (such as temperature or humidity).

14.5.14   Inspection, Test, and Operating Status

The QAPD should define the applicant's proposed provisions to control the inspection, test, and operating status of SSCs to prevent inadvertent use or bypassing of inspections and tests. The following are examples of areas/items that may be addressed to support implementation of the quality criteria:

a.   Measures to know the inspection and test status of items throughout fabrication.

b.   Measures to establish procedures to control the application and removal of inspection and welding stamps and operating status indicators (such as tags, markings, labels, and stamps).

c.   Measures to ensure procedures under the cognizance of the QA organization controls the bypassing of required inspections, tests, and other critical operations.

d.   Measures to specify the organization responsible for documenting the status of nonconforming, inoperative, or malfunctioning SSCs and identifying the item to prevent inadvertent use.

14.5.15   Nonconforming Materials, Parts, or Components

The QAPD should define the applicant's proposed provisions to control the use or disposition of nonconforming materials, parts, or components. The following are examples of areas/items that may be addressed to support implementation of the quality criteria:

a.   Measures to establish procedures to control the identification, documentation, tracking, segregation, review, disposition, and notification of affected organizations regarding nonconforming materials, parts, components, services, or activities.

b.   Measures to provide for adequate documentation to identify nonconforming items and describe the nonconformance, its disposition, and the related inspection requirements. The measures should also provide for adequate documentation and include signature approval of the disposition.

c.   Measures to establish provisions to identify those individuals or groups with the responsibility and authority for the disposition and closeout of nonconformance.

d.   Measures to ensure nonconforming items are segregated from acceptable items and identified as discrepant until properly dispositioned and closed out.

e.   Measures to verify the acceptability of reworked or repaired materials, parts, and SSCs by re-inspecting and retesting the item as originally inspected and tested or by using a method that is at least equal to the original inspection and testing method. In addition, the measures should provide for documentation of the relevant inspection, testing, rework, and repair procedures.

f.   Measures to ensure nonconformance reports designated "accept as is" or "repair" are made part of the inspection records and forwarded with the hardware to the customer for review and assessment.

g.   Measures to periodically analyze nonconformance reports to show quality trends and help identify root causes of nonconformance. Significant results should be reported to responsible management for review and assessment.

14.5.16   Corrective Action

The QAPD should define the applicant's proposed provisions to ensure that conditions adverse to quality are promptly identified and corrected, and that measures are taken to preclude recurrence. The following are examples of areas/items that may be addressed to support implementation of the quality criteria:

a.   Measures to evaluate conditions adverse to quality (such as nonconformance, failures, malfunctions, deficiencies, deviations, and defective material and equipment) in accordance with established procedures to assess the need for corrective action.

b.   Measures to initiate corrective action to preclude recurrence of a condition identified as adverse to quality.

c.   Measures to conduct follow-up activities to verify proper implementation of corrective actions and close out the corrective action documentation in a timely manner.

d.   Measures to document significant conditions adverse to quality, as well as the root causes of the conditions, and the corrective actions taken to remedy the and preclude recurrence of the conditions. In addition, this information should be reported to cognizant levels of management for review and assessment.

14.5.17   Quality Assurance Records

The SAR should define the applicant's proposed provisions for identifying, retaining, retrieving, and maintaining records that document evidence of the control of quality for activities and SSCs important to safety. The following are examples of areas/items that may be addressed to support implementation of the quality criteria:

a.   Measures to define the scope of the records program such that sufficient records will be maintained to provide documentary evidence of the quality of items and activities affecting quality. To minimize the retention of unnecessary records, the records program should list records to be retained by "type of data" rather than by record title.

b.   Measures to ensure that QA records include operating logs; results of reviews, inspections, tests, audits, and material analyses; monitoring of work performance; qualification of personnel, procedures, and equipment; and other documentation such as drawings, specifications, procurement documents, calibration procedures and reports, design review and peer review reports, nonconformance reports, and corrective action reports.

c.   Measures to ensure records are identified and retrievable.

d.   Measures to ensure requirements and responsibilities for record creation, transmittal, retention (such as duration, location, fire protection, and assigned responsibilities), and maintenance subsequent to completion of work are consistent with applicable codes, standards, and procurement documents.

e.   Measures to ensure inspection and test records contain the following information, where applicable:

*   A description of the type of observation.
*   The date and results of the inspection or test.
*   Information related to conditions adverse to quality.
*   Identification of the inspector or data recorder.

- Evidence as to the acceptability of the results.
- Action taken to resolve any noted discrepancies.

f. Measures to ensure record storage facilities are constructed, located, and secured to prevent destruction of the records by fire, flood, theft, and deterioration by environmental conditions (such as temperature or humidity). In addition, the facilities are to be maintained by, or under the control of, the licensee throughout the life of the DSS or the individual product.

14.5.18   Audits

The QAPD should define the applicant's proposed provisions for planning and scheduling audits to verify compliance with all aspects of the QA program, and to determine the effectiveness of the overall program. The following are examples of areas/items that may be addressed to support implementation of the quality criteria:

a. Measures to perform audits in accordance with written procedures or checklists; qualified personnel tasked with performing these audits should not have direct responsibility for the achievement of quality in the areas being audited.

b. Measures to ensure audit results are documented and reviewed with management having responsibility in the area audited.

c. Measures to establish provisions for responsible management to undertake appropriate corrective action as a follow-up to audit reports. In addition, the measures should ensure auditing organizations schedule and conduct appropriate follow-up to ensure that the corrective action is effectively accomplished.

d. Measures to perform both technical and QA programmatic audits to achieve the following objectives:

- Provide a comprehensive independent verification and evaluation of procedures and activities affecting quality.

- Verify and evaluate suppliers' QA programs, procedures, and activities.

e. Measures to ensure audits are led by appropriately qualified and certified audit personnel from the QA organization. The measures should also ensure that the audit team membership include personnel (not necessarily QA organization personnel) having technical expertise in the areas being audited.

f. Measures to schedule regular audits on the basis of the status and importance to safety of the activities being audited. The measures should also address that audits be initiated early enough to ensure effective QA during design, procurement, and contracting activities.

g. Measures to analyze and trend audit deficiency data as well as ensuring resultant reports, indicating quality trends and the effectiveness of the QA program, should be given to management for review, assessment, corrective action, and follow-up.

h. Measures to ensure that audits objectively assess the effectiveness and proper implementation of the QA program and should address the technical adequacy of the activities being conducted.

I. Measures to establish provisions requiring the performance of audits in all areas to which the requirements of the QA program apply.

14.6 Evaluation Findings

If the reviewer determines that the applicant's QAPD does not adequately address the Part 72 requirements, a request for additional information (RAI) must be prepared and submitted to the Project Manager to be forwarded to the applicant for resolution and response to the NRC. If the reviewer concludes that information provided with the application, along with additional information provided in response to NRC RAI(s), shows that the QA program description meets the acceptance requirements referenced in Section 14.4, findings of the following type should be included in the staff's SER or in a letter to the applicant, if the applicant's QA program description was submitted separate from a SAR.

(finding numbering is for convenience in referencing within the FSRP and SER):

F14.1 Based upon a review and evaluation of the QA program description contained in the Safety Analysis Report or applicant's submittal (identified by date and any other pertinent identifiers) for a DSS, the staff concludes that:

- The licensee's description of the QA program indicates requirements, procedures, and controls that, when properly implemented, should comply with the requirements of 10 CFR 72, Subpart G.

- The licensee's description of the QA program covers activities affecting SSCs important to safety as identified in the Safety Analysis Report.

- The licensee's description of the QA program describes organizations and persons performing QA functions indicating that sufficient independence and authority should exist to perform their functions without undue influence from those directly responsible for costs and schedules.

- The licensee's description of the QA program is in compliance with applicable NRC regulations and industry standards, and the acceptance of the QA program description by NRC allows implementation of the associated QA program for the (specify: design, fabrication and construction, operation, decommissioning) phases of the installation's life cycle.

APPENDIX A   CONSOLIDATED REFERENCES

A.1   U.S. Nuclear Regulatory Commission (NRC) Documents Cited

A.1.1   *U.S. Code of Federal Regulations* (CFR), Title 10, "Energy"

Part 2, "Rules of Practice for Domestic Licensing Proceedings and Issuance of Orders," August 15, 1991.

Part 20, "Standards for Protection Against Radiation," September 11, 1988.

Part 50, "Domestic Licensing of Production and Utilization Facilities," August 15, 1991.

Part 71, "Packaging and Transportation of Radioactive Material," Appendix H, Quality Assurance, September 28, 1995.

Part 72, "Licensing Requirements for the Independent Storage of Spent Nuclear Fuel, High-Level Radioactive Waste, and Reactor-Related Greater Than Class C Waste," January 1, 2001.

Part 73, "Physical Protection of Plants and Materials," December 28, 1973.

Part 100, "Reactor Site Criteria," January 10, 1997.

Part 961, "Standard Contract for Disposal of Spent Nuclear Fuel and/or High Level Radioactive Waste," April 18, 1983.

A.1.2   Regulatory Guides (RG)

RG 1.25, "Assumptions Used for Evaluating the Potential Radiological Consequences of a Fuel Handling Accident in the Fuel Handling and Storage Facility for Boiling and Pressurized Water Reactors," March 1972.

RG 1.26, "Quality Group Classification and Standards for Water-, Steam-, and Radioactive-Waste-Containing Components of Nuclear Power Plants," Revision 4, March 2007, ML070290283.

RG 1.29, "Seismic Design Classification," Revision 4, March 2007, ML070310052.

RG 1.33, "Quality Assurance Program Requirements (Operation)," Revision 3, February 1978, ML0037399.

RG 1.59, "Design Basis Floods for Nuclear Power Plants," Revision 2, August 1977 with Errata of 7/30/1980, ML003740388.

RG 1.60, "Design Response Spectra for Seismic Design of Nuclear Power Plants," Revision 1, December 1973. ML003740207.

RG 1.61, "Damping Values for Seismic Design of Nuclear Power Plants," Revision 1, March 2007, ML070260029.

RG 1.76, "Design Basis Tornado and Tornado Missiles for Nuclear Power Plants," Revision 1, March 2007, ML070360253.

RG 1.86, "Termination of Operating Licenses for Nuclear Reactors," June 1974, ML003740243.

RG 1.92, "Combining Modal Responses and Spatial Components in Seismic Response Analysis," Revision 2, July 2006, ML053250475.

RG 1.102, "Flood Protection for Nuclear Power Plants," Revision 1, September 1976, ML003740308.

RG 1.109, "Calculations of Annual Doses to Man from Routine Releases of Reactor Effluents for the Purpose of Evaluating Compliance with 10 CFR Part 50, Appendix I," Revision 1, October 1977, ML003740384.

RG 1.117, "Tornado Design Classification," Revision 1, April 1978, ML003739346.

RG 1.136, "Design Limits, Loading Combinations, Materials, Construction, and Testing of Concrete Containments," Revision 3, March 2007, ML070310045.

RG 1.142, "Safety-Related Concrete Structures for Nuclear Power Plants (Other than Reactor Vessels and Containments)," Revision 2, November 30, 2001, ML013100274.

RG 1.143, "Design Guidance for Radioactive Waste Management Systems, Structures, and Components Installed in Light-Water-Cooled Nuclear Power Plants," Revision 2, November 2001, ML013100305.

RG 1.145, "Atmospheric Dispersion Models for Potential Accident Consequence Assessments at Nuclear Power Plants," February 1989, ML003740205.

RG 1.183, "Alternative Radiological Source Terms for Evaluating Design Basis Accidents at Nuclear Power Plants," Revision 0, July 2000, ML003716792.

RG 1.193, "ASME Code Cases Not Approved for Use," Revision 2, October 2007, ML072470294.

RG 3.60, "Design of an Independent Spent Fuel Storage Installation (Dry Storage)," March 1997, ML003739501.

RG 3.61, "Standard Format and Content for a Topical Safety Analysis Report for a Spent Fuel Dry Storage Cask," February 1989, ML003739511.

RG 7.11, "Fracture Toughness Criteria of Base Material for Ferritic Steel Shipping Cask Containment Vessels with a Maximum Wall Thickness of 4 Inches," Revision 0, June 1991, ML003739413.

RG 7.12, "Fracture Toughness Criteria of Base Material for Ferritic Steel Shipping Cask Containment Vessels with a Wall Thickness Greater Than 4 Inches, But Not Exceeding 12 Inches," Revision 0, June 1991, ML003739424.

RG 8.5, "Criticality and Other Interior Evaluation Signals, Revision 1, March 1981, ML003739454.

RG 8.8, "Information Relevant to Ensuring that Occupational Radiation Exposures at Nuclear Power Stations Will Be as Low as Reasonably Achievable," June 2001.

RG 8.10, "Operating Philosophy for Maintaining Occupational Radiation Exposures as Low as is Reasonably Achievable," September 1975, ML003739563.

RG 8.25, "Air Sampling in the Workplace," Revision 1, June 1992, ML003739616.

RG 8.34, "Monitoring Criteria and Methods to Calculate Occupational Radiation Doses," July 1992, ML003739502.

RG 8.36, "Radiation Dose to the Embryo/Fetus," July 1992, ML003739548.

A.1.3 NUREG

NUREG-0612, "Control of Heavy Loads at Power Plants," July 1980.

NUREG-0800, "Standard Review Plan for the Review of Safety Analysis Reports for Nuclear Power Plants," March, 2007.

NUREG-1567, "Standard Review Plan for Spent Fuel Dry Storage Facilities," March 2000.

NUREG-1571, "Information Handbook on Independent Spent Fuel Storage Installations," Raddatz, M.G. and Waters, M.D., December 1995.

NUREG-1614, "Strategic Plan, FY2004 - FY2009," Volume 3, August 2004.

NUREG-1727, "NMSS Decommissioning Standard Review Plan," September 2000.

NUREG-1745, "Standard Format and Content for Technical Specifications for 10 CFR Part 72 Cask Certificates of Compliance," June 2001.

NUREG-1864, "A Pilot Probabilistic Risk Assessment of a Dry Cask Storage System at a Nuclear Power Plant," March 2007

A.1.4 NUREG/CR

NUREG/CR-1815, "Recommendations for Protecting Against Failure by Brittle Fracture in Ferritic Steel Shipping Containers Up to Four Inches Thick," LLNL, June 1981.

NUREG/CR-3826, "Recommendations for Protecting Against Failure by Brittle Fracture in Ferritic Steel Shipping Containers Greater than Four Inches Thick," LLNL, July 1994.

NUREG/CR-4554, "SCANS (Shipping Cask Analysis System): A Microcomputer Based Analysis System for Shipping Cask Design Review," LLNL, March 1998.

NUREG/CR-4775, "Guide for Preparing Operating Procedures for Shipping Packages," UCID-20820, July 1988.

NUREG/CR-5502, "Engineering Drawings for 10 CFR Part 71 Package Approval," Lawrence Livermore National Laboratory (LLNL), May 1998.

NUREG/CR-6007, "Stress Analysis of Closure Bolts for Shipping Casks," Kaiser Engineering, January 1993.

NUREG/CR-6242, "CASKS (Computer Analysis of Storage Casks): A Microcomputer-Based Analysis System for Storage Cask Design Review," Lawrence Livermore National Laboratory (LLNL), February 1995.

NUREG/CR-6322, "Buckling Analysis of Spent Fuel Basket", UCRL-ID-119697, LLNL, May 1995.

NUREG/CR-6328, "Adequacy of the 123-Group Cross-Section Library for Criticality Analyses of Water-Moderated Uranium Systems," ORNL/TM-12970, ORNL, August 1995.

NUREG/CR-6361, "Criticality Benchmark Guide for Light-Water-Reactor Fuel in Transportation and Storage Packages." ORNL/TM-13211, U.S. Nuclear Regulatory Commission (NRC), ORNL, March 1997.

NUREG/CR-6407, "Classification of Transportation Packaging and Dry Spent Fuel Storage System Components According to Importance to Safety," INEL-95/0551, Idaho National Engineering Laboratory (INEL), February 1996.

NUREG/CR-6487, "Containment Analysis for Type B Packages Used to Transport Various Contents," (LLNL), November 1996.

NUREG/CR-6608, "Summary and Evaluation of Low-Velocity Impact Tests of Solid Steel Billet onto Concrete Pad," Lawrence Livermore National Laboratory, February 1998.

NUREG/CR-6700, "Nuclide Importance to Criticality Safety, Decay Heating, and Source Terms Related to Transport and Interim Storage of High-Burnup LWR Fuel," ORNL/TM-2000/284, ORNL, January 2001.

NUREG/CR-6701, "Review of Technical Issues Related to Predicting Isotopic Compositions and Source Terms for High-Burnup LWR Fuel," ORNL/TM-2000/277, ORNL, January 2001.

NUREG/CR-6716, "Recommendations on Fuel Parameters for Standard Technical Specifications for Spent Fuel Storage Casks," ORNL/TM-2000/385, Oak Ridge National Laboratory (ORNL), March 2001.

NUREG/CR-6759, "Parametric Study of the Effect of Control Rods for PWR Burnup Credit," ORNL/TM-2001/69, ORNL, February 2002.

NUREG/CR-6760, "Study of the Effect of Integral Burnable Absorbers for PWR Burnup Credit," ORNL/TM-2000/321, ORNL, March 2002.

NUREG/CR-6761, "Parametric Study of the Effect of Burnable Poison Rods for PWR Burnup Credit," ORNL/TM-2000/373, ORNL, March 2002.

NUREG/CR-6798, "Isotopic Analysis of High-Burnup PWR Spent Fuel Samples From the Takahama-3 Reactor," ORNL/TM-2001/259, ORNL, January 2003.

NUREG/CR-6801, "Recommendations for Addressing Axial Burnup in PWR Burnup Credit Analyses," March 2003.

NUREG/CR-6802, "Recommendations for Shielding Evaluations for Transport & Storage Packages," May 2003.

NUREG/CR-6835, "Effects of Fuel Failure on Criticality Safety and Radiation Dose for Spent Fuel Casks," ORNL/TM-2002/255, ORNL, September 2003.

A.1.5 Other NRC Publications

"Movement of Heavy Loads Over Spent Fuel, Over Fuel in the Reactor Core, or Over Safety Related Equipment," NRC Bulletin 96-02, April 11, 1996.

"Chemical, Galvanic, or Other Reactions in Spent Fuel Storage and Transportation Casks," NRC Bulletin 96-04, July 1996.

Confirmatory Action Letter 97-7-001, July 22, 1998.

Information Notice No. 91-26, "Potential Nonconservative Errors in the Working Format Hansen-Roach Cross Section Set Provided with the KENO and SCALE Codes," April 15, 1991.

Tang, David T., et al., "NRC Staff Technical Approach for Spent Fuel Storage Cask Drop and Tipover Accident Analysis," Spent Fuel Project Office, 1997.

A.2 Codes, Standards, and Specifications

American Concrete Institute (ACI), "Code Requirements for Nuclear Safety-Related Concrete Structures and Commentary," ACI 349-06/349R-06, 2006.

– – – "Building Code Requirements for Structural Plain Concrete and Commentary," ACI 318-05/310R-05 with Errata of 0/11/05, 2005.

– – – "Building Code Requirements for Masonry Structures, and Commentary," ACI 530-05, 2005.

American Society of Mechanical Engineers (ASME), "Cases of ASME Boiler and Pressure Vessel Code," Code Case N-595-4, 2004.

American Society of Mechanical Engineers, Boiler and Pressure Vessel (B&PV) Code, "Specification for Welding Rods, Electrodes and Filler Metals," Section II, "Materials" - Part C, 2001.

– – – Section III, "Rules for Construction of Nuclear Facility Components," 2007.
    Division 1 - General Requirements for Division 1 and Division 2; Subsection NB through NH, and Appendices.
    Division 2 - Code for Concrete Containment (Also known as ACI 359-07).

Division 3 - Containment for Transportation and Storage of Spent Nuclear Fuel and High Level Radioactive Material and Waste.

– – – ASME B&PV Code, Section V, "Nondestructive Examination Specifications and Procedures."

– – – ASME B&PV Code, Section VIII, Division 3, "Alternative Rules for the Construction of High Pressure Vessels," 2001.

– – – ASME B&PV Code, Section IX, "Qualification Standard for Welding and Brazing Procedures, Welders, Brazers, and Welding and Brazing Operators."

– – – ASME B&PV Code, Section XI, "Rules for Inservice Inspection of Nuclear Power Plant Components," 2001.

American Institute of Steel Construction (AISC), "Code of Standard Practice for Steel Buildings and Bridges," March 2005.

– – – "Specification for Structural Steel Buildings," March 2005.

ANSI/American Nuclear Society (ANS), "Design Criteria for an Independent Spent Fuel Storage Installation (Dry Storage Type)," ANSI/ANS 57.9-1992-R2000, 2000.

– – – "Neutron and Gamma-Ray Flux-to-Dose Conversion Factors," ANSI/ANS-6.1.1, 1977.

– – – "Neutron and Gamma-Ray Flux-to-Dose Conversion Factors," ANSI/ANS-6.1.1, 1991.

– – – "Nuclear Criticality Safety in Operations with Fissionable Material Outside Reactors," ANSI/ANS-8.1, 1998.

– – – "Administrative Controls and Quality Assurance for the Operational Phase of Nuclear Power Plants," ANSI/ANS 3.2.

ANSI, Institute for Nuclear Materials Management, "American National Standard for Leakage Tests on Packages for Shipment of Radioactive Materials," ANSI N14.5, 1997.

– – – "American National Standards for Radioactive Materials-Special Lifting Devices for Shipping Containers Weighing 10,000 Pounds (4500 Kilograms) or More," ANSI N14.6-1986, 1986.

– – – "Characterizing Damaged Spent Nuclear Fuel for the Purpose of Storage and Transport," ANSI N14.33-2005.

ANSI/American Nuclear Society (ANS), "Requirements for Collection, Storage, and Maintenance of Quality Assurance for Nuclear Power Plants," ANSI/ASME N45.2.9-1979.

ANSI/ANS, "Nuclear Facilities – Steel Safety Related Structures for Design Fabrication and Erection," N690.

ANSI/ASME B16.34, "Valves Flanged, Threaded and Welding End."

ANSI/ASME B31.1, "Power Piping."

ANSI/ASME B96.1, "Specification for Welded Aluminum-Alloy Field-Erected Storage Tanks."

ANSI/ASME NQA-1, "Quality Assurance Program for Nuclear Facilities."

ANSI/ASME NQA-2, "Quality Assurance Requirements for Nuclear Facilities."

American Petroleum Institute (API), "Recommended Rules for Design and Construction of Large Welded, Low-Pressure Storage Tanks," API 620, February 2002.

American Society of Civil Engineers (ASCE), "Minimum Design Loads for Buildings and Other Structures," ASCE 7-05, 2005.

– – – "Seismic Analysis of Safety-Related Nuclear Structures," ASCE 4-98, 2002.

American Society for Testing and Materials International (ASTM), "Draft 17-Guide for Evaluation of Materials Used in Extended Service of Interim Spent Nuclear Fuel Dry Storage Systems," October 2002.

– – – "Standard Practice for Prediction of the Long-Term Behavior of Waste Package Materials Including Waste Forms Used in the Geologic Disposal of High-Level Nuclear Waste," C1174-97, 2003.

– – – "Standard Practice for Qualification and Acceptance of Boron Based Metal Neutron Absorbers for Nuclear Criticality Control for Dry Storage Systems and Transportation Packaging," C1671.

– – – "Standard Test Method for Dynamic Tear Testing of Metallic Materials," ASTM E604-83, 2002.

– – – "Standard Test Method of Conducting Drop-Weight Test to Determine Nil-Ductility Transition Temperature of Ferritic Steels," ASTM E208-95a, 2000.

– – – "Standard Specification for Concrete Aggregates," C 33, 2002.

American Water Works Association (AWWA), "Welded Steel Tanks for Water Storage," AWWA D100.

American Welding Society (AWS), "Standard Symbols for Welding, Brazing, and Nondestructive Examination," AWS A2.4 (Latest Edition).

– – – "Structural Welding Code-Steel," AWS D1.1/D1.1M-2002, 2002.

International Commission on Radiological Protection (ICRP), "Statement from the 1980 Meeting of the ICRO," ICRP Publication 26, Pergammon Press, New York, New York, 1980.

International Conference Council (ICC), "International Building Code (IBC)," 2006.

American Society for Nondestructive Testing (SNT), "Personnel Qualification and Certification in Nondestructive Testing," SNT-TC-1A.

A.3   Other Government Agencies

Environmental Protection Agency, "Manual of Protective Action Guides and Protective Actions for Nuclear Incidents," EPA 410R-92-001.

– – – "Manual of Protective Action Guides and Protective Actions for Nuclear Incidents," EPA 410-R-92-001, May 1992.

– – – "External Exposure to Radionuclides in Air, Water, and Soil," EPA Guidance Report No. 12, 1993.

Newman, L.W., "The Hot Cell Examination of Oconee Fuel Rods After Five Cycles of Irradiation," DOE/ET/34212-50, U.S. Department of Energy (DOE), 1986.

U.S. Department of Energy (DOE), "Criticality Safety Good Practices Program Guide for DOE Nonreactor Nuclear Facilities," DOE G 421.1-1, August 25, 1999.

Nuclear Science Committee, Nuclear Energy Agency, "International Handbook of Evaluated Criticality Safety Benchmark Experiments," NEA/NSC/DOC(95)03, September 2003. (This document is updated and published annually in CD-ROM format).

A.4   Technical Reports

AEA Technology, "MONK - A Monte Carlo Program for Nuclear Criticality Safety and Reactor Physics Analyses, User Guide for Version 8," ANSWERS/MONK(98)6, June 1991. Issued through the ANSWERS Software Service.

ANSYS, Inc., "ANSYS Basic Analysis Procedures Guide," Fourth Edition, ANSYS Release 5.6, November 1999.

Bechtel, "Commercial Spent Nuclear Fuel Handling in Air Study," 000-30R-MGR0-00700-000000, March 2005.

Beyer, C.E., Letter from C.E. Beyer, Pacific Northwest National Laboratory, to K. Gruss, 2001.

Boase, D.G. and T.T. Vandergraaf, "The Canadian Spent Fuel Storage Canister: Some Materials Aspects," Nucl. Techol., 32, 60, (1977).

Bjorkman & Moore, *Influence of ISFSI Design Parameters on the Seismic Response of Dry Storage Casks*, 2001.

Bjorkman, et al., *Seismic Analysis of Plant Hatch ISFSI Pad and Stability Assessment of Dry Casks*, 2000.

Broadhead, B.L., et al., "Evaluating of Shielding Analysis Methods in Spent Fuel Cask Environments," EPRI TR-104329, Electric Power Research Institute (EPRI), Palo Alto, California, May 1995.

Cacciapouti, R.J., and S. Van Volkinburg. "Axial Burnup Profile Database for Pressurized Water Reactors." YAEC-1937. May 1997. Available as Data Package DLC-201 from the Radiation

Safety Information Computational Center at Oak Ridge National Laboratory (ORNL). http://www-rsicc.ornl.gov/ORDER.html.

Cappelaere, R. Limon, T. Bredel, P. Herter, D. Gilbon, S. Allegre, P. Bouffioux and J.P. Mardon. "Long Term Behaviour of the Spent Fuel Cladding in Dry Storage Conditions." 8th International Conference on Radioactive Waste Management and Environmental Remediation. October 2001. Bruges, Belgium.

Chung, H.M. and T.F. Kassner. "Cladding Metallurgy and Fracture Behavior During Reactivity-Initiated Accidents at High Burnup." Proceedings of the International Topical Meeting on Light Water Reactor Fuel Performance. American Nuclear Society. March 2-6, 1997. Portland, Oregon. 1997.

Chung, H.M. "Fundamental Metallurgical Aspects of Axial Splitting in Zircaloy Cladding." Proceedings of the International Topical Meeting on Light Water Reactor Fuel Performance. American Nuclear Society. April 10-13, 2000. Park City, UT. 2000.

Chun, R., Witte, M., and Schartz, M., "Dynamic Impact Effects on Spent Fuel Assemblies," UCID-21246, LLNL, October 20, 1987.

Cottrell, W.B., and Savolainen, A.W., "U.S. Reactor Containment Technology," ORNL-NSIC-5, Volume 1, Chapter 6, ORNL, August 1965.

Cunningham, M.E., E.R. Gilbert, A.B. Johnson, and M.A. McKinnon, "Evaluation of Expected Behavior of LWR Stainless Steel-Clad Fuel in Long-Term Dry Storage," EPRI TR-106440, April 1996.

DeHart, M.D. and O.W. Hermann, "An Extension of the Validation of SCALE (SAS2H) Isotopic Prediction for PWR Spent Fuel," ORNL/TM-13317, ORNL, September 1996.

Eckerman, K.F. and J.C. Ryman, "External Exposure to Radionuclides in Air, Water, and Soil," Federal Guidance Report No. 12, EPA 402-R-93-081, ORNL, September 1993.

Einziger, R.E., et al., "Examination of Spent Fuel Rods After 15 Years in Dry Storage," Argonne National Laboratory (ANL), 2002.

Einziger, R. E., et al., "High Temperature Postirradiation Materials Performance of Spent Pressurized Water Reactor Fuel Rods Under Dry Storage Conditions," Nuclear Technology, v. 57, p. 65, 1982.

Einziger, R.E. and R. Kohli, "Low Temperature Rupture Behavior of Zircaloy-Clad Pressurized Water Reactor Spent Fuel Rods under Dry Storage Conditions," Nuclear Technology, v. 67, p. 107, 1984.

Einziger, R.E. and J.A. Cook, "LWR Spent Fuel Dry Storage Behavior at 229°C," HEDLTME 84-17, NUREG/CR-3708, Hanford Engineering Development Laboratory (Aug 1984).

Einziger, R.E. and R.V. Strain, "Oxidation of Spent Fuel at Between 250° and 360°C," EPRI Report NP-4524, 1986.

Einziger, R.E., L.E. Thomas, H.V. Buchanan, and R.B. Stout, "Oxidation of Spent Fuel in Air at 175 to 195°C," J Nucl. Mater., 190, p53., (1992).

Federal Registry (FR), "List of Approved Spent Fuel Storage Casks: Holtec HI–STORM 100 Addition," Vol. 65, No. 84, pg. 25241, May 1, 2000.

Ferry, C, et al. - Synthesis on the Spent Fuel Long Term Evolution, Rapport CEA-R6084, (2005).

Fontana, M.G. and N.D. Greene, *Corrosion Engineering*, McGraw Hill, 1978.

Gao, J., "Modeling of Neutron Attenuation Properties of Boron-Aluminum Shielding Materials," Masters Dissertation, University of Virginia, August 1997.

Garde, A.M., et al., "Effects of Hydride Precipitate Localization and Neutron Fluence on the Ductility of Irradiated Zircaloy-4," Zirconium in the Nuclear Industry: Eleventh International Symposium, ASTM STP 1295, American Society for Testing and Materials (ASTM), 1996.

Goll, W., et al., "Short-Term Creep and Rupture Tests on High Burnup Fuel Rod Cladding," Journal of Nuclear Materials," v. 289, p. 247, 2001.

Hanson, B.D., 1998, "The Burnup Dependence of Light Water Reactor Spent Fuel Oxidation," PNNL-11929, Richland, Washington, Pacific Northwest National Laboratory. TIC: 238459.

Hermann, O.W. and M.D. DeHart, "Validation of SCALE (SAS2H) Isotopic Predictions for BWR Spent Fuel," ORNL/TM-13315, ORNL, September 1998.

Hoerner, S.F., *Fluid-Dynamics Drag*, Hoerner Fluid Dynamics, 1965.

Johnson, A.B., et al., "Exposure of Breached BWR Fuel Rods at 325°C to Air and Argon," Proc. NRC Workshop on Spent Fuel/Cladding Reaction During Dry Storage, Gaithersburg, Maryland, Aug 1983, NUREG/CR-0049, D. REISENWEAVER, Ed., S. Nuclear Regulatory Commission (1984).

Kammenzind, B.F., et al., "The Long-Range Migration of Hydrogen Through Zircaloy in Response to Tensile and Compressive Stress Gradients," Zirconium in the Nuclear Industry: Twelfth International Symposium, ASTM STP 1354, G.P. Sabol and G.D. Moan, Eds., American Society for Testing and Materials, pp. 196-233, 2000.

Kennedy, R.P., *Review of Procedures for the Analysis and Design of Concrete Structures to Resist Missile Impact Effects*, Holmes and Narver, Inc., September 1975.

Kese, K., "Hydride Re-Orientation in Zircaloy and its Effect on the Tensile Properties," SKI Report 98:32, 1998.

Knoll, R.W., et al., "Evaluation of Cover Gas Impurities and Their Effects on the Dry Storage of LWR Spent Fuel," PNL-6365, DE88 003983, PNNL, November 1987.

Lloyd, W.R., "Determination and Application of Bias Values in the Criticality Evaluation of Storage Cask Designs," UCID-21830, LLNL, January 1990.

Manteufel, R.D. and Todreas, N.E., "Effective Thermal Conductivity and Edge Configuration Model for Spent Fuel Assembly," *Nuclear Technology*, Vol. 105, pp. 421-440, March 1994.

Machiels, "Regulatory Applications Lessons Learned -- Industry Perspective." NEI Dry Storage Information Forum. Naples, FL. May 15-16, 2002.

MCNP5, "MCNP – A General Monte Carlo N-Particle Transport Code, Version 5; Volume II: User's Guide," LA-CP-03-0245, Los Alamos National Laboratory, April 2003.

Nakamura, J., T. Otomo, T. Kikuchi, and S. Kawasaki, "Oxidation of Fuel Rods under Dry Storage Condition," J Nuc. Sci. Tech., 32, [4], p321, (April 1995).

National Association of Corrosion Engineers (NACE), *Corrosion Data Survey*, 1985.

Novak, J., and I.J. Hastings, "Post-Irradiation Behavior of Defected UO2 in Air at 220250°C," Proc. NRC Workshop on Spent Fuel/Cladding Reaction During Dry Storage, Gaithersburg, Maryland, Aug. 1983, NUREG/CR-0049, D. REISENWEAVER, Ed., S. Nuclear Regulatory Commission (1984).

NRC. Subject: Transmittal of "Update of CSFM Methodology for Determining Temperature Limits for Spent Fuel Dry Storage in Inert Gas," November 27, 2001.

NRC Inspection Manual, Inspection Procedure 60851, "Design Control for ISFSI Components," ML0037287650.

Oak Ridge National Laboratory, "SCALE: A Modular Code System for Performing Standardized Computer Analyses for Licensing Evaluation," ORNL/TM-2005/39, Version 5, Vols. I-III, April 2005. Available from Radiation Safety Information Computational Center at Oak Ridge National Laboratory as CCC-725.

Pacific Northwest Laboratory (PNL), "Evaluation of Cover Gas Impurities and Their Effects on the Dry Storage of LWR Spent Fuel," PNL-6365, November 1987.

Parks, C.V., et al., "Assessment of Shielding Analysis Methods, Codes, and Data for Spent Fuel Transport/Storage Applications," ORNL/CSD/TM-246, ORNL, July 1988.

Rashid, Y.R. and R.S. Dunham, "Creep Modeling and Analysis Methodology for Spent Fuel in Dry Storage," TR-1003135, EPRI, 2000.

Rashid, Y.R., et al., "Creep as the Limiting Mechanism for Spent Fuel Dry Storage-Progress Report," EPRI TR-1001207, EPRI, 2000.

Roark, R.J., *Formulas for Stress and Strain*, McGraw Hill, 1965.

Sandoval, R.P., et al., "Estimate of CRUD Contribution to Shipping Cask Containment Requirements," SAND88-1358, TTC-0811, UC-71, SNL, January 1991.

Stokley, J.R., and D.H. Williamson, "Structural Integrity of Spent Nuclear Fuel Storage Casks Subjected to Drop," Nuclear Technology, Volume 114, Number 1, April 1996.

TRW Environmental Safety Systems, Inc. (TRW), "DOE Characteristics Database, User Manual for the CDB-R," November 16, 1992.

Uhlig, H.H., *Corrosion and Corrosion Control*, Wiley & Sons, Inc., 1985.

Wilson, D.W., et al., "Creep-Rupture Testing of Aluminum Alloys to 100,000 Hours, First Progress Report," Prepared for the Metal Properties Council, New York, November, 1969.

A.5   Correspondence

Beyer, C.E., PNNL, letter to K. Gruss, NRC, November 27, 2001, Subject: Transmittal of "Update of CSFM Methodology for Determining Temperature Limits for Spent Fuel Dry Storage Inert Gas," November 27, 2001.

Hendricks, L., Nuclear Energy Institute (NEI), letter to M.W. Hodges, NRC, Subject: Transmittal of Responses to the NRC Request for Additional Information on Storage of High Burnup Fuel, August 16, 2001.

NRC Confirmatory Action Letter 97-7-001, 1998 (ADAMS ML060620420).

Tsai, H.C. letter to K. Gruss, NRC, Subject: "A Recent Result on Thermal Creep of Surry Cladding after 15-y Dry Cask Storage," ANL, July 11, 2002.

Transnuclear (TN) Standardized NUHOMS Amendment 10 RAI Response, Docket No. 72-1004, November 7, 2007.

A.6   Conference Proceedings

Cappelaere, C., R. Limon, T. Bredel, P. Herter, D. Gilbon, S. Allegre, P. Bouffioux and J.P. Mardon. 2001, "Long Term Behavior of the Spent Fuel Cladding in Dry Storage Conditions," 8th International Conference on Radioactive Waste Management and Environmental Remediation, Bruges, Belgium, October 2001.

Chung, H.M. and T.F. Kassner, "Cladding Metallurgy and Fracture Behavior During Reactivity-Initiated Accidents at High Burnup," Proceedings of the International Topical Meeting on Light Water Reactor Fuel Performance. American Nuclear Society, Portland, OR, March 2-6, 1997.

Chung, H.M., "Fundamental Metallurgical Aspects of Axial Splitting in Zircaloy Cladding," Park City, Utah, Proceedings of the International Topical Meeting on Light Water Reactor Fuel Performance, American Nuclear Society, Park City, Utah, April 10-13, 2000.

Machiels, A., "Regulatory Applications Lessons Learned – Industry Perspective," NEI Dry Storage Information Forum, Naples, Florida, May 15-16, 2002.

APPENDIX B PROCESS FOR PRIORITIZING THE STANDARD REVIEW PLAN FOR DRY STORAGE SYSTEMS

B.1 Introduction

The purpose of this appendix is to describe the process used for prioritizing the review procedures contained in this NUREG. The application of this process, which is based upon determining relative importance, has resulted in assigning priorities of HIGH, MEDIUM or LOW to each of the review procedures in the SRPs. These priorities are intended to help focus staff review resources on those review procedures which are considered to be the most effective and important to worker and public safety. They are not, however, intended to relieve applicants of responsibility to comply with all requirements associated with dry cask storage licensing.

In 1995 the Commission issued a policy statement on the use of probabilistic risk assessment methods in all regulatory activities (60 FRN 42622, dated August 16, 1995). This policy statement has led to the development and application of "risk-informed" approaches in various regulatory areas. Specifically, a "risk-informed" approach represents a philosophy where risk insights are considered together with other factors to establish requirements that better focus licensee and regulatory attention on design and operational issues commensurate with their importance to safety. In general, "Risk-informed" approaches lie between "risk-based" and purely deterministic approaches, and are intended to:

- Allow consideration of a broader set of challenges to safety;

- Provide a means for prioritizing these challenges based on risk significance, operating experience and / or engineering judgment;

- Facilitate an integrated consideration of a broader set of factors (i.e., defense-in-depth, human reliability) to defend against these challenges;

- Explicitly identify and quantify sources of uncertainty in the analysis; and

- Provide a means to test the sensitivity of the results to key assumptions.

Where appropriate, a risk-informed regulatory approach can also be used to reduce unnecessary conservatism in purely deterministic approaches, or can be used to identify areas with insufficient conservatism in deterministic analyses and provide the basis for additional requirements or regulatory actions.

Prioritizing the various elements of the licensing review of an applicant's submittal, by noting areas in the SRP review procedures of higher and lower importance, can also be viewed as an identification of the review areas that have more or less value (i.e., effectiveness and importance to safety). Therefore, by focusing review resources on areas of the review that are the most effective and safety significant, efficiency can also be improved.

B.2 Scope, Approach and Process Description

B.2.1 Scope

The scope of the SRP prioritizing effort includes all SRP chapters. Within each of these chapters, only the review procedures were prioritized. The regulatory requirements and their acceptance criteria contained in each chapter were not prioritized, since these need to be met regardless of the priority of its corresponding review procedure.

B.2.2 Approach

The approach used in developing the prioritization process is a graded approach that combines likelihood or consequence insights with deterministic considerations and operating experience. It is directed to assess the relative value of performing each review procedure and results in a qualitative prioritization considering:

1) The likelihood of the applicant's non-compliance with a review procedure in the SRP.

2) The perceived "value added" provided by the NRC review of a given SRP procedural step.

3) The potential consequence if the non-compliance were to remain undetected and uncorrected.

4) The impact on defense-in-depth if the non-compliance remains undetected, assuming the review procedure being prioritized was related to a defense-in-depth item. Likelihood or consequence insights are those associated with the perceived radiological exposure to workers as well as to the public.

The prioritization was done on a generic basis (i.e., no specific dry cask design being considered) using the SRP review procedures identified for prioritization. However, it is always possible that a design being reviewed will have such unique features (e.g., new material, new configurations) that the prioritization needs to be revisited. This can be done on a case-by-case basis by reapplying this process on an actual application.

Finally, in developing the prioritization approach and process, certain assumptions were developed . These assumptions included:

- The cost of correcting a noncompliance was not a factor included in the process.

- The time and resources required to perform a review procedure were not factors included in the process.

- Dose thresholds used in this process were consistent with thresholds established in 10 CFR 20 and 10 CFR 72.104.

- The "value added" by the review was consistent with the current review level of effort and staff experience.

- Items to be prioritized were chosen such that overlap between them is minimized.

- All other requirements, except those included in the specific SRP review procedure being prioritized, were assumed to be satisfied.

B.2.3 Process Description

The process was applied to each technical discipline area in the SRP. The process was implemented by the NRC staff reviewers responsible for that discipline (i.e., multiple reviewers participated in the prioritization of each review procedure, and the final priority was developed based upon a consensus among the reviewers). The process involved looking at each SRP review procedure paragraph (or group of paragraphs) in each technical discipline area, and asking a structured set of questions. These questions addressed:

- What is the likelihood of the applicant not meeting the requirement(s) contained in the SRP review procedure being prioritized (need for staff review)?

- What is perceived value added by the staff review (i.e. likelihood of identifying a non-compliance for a given review procedure).

- What is the potential consequence to public and/or worker radiological safety if the requirement(s) remain unmet?

- What is the impact on defense-in-depth, if any, if the review procedure remains unmet?

The answers to the above questions were based upon the judgment of the NRC staff reviewers who participated in the prioritization process. This judgment reflected the reviewer's experience with current and previous applications and their views regarding potential future problems.

NUREG-1864, "A Pilot Probabilistic Risk Assessment of a Dry Cask Storage System at a Nuclear Power Plant" was previously developed to assess the risk to the public of a specific dry storage system at a boiling water reactor site to postulated events. The PRA information was not explicitly used in this SRP prioritization because it was limited in scope and assumed that the cask was properly designed, constructed and tested. Furthermore, the PRA did not address the factors listed in Table B-1 and B-2. It only assessed the risk during cask use from external hazards (e.g., fire) and operational errors (e.g., cask drop). Some of these accident sequences were also outside the scope of regulatory accidents typically evaluated under Part 72 for certified cask systems. In summary, the prioritized review procedures in the SRP address cask design, construction and testing, operations, and performance under normal and accident conditions to verify compliance with 10 CFR Part 72.

The steps the reviewers took in prioritizing each SRP review procedure were the following. First, the answers to the first two questions were qualitatively determined using a 5 tier qualitative ranking. Second, the answer to the third question was qualitatively determined using a 3-tier qualitative ranking system. The ranking systems are defined in Tables B-1, B-2 and B-3. The quantitative values used in Tables B-1, B-2 and B-3 are intended to serve as guidance in the selection of the appropriate qualitative ranking and reflect conservative estimates so as to provide a margin to account for uncertainties. The qualitative rankings resulting from Tables B-1, B-2 and B-3 were then assigned point values as shown in Table B-4. The point values corresponding to the qualitative rankings from Tables B-1, B-2 and B-3 were added together and, using the guidance described in Table B-4, an overall qualitative risk component of the prioritization (High, Medium or Low) was determined. The reason the scores from Tables B-1, B-2 and B-3 were added is that each is a reflection of the importance of the NRC staff performing the review procedure being prioritized. Finally, the answer to the last question (defense-in-depth) was qualitatively determined using a 3-tier scale (High, Medium or Low) following the guidance contained in Table B-5 and Attachment 2 and the reviewer's expert opinion.

At the end of this prioritization process the scoring of all Question 2 were revisited. If the score was low or very low (likelihood of staff finding applicants error) it was compared to the LOC score (Question 3). If the LOC score was medium or high, then the staff would reevaluate the appropriateness of original scoring for Question 2 and would consider raising the score to be at least commensurate with the LOC score (Question 3). If the LOC score was low, then the significance of the staff failing to discover the mistake is of little consequence.

The result was a prioritization based on likelihood or consequence and, if applicable, a defense-in-depth prioritization ranking. The final prioritization for the SRP review procedure was the overall LOC ranking and, if also related to defense-in-depth, a weighed combination of these two, with the weights determined by the NRC staff. These weights were determined for each review procedure prioritized and used only for that respective item (i.e., the importance of LOC versus defense-in-depth may vary from item to item). Attachment 1 to this appendix lists the detailed steps associated with implementing the prioritization process that was used in assessing the priority of each SRP review procedure. Attachment 2 provides a more detailed discussion on defense-in-depth. Attachment 3 provides an example of the documentation and major considerations associated with implementation of the process for one specific review procedure.

B.3   SRP Priority Designation and Implications

Upon completion of the prioritization process, the priority (HIGH, MEDIUM or LOW) associated with each review procedure has been indicated in the SRP at the beginning of each paragraph in the review procedures.

The prioritized procedures are intended to ensure that reviews are adequately focused on areas that have the most significant impact on safety and compliance with regulatory limits. It is important to remember that the priority designations were developed on a generic basis and may need to be adjusted depending upon the characteristics of specific applications. It is the responsibility of the individual reviewer to assess the design and determine the ultimate rigor needed to make a safety determination, with reasonable assurance, in each review area.

Finally it should be noted that a low or medium priority review procedure does not mean an application is exempted from any associated regulatory requirement, design requirement, or safety analyses that is expected within the review objectives and acceptance criteria.

Table B-1 Likelihood of Applicant's Non-Compliance with the SRP Review Procedure

Likelihood of Not Meeting the Requirements	Description
Very High	**Qualitative:** Likely to occur. **Quantitative:** $P > 0.5$
High	**Qualitative:** Probably will occur. **Quantitative:** $0.1 < P < 0.5$
Medium	**Qualitative:** May occur. **Quantitative:** $0.03 < P < 0.1$
Low	**Qualitative:** Unlikely to occur. **Quantitative:** $0.01 < P < 0.03$
Very Low	**Qualitative:** Occurrence improbable. **Quantitative:** $P < 0.01$

P = Probability

Table B-2 Potential "Value Added" Through the NRC Review Process

Likelihood that the NRC Review of a Specific Review Procedure Step Will Identify a Non-Compliance	Description
Very High	**Qualitative:** Likely to occur. **Quantitative:** $P > 0.5$
High	**Qualitative:** Probably will occur. **Quantitative:** $0.1 < P < 0.5$
Medium	**Qualitative:** May occur. **Quantitative:** $0.03 < P < 0.1$
Low	**Qualitative:** Unlikely to occur. **Quantitative:** $0.01 < P < 0.03$
Very Low	**Qualitative:** Not probable. **Quantitative:** $P < 0.01$

P = Probability

Table B-3 Potential Impact if the Noncompliance Were to Remain Uncorrected

Increase in Likelihood or Consequence (LOC) if Requirements Remain Unmet	Description
High	**Qualitative:** Likely to occur or significant consequences.
	Quantitative: >10^{-3}/yr* or >25 rem to worker or > 1 rem to public.
Medium	**Qualitative:** May occur or moderate consequences.
	Quantitative: <10^{-3}/yr but >10^{-5}/yr** or 5 -25 rem to worker or 0.1 rem - 1 rem to public.
Low	**Qualitative:** Occurrence improbable or minimal consequences.
	Quantitative: < 10^{-5}/yr or less than 10 CFR 20 dose limits for workers and the public.

* 10^{-3}/yr corresponds to the likelihood of an event that could occur in one or more casks over a 20 year life of 50 casks.

** 10^{-5}/yr corresponds to the likelihood of an event that could occur in one or more casks over a 20 year life of 5000 casks (i.e., 50 at each of 100 operating reactors).

Table B-4 Overall LOC Ranking

Numerical values for Tables B-1, B-2 and B-3 are assigned as follows (note that Table B-3 only assigns values of 1 through 3):

Very High	4
High	3
Medium	2
Low	1
Very Low	0

For each SRP review procedure, the qualitative scores from Tables B-1, B-2 and B-3 are added and a combined qualitative score is determined as follows:

High	9 - 11
Medium	6 -8
Low	1 - 5

Table B-5 Defense-in-Depth Ranking

Defense-in-depth has long been a key element of the NRC's safety philosophy. It is intended to ensure that the accomplishment of key safety functions is not dependent upon a single element of design, construction, maintenance or operation. In effect, defense-in-depth is used to provide one or more additional measures to back up the front line safety measures, to provide additional assurance that key safety functions will be accomplished. Traditional defense-in-depth measures for reactors have included items such as confinement, containment, redundant and diverse means of decay heat removal and emergency evacuation plans. For DSS, examples of measures associated with defense-in-depth are discussed in Attachment B-2. Defense-in-depth measures are generally decided upon using deterministic considerations (i.e., engineering judgment) regarding the importance of the safety function and the potential uncertainties that could affect its performance.

With respect to prioritizing the review procedures in this SRP, a review procedure can be considered associated with defense-in-depth if it is related to providing a backup to the front line of defense (e.g., confinement is generally considered a defense-in-depth measure since it provides a backup to cladding integrity).

Defense-in-depth measures are not intended to detract from the importance of front line safety measures. Defense-in-depth measures are intended to provide additional assurance so the safety function can be accomplished. It is not the intent of defense-in-depth to reduce the importance of the front line safety measures since, if their importance were reduced, the importance of the NRC staff review associated with those measures could also be reduced, which could affect the reliability or performance of the front line safety measures. This could leave the defense-in-depth measures as the primary means of performing the safety functions, instead of being the backup.

If failure to perform the review procedure could impact defense-in-depth (assuming the front line safety measure has failed) and has:

- a low likelihood and/or consequence, then the paragraph should be prioritized as "LOW."

- a medium likelihood and/or consequence, then the paragraph should be prioritized as "MEDIUM."

- a high likelihood and/or consequence, then the paragraph should be prioritized "HIGH."

Likelihood and consequence are defined in Table B-3.

Attachment B-1

Process Steps to Prioritize SRP Review Procedures

The following steps should be followed in prioritizing each review procedure. Multiple staff reviewers in each technical area should participate in the prioritization so as to arrive at a consensus on the priority. The checklist at the end of this attachment can be used to document each step.

1.   Identify the SRP review procedures to be prioritized, with a focus on the requirements that the review procedure is checking. This will result in individual paragraphs (or groups of paragraphs) being prioritized as separate items.

2.   Estimate the likelihood that the requirement related to the SRP review procedure will not be met by the applicant by choosing the appropriate likelihood range from Table B-1 (Likelihood of Applicant's Non-Compliance with the SRP Review Procedure). This estimate can be affected by several factors, including the experience of the applicant, the novelty of the technology used in the application, the difficulty level of meeting the requirement, the applicant's quality assurance program, etc.

The rankings listed in Table B-1 are arranged to provide more staff review effort where it is determined that the applicant is less likely to meet the review procedure. Conversely, where it is felt that the applicant will meet the review procedure, less staff effort would be required.

3.   Estimate the likelihood that if the requirement is not met, this fact will be discovered by performing the SRP review procedure. This is done by choosing the appropriate likelihood range from Table B-2 (Likelihood that the NRC Review Would Identify the Non-Compliance, Given that it Exists). This factor may be relatively high, however, there may be review procedures that have varying degrees of implementation.

The rankings listed in Table B-2 are arranged to continue to provide a high level of staff effort in areas where the staff review has typically identified problems. Conversely, where historical staff review efforts have not identified problems, that level of staff effort is minimized.

4.   Estimate the potential radiological likelihood or consequence to public and worker safety if the requirement were to remain unmet. It is recognized that this is not a trivial task and that no complete probabilistic risk assessment (PRA) is available for dry casks or ISFSIs. The following was intended to aid the prioritizer with this assessment:

- Consider potential event sequences or a set of event sequences, such that the dose to the most exposed person from these sequences includes the bulk of the dose from all possible sequences. The premise here is that every possible sequence of events has some likelihood of occurring and results in some dose to workers and the public. Some sequences are very likely and result in very little dose, others are very unlikely and result in very large dose, etc. The prioritizer should use experience in considering the sequence(s) that have the highest radiological likelihood or consequence to the most exposed person. This is equivalent to answering the following questions:

- What can happen? (i.e., what can go wrong?)
- How likely is it that that will happen?
- If it does happen, what are the consequences?

- Using Table B-3 Potential Impact if a Noncompliance is not identified, determine the corresponding range of increased likelihood or dose. This range corresponds with the likelihoods or consequences for the dominant sequences.

The rankings listed in Table B-3 are weighted to devote more staff resources to the review procedures that are viewed to be more likely or consequence significant and less staff resources to those that are viewed to be less likely or consequence significant.

5. The prioritizer now has three qualitative rankings corresponding to:

    - Likelihood of the applicant not meeting the requirements.
    - Likelihood that the NRC Review would find the discrepancy, given that it exists.
    - Potential consequences if the requirements remain unmet.

    Using these three rankings, determine the overall qualitative LOC (Likelihood or Consequence) ranking (High, Medium or Low) for this review procedure by adding the numerical values assigned to each qualitative ranking and the guidance in Table B-4.

6. Using Table B-5, assess the applicability and impact on defense-in-depth, if any, if the SRP review procedure is not met. Defense-in-depth consists of a number of elements as discussed in Attachment 2 and will not be applicable to all review procedures. If applicable, this step results in a High / Medium / Low qualitative ranking.

7. There is now a qualitative ranking and, if applicable, a qualitative defense-in-depth ranking. The method of combining these scores reflects the relative importance given to risk versus defense-in-depth. Judgment must be used to integrate these two rankings into a single ranking applicable to the SRP review procedure. This integration is done by weighing the two rankings using weights determined by the NRC reviewers. The weights are determined for each review procedure being prioritized and used for that procedure only.

8. A prioritization process checklist is to be filled out for each paragraph (or group of paragraphs) prioritized, so as to document the basis for the priorities assigned to each review procedure. This checklist is shown on the following page and Attachment B-3 provides an example of a completed checklist for a specific review procedure.

Prioritization Process Checklist

Chapter: Paragraph Number:

STEP	SCORE	COMMENTS
1. Identify the SRP procedure to be prioritized.	N/A	
2. Likelihood that requirement will not be met (Table B-1).		
3. Likelihood that staff reviews will find discrepancy (Table B-2).		
4. LOC if requirement is not met (Table B-3).		
5. Determine combined LOC value (Table B-4).		
6. Determine defense-in-depth value (Table B-5), if applicable.		
7. Determine relative weight of risk and defense-in-depth values determined in (steps 5 and 6 above).		
8. Overall priority (Combine LOC and defense-in-depth values).		

Attachment B-2

Defense-in-Depth (DID)

Defense-in-depth has long been a key element of NRC's safety philosophy. It is intended to ensure that the accomplishment of key safety functions is not dependent upon a single element of design, construction, maintenance or operation. In effect, defense-in-depth is used to compensate for uncertainties by employing one or more additional measures to back up the front line safety measures, thus providing additional assurance that key safety functions will be performed. Traditional defense-in-depth measures for reactors have included items such as confinement, containment, redundant and diverse means of decay heat removal and emergency evacuation plans. Defense-in-depth measures are generally decided upon using deterministic considerations (i.e., engineering judgment) regarding the importance of the safety function and the potential uncertainties that could affect its performance.

In the dry cask SRP prioritization, each paragraph (or group of paragraphs) to be prioritized, would be examined individually from a DID perspective to determine if that paragraph (or group of paragraphs) is related to defense-in-depth. If so, and if the paragraph is not met, a determination would then be made as to whether or not a defense-in-depth measure could be compromised and the risk significance.

To determine if a defense-in-depth measure could be compromised, it is first necessary to decide what are defense-in-depth measures? To help make this decision, the following guidance was used.

- A defense-in-depth measure is any design feature or action that is required by the SRP as a backup measure to the front line safety measures. This ensures that, if the front line safety measure is lost, the backup measure is present to perform that safety function.

DSS defense-in-depth measures may include:

- Confinement System (2nd barrier to fuel clad integrity);

- Operating Controls and Monitoring

- Non-mechanistic and bounding event analyses (to mitigate site-specific uncertainties).

SRP review procedures that relate to items that can be considered defense-in-depth should receive a DID ranking.

If the SRP paragraph (or group of paragraphs) being prioritized is related to a measure that meets the above guidance, then it would be evaluated as a defense-in-depth measure and prioritized as follows:

- If the failure of the front line and DID measures *relative to the issue identified in the SRP review procedure* would result in a low likelihood and / or consequence, then the paragraph should be prioritized as "LOW."

B-11

- If the failure of the front line and DID measures *relative to the issue identified in the SRP review procedure* would result in a medium likelihood and / or consequence, then the paragraph should be prioritized as "MEDIUM."

- If the failure of the front line and DID measures *relative to the issue identified in the SRP review procedure* would result in a high likelihood and / or consequence, then the paragraph should be prioritized "HIGH."

Risk and consequence are defined in Table B-3.

It should be noted that defense-in-depth measures are not intended to detract from the importance of front line safety measures. Defense-in-depth measures are intended to provide additional assurance so the safety function can be accomplished.

Attachment B-3

This attachment provides an example of a completed prioritization checklist to illustrate the level of documentation and major considerations associated with the prioritization of each specific review procedure. The review procedure used in the example is Section 4.5.4.7 "Confirmatory Analysis" in Chapter 4 "Thermal Evaluation" of NUREG-1536. A total of three staff reviewers participated in the prioritization of Chapter 4 and the prioritization input and outcome reflects a consensus among the reviewers.

Prioritization Process Checklist

Chapter: *4 - "Thermal Evaluation"*   Paragraph Number: *4.5.4.7*

STEP	SCORE	COMMENTS
1. Identify the SRP procedure to be prioritized.	N/A	Done by reviewers.
2. Likelihood that requirement will not be met (Table B-1).	L	Applicant provides calculations using generally accepted analytical tools.
3. Likelihood that staff reviews will find discrepancy (Table B-2).	H	Staff provides a thorough review.
4. Risk if requirement is not met (Table B-3).	H	Fuel cladding (i.e., first line-of-defense for fission product retention) could fail if thermal analysis is incorrect.
5. Determine combined risk value (Table B-4).	M	L (1) + H (3) + H (3) = 7 (MEDIUM)
6. Determine defense-in-depth value (Table B-5), if applicable.	H	Provides independent check (i.e., second line-of-defense) as backup to front line staff review of applicant's submittal.
7. Determine relative weight of risk and defense-in-depth values determined in (steps 5 and 6 above).	DID > Risk	DID is more important than risk since it has the potential to uncover applicant or staff review errors and can provide additional insights for probing the validity of the applicant's analysis.

STEP	SCORE	COMMENTS
8. Overall priority (Combine risk and defense-in-depth values).	*H*	*DID controls final priority.*

APPENDIX C
INTERIM STAFF GUIDANCE (ISG) INCORPORATED INTO NUREG-1536 Revision 1

ISG # & Rev.	Title	NUREG 1536 Revision 1 Status
ISG 1 Rev. 2	Damaged Fuel	Added
ISG 2 Rev. 1	Fuel Retrievability	Added
ISG 3	Post Accident Recovery and Compliance with 10 CFR 72.122(l)	Added
ISG 4 Rev. 1	Cask Closure Weld Inspections	Superseded by ISGs 15 and 18
ISG 5 Rev. 1	Confinement Evaluation	Added
ISG 6	Establishing Minimum Initial Enrichment for the Bounding Design Basis Fuel Assembly(s)	Added
ISG 7	Potential Generic Issue Concerning Cask Heat Transfer in a Transportation Accident	Added
ISG 8 Rev. 2	Burnup Credit in the Criticality Safety Analyses of PWR Spent Fuel in Transport and Storage Casks	Not Added, but referenced
ISG 9 Rev. 1	Storage of Components Associated with Fuel Assemblies	Added
ISG 10 Rev. 1	Alternatives to the ASME Code	Added
ISG 11 Rev. 3	Cladding Considerations for the Transportation and Storage of Spent Fuel	Added
ISG 12 Rev. 1	Buckling of Irradiated Fuel Under Bottom End Drop Conditions	Added
ISG 13	Real Individual	Added
ISG 14	Supplemental Shielding	Added

ISG # & Rev.	Title	NUREG 1536 Revision 1 Status
ISG 15	Materials Evaluation	Added
ISG 16	Emergency Planning	NA
ISG 17	Interim Storage of Greater Than Class C Waste	NA
ISG 18 Rev. 1	The Design & Testing of Lid Welds on Austenitic Stainless Steel Canisters as Confinement Boundary for Spent Fuel Storage	Added
ISG 19	Moderator Exclusion Under Hypothetical Accident Conditions and Demonstrating Subcriticality of Spent Fuel Under the Requirements of 10 CFR 71.55(e)	NA
ISG 20	Transportation Package Design Changes Authorized Under 10 CFR Part 71 Without Prior NRC Approval	NA
ISG 21	Use of Computational Modeling Software	Added
ISG 22	Potential Rod Splitting Due to Exposure to an Oxidizing Atmosphere During Short-Term Cask Loading Operations in LWR or Other Uranium Oxide Based Fuel	Added
ISG 23 (Draft)	Draft - Application of ASTM Standard Practice C1671-07 when performing technical reviews of spent fuel storage and transportation packaging licensing actions	Not Added
ISG 24 (Draft)	Reserved	N/A
ISG 25 (Draft)	Draft - Pressure and Helium Leakage Testing of the Confinement Boundary of Spent Fuel Storage Casks	Added
ISG 26 (Draft)	Reserved	N/A

APPENDIX D   PUBLIC COMMENTS RECEIVED AND THEIR DISPOSITION

The purpose of this appendix is to list all the public comments received on NUREG-1536 "Standard Review Plan for Spent Fuel Storage Systems at a General License Facility," Revision 1A.   The NRC issued NUREG-1536, Revision 1A (ML 090500630) for public comment on April 15, 2009 for a 90 day period and received comments from the following two sources:

- NEI, Nuclear Energy Institute, letter to Mr. Ron Parkhill, USNRC, dated July 14, 2009 (ML 091970430)
- NAC International, email from Mr. Tony Patko to Mr. Ron Parkhill, USNRC, dated July 15, 2009 (ML 092020356)

The staff's resolution and any associated changes to the standard review plan are listed for each comment.   Note that all line numbers listed in the attached table refer to the line numbering of Revision 1A of NUREG-1536.

Comment	SRP Location	Public Comment	Resolution	Changes to SRP
NEI 1	General	The SRP discusses the content of the Technical Specifications in numerous locations. While the NRC does not have a policy statement on technical specifications for dry cask storage systems, the NRC Final Policy Statement on Technical Specification Improvements for Nuclear Power Reactors, as published in the Federal Register at 58 FR 39132, July 22, 1993, provides useful guidance. The Final Policy Statement discusses, in the Background, the trend towards adding information to the Technical Specifications by stating: "… since 1969 there has been a trend towards including in technical specifications not only those requirements derived from the analyses and evaluation included in the plant's safety analysis report but also essentially all other NRC requirements governing the operation of nuclear power plants. … In the Commission's view, this has diverted both NRC	The staff recognizes the policy statement for operating reactors. The staff also notes that site-specific operating reactors are different than certified dry cask systems. Reactors represent an inherently higher risk from accidents, and maintain several active systems and active monitoring of key performance parameters during operations, Reactor technical specifications (TS) are established to ensure these functions are maintained in order to ensure adequate containment, reactivity control and thermal hydraulic control of the system during operations. Dry storage casks are passive in nature, and do not typically rely upon multiple active safety systems to mitigate events during storage operations. Instead they rely on passive design features and administrative controls to assure criticality safety, confinement safety, and cladding protection during normal, off-normal, and accident conditions. The staff considers TS to be valuable, in part, to assure the most important fabrication, design features,	Chapter 13 is clarified to state: If a reviewer determines that a design feature, content specification, analytical assumption, operating assumption, limiting condition of operation, elements of reactor programmatic controls, or other SAR item is important and should not be changed without NRC staff approval, then it should be further evaluated and considered as a potential CoC condition or technical specification. The reviewer should consider, in part, risk-insights, safety margins, operational experience, defense-in-depth considerations, design novelty, and other issues that are unique to each proposed design. The reviewer should also implement the guidance in this chapter for establishing such conditions and technical specifications in the CoC.

Comment	SRP Location	Public Comment	Resolution	Changes to SRP
		staff and licensee attention from the more important requirements in these documents to the extent that it has resulted in an adverse but unquantifiable impact on safety." The Final Policy Statement also stated: "The purpose of Technical Specifications is to impose those conditions or limitations upon reactor operation necessary to obviate the possibility of an abnormal situation or event giving rise to an immediate threat to the public health and safety by identifying those features that are of controlling importance to safety and establishing on them certain conditions of operation which cannot be changed without prior Commission approval." A similar philosophy where only those items that have a direct nexus to the protection of the public health and safety from an immediate threat are included in the Technical specifications should be	contents, and operations of the system are appropriately controlled among the diversity of site users, to mitigate the likelihood and consequences of potential off-normal and accident conditions. The dry cask storage certificate includes a condition which specifies TS. These control the fabrication, safe use, and operation of the dry cask system during loading, transfer, and passive storage. This is consistent with the Commission's policy statement published in Federal Register, 58 FR 39132, July 22, 1993. The staff also believes the format typically employed for DCSS TS is amenable to the use by general licensees into assuring safe use and operation. Several factors may influence the content of TS. Chapter 13 is revised to clarify these factors. The staff finally notes that NEI has identified this as a future issue to be discussed between the NEI dry storage task force and NRC requested to discuss concerns with cask TS in a separate interaction with NRC (see ML093310122). If a new philosophy were adopted,	

Comment	SRP Location	Public Comment	Resolution	Changes to SRP
		adopted. The guidance to the staff in the draft SRP in regards to Technical Specifications should be revised accordingly.	or generic changes were made to dry cask technical specifications, then the standard review plan may be subject to future revision to implement any associated guidance with the changes.	
NEI 2	General	The document should state throughout that for canister-based systems the "confinement cask" is the welded canister assembly.	The SRP is written to apply to both bolted casks as well as to welded canister systems. The term confinement cask applies to both the bolted cask, as well as, welded canister. SRP Section 5.5.1.2. specifies the design and qualification guidance for a welded canister to qualify as a confinement boundary	No Change
NEI 3	General	In a number of locations, the guidance gets into specifying the details of the ASME Code and other codes. Unless NRC does not accept what the codes require, the guidance should avoid repeating the code details and simply refer to the code at a higher level (e.g., "Section III, Subsection NB").	Specific guidance is provided in certain instances to avoid misunderstanding and possible conflicting interpretations. The specific guidance also assists reviewers in focusing on important elements of the ASME Code with respect to the associated review objectives.	No Change
NEI 4	General	Renumbering the chapters in the SRP may create confusion during future licensing actions where the SRP chapters will not coincide with the SAR chapters. Please consider restoring the current SRP revision chapter numbering	This revision of the SRP included the addition of a new Materials chapter and the deletion of the Decommissioning chapter which affected the numbering of many of the chapters. New certificate requests may follow the new format, and amendments to an	No Change

Comment	SRP Location	Public Comment	Resolution	Changes to SRP
		sequence.	existing certificate may follow the format as licensed. However, NRC intends to revise associated Regulatory Guides that specify acceptable content and format of certificate applications. The revised draft regulatory guides will also be issued for public comment.	
NEI 5	34	The statement that ISGs were developed to address changes in requirements differs from the definition of ISGs provided at line 660. This statement should be consistent with line 660 to avoid implying that ISGs impose new requirements as could be interpreted by the current wording.	The staff agrees the wording should be more consistent and the SRP is revised as appropriate. The statements were not intended to imply that ISGs impose new regulatory requirements, because the SRP is only for guidance to staff.	Changed abstract to state; "These ISGs were developed to clarify important aspects of regulatory requirements, reflect lessons learned and evolving technology, and document detailed technical positions." Also changed the second sentence of the ISG definition accordingly.
NEI 7	542	Editorial: Change "term" to "terms."	Agree with comment	Changed to "terms".
NEI 8	542 791 1259 4522 4993 6853	Change "containment" to "confinement" to use more storage-specific language.	Agree with comment	Changed these lines to state confinement in lieu of containment.
NEI 9	541-542 8319-8320	It is not clear why peak rod average burnup is included in this definition and later in the SRP. Assembly average burnup is typically used for specifying allowable contents and should be sufficient	Agree with comment that assembly average is typically used for specifying allowable contents. In addition, the peak rod average burnup is a parameter considered in the fuel integrity analyses in the materials review.	Definition changed to indicate that assembly average burnup is used for assessing allowable content, and that peak rod average burnup is specified for assessing fuel cladding integrity in the materials review. Similar exception added to Section

Comment	SRP Location	Public Comment	Resolution	Changes to SRP
				8.4.17.
NEI 10	605-606	Revise definition to account for a DFC that could contain less than one assembly (e.g. failed rod basket with 50 rods vs. 264 for an assembly) or more than one assembly for a consolidated rod can. Suggest "A metal enclosure to confine damaged spent fuel. A damaged fuel can with its damaged spent fuel contents must satisfy …"	This is the definition currently in ISG-1 Rev 2. Because the ISG also applies to the transportation SRPs, the definition will be maintained for consistency at this time. However, this does not preclude an applicant proposing the use and evaluation of a DFC that may contain fuel rods that are more or less than that associated with one fuel assembly.	No change
NEI 11	667-669	a) M.O.S. is not "identical" to F.O.S. b) "M.O.S" in the first set of parentheses should be "F.O.S." c) Line 669: delete the first occurrence "-1" ,	Agree with comment. Margin of safety was restated in the document as suggested.	Text changed to: "This term may be defined, through a factor of safety, f.s. = capacity/demand, as MofS = F.S. (capacity /demand) -1 (with minimum acceptable MofS $\geq$ 0.0
NEI 12	684	In the 2nd sentence, add "neutron" between "high" and "absorption."	Agree with comment.	Revised to "high neutron absorption"
NEI 13	687	Suggest deleting "and transporting" because this SRP is exclusively for storage.	Agree with comment.	Definition changed to eliminate reference to "transportation" or "storage".
NEI 14	720	A definition is provided for BPRA at line 532 but definitions are not provided for control element assemblies (CEAs) or thimble plug assemblies (TPAs).	Agree with Comment	The definition section was revised to include Added definitions for CEA and TPA.
NEI 15	740	While preferential loading is currently used for thermal	Agree with comment.	Definition changed to: A non-uniform loading

Comment	SRP Location	Public Comment	Resolution	Changes to SRP
		loading, it is also used for dose reduction and could be used in the future for other reasons (e.g., criticality control). This definition should be more flexible.		configuration of spent fuel assemblies within a dry storage system that is typically specified by assigning a fuel zone designation to each basket cell, and specifying limiting nuclear and physical parameters of SNF assemblies that can be loaded into each zone. Preferential loading is often used as a means to optimize allowable SNF parameters (e.g. burnup, cooling time, decay heat), while satisfying the shielding, criticality, and thermal performance objectives of the cask system.
NEI 16	748-752	The definition of "Ready Retrievability" is incorrect and inconsistent with Section 12.4.5 (lines 11208 – 11219) of the SRP and draft ISG-2 Rev 1 which has been issued by the NRC for comment. The first sentence of this definition is the definition of recovery not retrievability. This definition should be revised and a definition for "recovery" should be added.	Agree with comment. ISG-2, Rev 1 was issued as final on February 22, 2010 (ML100550817). The ISG considered and addressed public stakeholder comments. This SRP has been administratively updated to incorporate Rev 1 of ISG-2 Definitions for "ready retrieval" and "normal means" have been added in accordance with the ISG-2 Rev. 1 guidance. The definition for "retrievability" is changed to be consistent with the language of 10 CFR 72.122(l). A definition for "recovery" has been added. The SRP is also revised to use	Changed definitions to include Retrievability, Ready Retrieval Normal Means, and Recovery: "Retrievability - In accordance with 10 CFR 72.122(l), storage systems must be designed to allow ready retrieval of spent fuel, high-level radioactive waste, and reactor-related GTCC waste for further processing or disposal. Ready retrieval -The ability to move a canister containing spent fuel to either a

Comment	SRP Location	Public Comment	Resolution	Changes to SRP
			consistent terminology for retrievability and ready retrieval in sections 1, 2, 8, and 12. The term "ready retrievability" has been change to "retrievability" or "ready retrieval" as appropriate..	transportation package or to a location where the spent fuel can be removed. Ready retrieval also means maintaining the ability to handle individual or canned spent fuel assemblies by the use of normal means. Normal means - The ability to move a fuel assembly and its contents by the use of a crane and grapple used to move undamaged assemblies at the point of cask loading. The addition of special tooling or modifications to the assembly to make the assembly suitable for lifting by crane and grapple does not preclude the assembly as being considered moveable by normal means Recovery - The capability to return the stored radioactive material to a safe condition after an accident event without endangering public health and safety. This generally means ensuring that any potential release of radioactive materials to the environment or radiation exposures is not in excess of the limits in 10 CFR Part 20 during post-accident recovery

Comment	SRP Location	Public Comment	Resolution	Changes to SRP
				operations."
NEI 17	810	Clarify this definition to say that the supplemental shielding is only ITS if it is credited in the 72.104 dose analysis.	Agree with comment.	Changed to: Supplemental shielding shall be deemed as component(s) important to safety and be specified in the Technical Specifications as a condition for use of the system as designed, if credited in the shielding and radiation protection analyses for meeting 72.104(a) or 72.106(b) requirements.
NEI 18	892-895 2358-2359 6623-6625 7261-7266 7286 7318-7319 7413-7414 7456-7457 11350-11351 11366 11454-11461 11513-11516 11521-11523 11527-11597	The bases for what requirements should be in the CoC or TS provided in these sections are vague, subjective, not risk-informed, and not consistent with practice in NRR (i.e., Part 50 TS). Examples: a) "Any aspect of the design or procedures that the NRC determines should not be changed" (892-895) b) "preclude the possibility of damage to the structure or damage to the confined nuclear material" (2358-2359) c) "any technical aspect of the design which is deemed critical to nuclear safety" (7318-7319) d) whatever "the staff deems necessary" (11350 – 11351) e) "a reviewer deems an item	See resolution to NEI1 Technical specifications are part of the CoC, and specific guidance remains important for limiting parameters or procedures in these areas to ensure safety of the system during normal operations and accident conditions. There is a diversity of dry cask storage technologies, which employ different types of design features and analytical methodologies to ensure safety with different safety margins calculated within each discipline. Cask technologies continue to evolve with innovative, first-of-a-kind approaches to ensure confinement, shielding, and criticality safety. In addition, vendors have proposed a diversity	See changes described for NEI1. In addition, the only changes made as a result of this comment include the moving of technical specification items identified via an asterisk from Section 8.2 of the Public Comment version to Section 8.4.1. Also, the discussion of technical specifications in Section 8.4, Review Procedures and Acceptance Criteria, of the Public Comment version was removed because it was not applicable to this section, not because it was incorrect.

Comment	SRP Location	Public Comment	Resolution	Changes to SRP
		so important" (11366) Given that these casks are loaded and operated at NRC-licensed Part 50 facilities, we suggest SFST adopt a function-based, risk-informed set of criteria for what information belongs in the CoC and TS, similar to 10 CFR 50.36(c) for power reactors, recognizing the passive design and operation of storage casks and modules. In general, the TS should only cover operational items under the user's control for implementation, and only critical design features under the control of the CoC holder, similar to those in the "Design Features" section of Part 50 TS. Examples of information not appropriate for inclusion in TS: fuel basket dimensions (line 6624); alternate materials and other material requirements (7261-7266, 7456-7457); QA/QC documents, procedures, and test protocols for neutron absorbers (7413-7414); ASME Code information (11454-11461), and training (11521-11523). Including this information only	of TS in terms of scope and format to both assure that safety is maintained and to satisfy specific operational needs of general licensees. All of these factors have contributed to the diversity of TS formats; as well as some of the generality specified for TS in the SRP guidance. The staff recognizes that if applied correctly 72.48 may be used to evaluate if NRC approval is needed for changes. However, 72.48(c)(1)(B) itself, recognizes the role of certificate conditions and TS in limiting design changes without NRC approval. These certificate conditions and TS areas established at the discretion of NRC. The staff finally notes that NEI has identified this as a future issue to be discussed between the NEI dry storage task force and NRC (see ML093310122). If a new philosophy were adopted or generic changes were made to dry cask technical specifications, then the standard review plan may be subject to future revision to implement any associated	

Comment	SRP Location	Public Comment	Resolution	Changes to SRP
		in the FSAR is appropriate based on risk. 72.48 provides adequate controls for determining whether prior NRC approval is required for changes to these items, and the QA program adequately addresses training and manufacturing. It is also a poor practice from a human factors standpoint to incorporate portions of the FSAR into the CoC by reference.	guidance with any change. Further resolution of this issue through the SRP comment resolution process would not be practical at this time.	
NEI 19	1255	Suggest the word "removed" instead of "retrieved". The damaged fuel container is used to assist in placing and removing damaged fuel from the canister.	Agree with comment.	The last sentence in Section 1.5.1 was modified to "Therefore, the reviewer should verify that the application contains a description of how the damaged fuel would be canned, the characteristics of the can, and the means by which the can would be inserted into and removed from the cask."
NEI 20	1259 11524 11527	Editorial: Add "and Limits" to the title of Chapter 13.	Agree with comment	Changed these lines to: "and Limits"
NEI 21	1540-1541	The operational history parameters need to be reasonable values assumed in the depletion calculations and not bounding values the user must verify that their reactor history meets	When using burnup credit, the fuel must be confirmed to meet the bounds of the operational history parameters assumed in the analysis or these parameters must be shown to be sufficiently bounding over the full range of fuel to be authorized for loading.	No change.

Comment	SRP Location	Public Comment	Resolution	Changes to SRP
			NUREG/CR-6716 provides-results on a study of the importance of and the sensitivity of K-effective to changes in some major parameters.

Section 7.5.5.3 of the SRP provides additional guidance regarding bounding assumptions. | |
| NEI 22 | 1552-1553 | Delete this bullet. "Inerting atmosphere requirements" is not an SNF specification and the maximum number of fuel assemblies is specified two bullets prior. | Agree with comment. | Removed bullet |
| NEI 23 | 1154-1155 | Based on the elimination of the SAR chapter on decommissioning, consider deleting the sentence regarding the planned decommissioning process. | Agree with comment. | Deleted the following sentence: "Additionally, a discussion should be included of the planned decommissioning process." |
| NEI 24 | 1253
1600-1601
2139
2204
2337-2338
2401 (flowchart box for Chapter 12)
2508-2512
3037
3053
8803
9075
11314 | These lines are inconsistent with Section 12.4.5. of the draft SRP (lines 11215-11219) and other portions of the SRP which state that retrievability in 10CFR72.122(10CFR72.122 (I) applies only to normal and off-normal conditions and not accident conditions. These lines are also inconsistent ISG-2 Revision 0 and draft ISG-2 Revision 1. Reference to retrievability should be removed in discussions of | Agree with Comment.

See response to NEI 16. Section 12.4.5 is also clarified to discuss the applicability of recovery and retrievability. Retrievability applies to normal and off-normal events, and not design basis accidents. The meaning of normal condition, off-normal events and design-basis accidents, in this context, are further clarified in Section 12.1 | See changes in NEI 16.

The SRP is revised to use consistent terminology for retrievability and ready retrieval in sections 1, 2, 8, and 12. The term "ready retrievability" has been change to "retrievability" or "ready retrieval," as appropriate. Other referenced guidance for off-normal and design-basis events have been clarified with distinguishing terminology such |

Comment	SRP Location	Public Comment	Resolution	Changes to SRP
		accident conditions throughout the SRP.		as "retrievability or "recovery", as applicable. Section 12.1 has been revised to include "Normal conditions are the intended operations, planned events, and environmental conditions, that are known or reasonably expected to occur with high frequency during storage operations. Off-normal events are those man-made events or natural phenomena expected to occur with moderate frequency or once per calendar year. ANSI/ANS 57.9 refers to these events as Design Event II."
NEI 25	1374	Since the NRC is currently working on rulemaking that would change the licensed lifetime of a cask, it is suggested that a reference to the 20 year limit be removed here and throughout the document and that a reference to the regulation be provided instead.	Agree with Comment	Changed Section 1.5.5 to remove the specific time period and referenced the regulation where appropriate. Also similarly changed Section 2.4.3.1, Section 3.6 F3.6, and Section 8.5 F8.6.
NEI 26	1704	Identifying the fuel vendor is not pertinent to the review and should be deleted.	It is necessary to distinguish the fuel vendor so that staff can distinguish between the many different types of fuel assembly	No change.

Comment	SRP Location	Public Comment	Resolution	Changes to SRP
			variations that exist and whose materials properties are not identical.	
NEI 28	1913-1917	This paragraph is inconsistent with ISG-5 (for metal casks) and ISG-18 for welded canisters. Non-mechanistic confinement boundary failures are no longer part of the cask design and licensing basis.	Agree with comment	Removed the following Sentence: "Nevertheless, for assessment purposes and to demonstrate the overall safety of the storage cask system, the NRC staff considers that the DSS should be evaluated for the effects of a confinement boundary failure.
NEI 29	1992	Editorial: Change ".." to "."	Agree with comment	Changed as stated in comment
NEI 30	2041	Change "SNF retrieval" to "retrievability".	Agree with comment	Changed as stated in comment
NEI 31	2110	Change "retrieval capability" to "retrievability".	Agree with comment	Changed as stated in comment
NEI 32	2271-2280	ANSI/ANS-57.9 is outdated and not germane to many of today's commercial spent fuel systems. Other than the design event classifications, care should be used in referring to this standard for today's DSS designs.	The use of ANSI/ANS-57.9 is broader than just event classifications. Each applicant should evaluate and justify the applicability of ANSI/ANS-57.9 to its proposed DSS. The reference to the review standard is reworded to be consistent with current terminology in the structural review chapter.	Revised the words, "the cask system structures," to read, "the ISFSI dry storage systems" in Section 3.4.2.
NEI 33	2278-2280	Editorial: The last sentence of this paragraph does not appear to be grammatically correct.	Agree with comment	Changed sentence to read: The loadings defined in American Society of Civil Engineers, "Minimum Design Loads for Buildings and Other Structures," (ASCE 7) can be used when load combinations

Comment	SRP Location	Public Comment	Resolution	Changes to SRP
				are considered on the basis of ANSI/ANS-57.9.
NEI 34	2308, 2626 3085, 3501	Some inconsistency is noted regarding the specified Code years When referring to the ASME code, no code year was mentioned. However, when referring to a non-ASME code, a code year was mentioned. For example, line 2308, IBC code (2006), line 2626, ASTM C33 (2002), line 3085, ANSI/ANS-57.9 (1992), line 3501, ACI 349 (2006). To avoid confusion and permit appropriate flexibility for the applicant, the code year should not be mentioned in the review plan	Specific guidance regarding codes year is provided in certain instances to avoid misunderstanding and possible conflicting interpretations. However, this does not preclude an applicant from proposing the use of alternate codes or code years with appropriate justification.	No Change
NEI 35	2340, 2713	Regarding Line 2340, "This position does not necessarily require that all confinement system and other structures important to safety survive all design-basis accident and extreme natural phenomena without any permanent deformation or other damage" and Line 2713, "The system should not experience any permanent deformation or loss of safety function capability during normal or off-normal operation conditions. However, the system may	The NRC staff agrees that the design analysis, in accordance with provisions in Section III of the ASME Code, does not restrict use of linear-material properties. To ensure clarity, the paragraph beginning in Line 3168 was rewritten to recognize the potential inelastic structural behavior for the ASME Code, Section III, Appendix F accident load conditions.	The entire paragraph beginning in the second paragraph in Section 3.5.1.4 ii.(1) was changed to, "Consistent with the provisions of ASME Code, Section III, Appendix F, inelastic material properties may be used for the storage cask design analysis evaluation for accident loads. The SAR should identify the sources used for the inelastic material properties."

Comment	SRP Location	Public Comment	Resolution	Changes to SRP
		experience some permanent deformation, but no loss of safety function capability, in response to an accident" please consider the following: Based on the above discussion, elastic-plastic analysis should be allowed to analyze the accident load; however, Line 3168, "to be consistent with the provision in Section III of the ASME code, the analysis should use linear material properties. For materials that do not serve in structural capacity (such as shielding materials), inelastic material properties may be used for cask components that are not stress-limited and respond inelastically to the load conditions for storage casks" implies that only elastic analysis can be used unless you use strain limited criteria. In the past NRC has accepted the use of elastic-plastic properties for all the accident load analyses and stress limited criteria are used per ASME Appendix F.		

Comment	SRP Location	Public Comment	Resolution	Changes to SRP
NEI 36	2357-2362	The first sentence of this paragraph seems to indicate that TSs should be in place to preclude possibility of damage to the structure or the confined material during cask handling and operations. The second sentence of the same paragraph seems to indicate that TS should describe the actions and inspections to be conducted upon occurrence of "events" that may cause such damage. These two statements appear to be contradicting each other.	In the unlikely event of cask damage resulting from cask handling and/or operation, the second sentence discusses actions and inspections that should be conducted to ensure that the cask is secured in a safe configuration.	No change
NEI 37	2380	Editorial: Add a blank line between lines 2379 and 2380	Agree with comment	Changed as stated in comment
NEI 38	2612-2640	This section seems to imply that the alternate concrete temperatures described apply only to the steel-lined concrete confinement cask system designed to ACI 359. Similar concrete temperature provisions have been accepted for the NUHOMS HSM type concrete structure designed to ACI 349.	The text explicitly states that the temperature limits presented as an exception to Section CC-3340 of ACI 359 are temperature limits that apply as an alternative to ACI 349, A.4. The inclusion of steel-lined concrete confinement cask systems is an additional configuration in which alternative temperature requirements can be employed, not the sole configuration for their use.	No Change
NEI 39	2621, 8210	Add ASTM C150 as the standard specification for Type II cement.	Agree with comment	Changed to the following: Satisfy ASTM C33, … Satisfy ASTM C150, ("Standard

Comment	SRP Location	Public Comment	Resolution	Changes to SRP
				Specfication for Portland Cement") requirements and other requirements referenced in ACI 349 for cement. Have demonstrated...
NEI 40	2627	Delete "2002" (edition year of ASTM C33)	Agree with comment	Rewrite "Satisfy ASTM C33, ("Standard Specification for Concrete Aggregates") requirements...
NEI 41	2729 2735 5908 8889	Change "retrieval" to "unloading" or "removal" as applicable.	These lines discuss normal condition operations, so retrieval is the correct word choice for each case, and is essentially synonymous with "unloading" or "removal" in the context it is used.	No change
NEI 42	2879-2882	The passage: "The SAR should identify the maximum response determined. That response should be sufficiently low such that while damage may occur, it would not impair the capability of the component to perform its safety functions" is not clear. What, specifically, is meant by "maximum response"?	In the context of structural analysis for the explosive overpressure, the generalized term, "maximum response," generally means to include pressure induced maximum stresses at critical cask locations and governing structural performance modes for the cask components important to safety. This is added to the SRP to provide clarity.	Add the following to the end of second paragraph in Section 3.5.1.4 i. (3) (e): The maximum response includes pressure-induced maximum stresses at critical cask locations and governing structural performance modes for the cask components important to safety.
NEI 43	2885	The third paragraph of the current SRP version has been deleted in this proposed revision to NUREG 1536. The deleted paragraph accepted the fire parameters from Part 71 as a basis for characterizing the fire during storage. Additionally, it	Partially agree with comment. The deleted paragraph contains the lead sentence, "The NRC has accepted the fire parameters included in 10 CFR Part 71 as the basis for characterizing the heat transfer associated with fire during storage." This may or may not be conservative for the fire accident	At the end of the second last paragraph of Section 3.5.1.4 i. (3) (c), reinstated the following sentences: "Spalling of concrete that may result from a fire is generally considered acceptable and need not be estimated or evaluated. Such damage is readily detectable,

Comment	SRP Location	Public Comment	Resolution	Changes to SRP
		accepted spalling of concrete due to fire without further evaluation. It also accepted concrete temperatures that exceeded ACI 349 limits as long as corrective actions are taken for continued safe storage. The revised version does not provide guidance on the structural assessment to fire event. Suggest restoring this paragraph.	evaluation in a licensee's Part 72.212 site parameter report. To preserve the evaluation bases discussed in the original paragraph, only the lead sentence of the paragraph, which refers to Part 71 transportation provisions, will be deleted.	and appropriate recovery or corrective measures may be presumed. The NRC accepts concrete temperatures that exceed the temperature limits of ACI349 for accidents, providing that the temperatures result from a fire. However, corrective actions may need to be taken for continued safe storage.
NEI 44	2962	Line 2962 states that consequences of floods such as damage to access routes, temporary blockage of ventilation passages, etc. "should be identified in the CoC so that a general licensee will be able to consider these factors when sitting an ISFSI". This is a general site characterization issue more appropriate to be addressed in the 10 CFR 72.212 Report. Generic flooding depth and moving water limits the DSS is designed for should be described in the SAR and the CoC.	Flood consequences, such as temporary but prolonged blockage of ventilation passages, may adversely affect thermal performance of the cask system. The staff agrees with the comment that evaluation of whether design-basis floods are bounded by floods analyzed in the certified cask system is a site-specific 72.212 characterization issue. Evaluation of additional flood consequences is generally at the discretion of cask vendors.	Replace the word, "should" with the word, "may."
NEI 45	3083-3103	Lines 3085-3103 deal with the response of the storage system sitting on a flexible pad and subjected to earthquake loads. It requires that the	The section of the document provides guidance to staff regarding the ISFSI seismic analysis and for reviewing calculations that show a cask will	No change

Comment	SRP Location	Public Comment	Resolution	Changes to SRP
		flexibility of the pad be taken into consideration in the seismic analysis. This is not an appropriate requirement for a system that is licensed to be used under a general license where the system design is based on a design response spectra (e.g. a RG 1.60 response spectra) anchored to a defined maximum acceleration for the horizontal directions and a maximum acceleration in the vertical direction. Each particular user is to ensure as part of their 72.212 evaluation that the system as qualified is adequate for each particular site considering the characteristics of the pad and its response when coupled with the underlying supporting media.	not tipover or drop during a seismic event.	
NEI 46	3106, 12537	RG 1.60 imposes excessive conservatism for seismic evaluations. RG 1.60 should be replaced by NUREG/CR-6728 and also NUREG/CR-6865.	RG 1.60 provides general guidance for generating design response spectra and has not been replaced by the cited NUREG reports.	No change
NEI 47	3139-3140	The term "confinement casks" is confusing. Should this be "confinement boundary"?	Agree with comment	Changed "confinement casks" to "the confinement boundary of the cask".
NEI 48	3153	In the previous paragraph, Subsection NB is used to	Agree with comment. To provide clarity, a sentence to recognize the	Add a 2nd sentence to Section 3.5.1.4 ii, 3rd paragraph which

Comment	SRP Location	Public Comment	Resolution	Changes to SRP
		define stress qualification for the confinement boundary, which is a pressure retaining boundary. In the paragraph including line 3153 it does not clearly state that the basket is a non pressure retaining boundary, and that the applicant should use Subsection NG. Need to state that Subsection NG is acceptable, or the reader is left to believe that Subsection NB applies to non pressure boundary baskets. It should also confirm that Appendix F is applicable for use with Subsection NG.	code requirements for the basket is added.	reads: "For the fuel basket, Subsection NG of the Code applies."
NEI 49	3168	Although not a change from the existing version of NUREG 1536, this paragraph appears to imply that Section III analysis should be only linear elastic. This section should be clarified to allow elastic-plastic and other non-linear analysis as permitted by the Code. It should state that Subsection NB and Subsection NG do permit the use of Appendix F which does permit the use of inelastic properties for components which serve as the pressure boundary or also non-pressure boundary	See response to NEI Comment 35 regarding use of inelastic material properties.

The strain-based criteria are not recognized by the ASME Code or other applicable standards. Recognizing that the SRP provides review guidance for broad base, common subjects. An applicant, however, may propose, the NRC staff may consider the use of other acceptance criteria, such as strain-based criteria, only on a case by case basis with appropriate justification. The staff may review alternate strain-based proposals in | See response to NEI Comment 35 for the first part of the comment on using inelastic properties.

No change with respect to strain-based criteria. |

Comment	SRP Location	Public Comment	Resolution	Changes to SRP
		applications, such as baskets. It should also state that strain-based criteria can be employed for energy-limited accident conditions, provided the applicant provides such basis for its use.	greater depth depending on the applicability and experience with the criteria to the proposed DSS design.	
NEI 50	3171	In many applications for drop conditions, it should be acceptable to use strain-rate-sensitive properties. Appendix F permits their use. Need to include "strain rate properties, which needs the appropriate references."	As worded, the SRP does not preclude use of strain-rate-sensitive material properties for design analysis of cask drop conditions.	No change
NEI 51	3315	Editorial: Delete either "for" or "of."	Agree with the comment.	Deleted the word, "of."
NEI 52	3321-3338	Please clarify the trunnion design stress criteria used to compare the stress at the trunnion connection with the cask body at that interface. Regarding Line 3338, "the applicant should evaluate the stresses and forces in the trunnion connections with cask body...", since the cask body is typically designed per ASME Code Section NB, the NB stress criteria should be used instead of yield and tensile strength. Please clarify.	The SRP provides guidance for implementing the ANSI/ANS N14.6 stress design factors evaluation by recognizing that the maximum bending stress occurs at the base of the trunnion. Implicit in this evaluation is a classical strength of materials approach to calculate the maximum average shear over the trunnion cross section. In the case of a loaded cask consisting of the transfer cask, as a special lifting device, and the loaded enclosure vessel in which the basket and its fuel assemblies are emplaced, the SRP provides that the applicant should evaluate the stresses and forces in the trunnion connections	Changed second to last sentence in Section 3.5.1.4.ii.(3)(c) to read, "If other assumptions, including ASME Section III stress limits by the finite element design analysis and slight material yielding at localized regions, are considered, the applicant should provide adequate justifications"

Comment	SRP Location	Public Comment	Resolution	Changes to SRP
			with the cask body and in the cask body near the trunnions. The Line 3336 statement will be revised to recognize potential localized materials yielding also in accordance with the ANSI/ANS N14.6 provisions for stress design factors.	
NEI 53	3380	Section 3.5.2 "Other System Components Important to Safety" does not contain the alternate concrete temperatures as listed in Lines 2612-2640 for the steel-lined concrete confinement cask structure.	Agree with comment. The temperature limit alternative listed from 2607-2640 and 8196-8227 is added to Section 3.5.2 for consistency	Incorporated text from the referenced sections as well as the revision from NEI Comment 39 into Section 3.5.2
NEI 54	3747	"Appendix C" should read Appendix F for the version year of the ACI 349 that is described in Line 3501.	Agree with comment	Changed as stated in comment
NEI 55	3758	Editorial: "30 ksi" should be "3 ksi." Also, should the example list include a maximum compressive strength because that value is a limit for drop and tipover analyses?	Agree with comment. The NRC staff notes that the cask tipover analysis generally places a limit on the maximum concrete compressive strength of the cask storage pad. To ensure that the analyzed configuration remains applicable in a general licensee's Part 72.212 site parameters evaluation report, an upper limit on the maximum compressive strength should be reported.	In Section 3.5.2.3 i. (3) changed fifth bullet, "30 ksi," to read, "3 ksi." Additionally, revise the entire sentence to read, "Upper limit (60 ksi, 4219 kgf/cm2) on the specified yield strength of reinforcement, lower limit (3 ksi, 211 kgf/cm2) on concrete specified compressive strength (f'c), and upper limit on concrete strength, as analyzed and specified for the ISFSI cask storage pads.
NEI 56	4182-4184/JS	The sentence regarding delivery of electronic media is	Agree with comment.	Deleted sentence: "It should be noted that electronic media

Comment	SRP Location	Public Comment	Resolution	Changes to SRP
		guidance for the applicant rather than the staff and as such should it may be more appropriate in another document.		should be delivered to the appropriate SFST staff directly, if possible, as electronic media sent to the NRC Document Control Desk may be damaged during security screening."
NEI 57	4302	The discussion about annotation of input files is too broad. It may be important for the reviewer to see and perhaps use the applicant's files, but it is not necessary to understand all aspects of the input files. Some of these files come from Journal files or Log files which are generated by the program. It is not feasible to add comments to these files. Open ended statements such as adding "annotation" leads to vague expectations by the reviewer for the need of such documentation.	Well documented input files expedite the review since it is easier for the reviewer to verify that analysis files are consistent with the design information provided in the SAR. If it is not feasible to add comments to some files, then, as indicated in the SRP, "the applicant should provide an adequate explanation of how computer models were assembled using the CMS in the appropriate SAR chapters or related documents."	No change
NEI 58	4313-4315	Delete these lines. The level of review described here seems to be beyond an audit review and more like a third party validation of the computer analysis. It is the responsibility of the applicant's QA program to ensure that the analyses are performed correctly.	The level of review depends on the complexity of the application, including the uniqueness of new designs and safety margins. The guidance also reflects previous NRC licensing review experience in identifying insufficient analyses, (performed under applicant's QA program) in these areas of computational analyses.	No change
NEI 59	4332	Clarify or delete "mesh type."	Agree with comment.	Changed first bullet, first two sentences of Appendix 3A,

Comment	SRP Location	Public Comment	Resolution	Changes to SRP
				under Sensitivity Studies to: "The reviewer should verify that the applicant has completed sensitivity studies for relevant CMS modeling parameters. This includes element type and mesh density, load…."
NEI 60	4335-4336	A mesh sensitivity study is not required when stress linearization is being used for primary loading. Such detailed studies should be restricted to fatigue evaluations at stress discontinuities.	A mesh sensitivity is required to make sure the analysis results are mesh independent.	Added sentence at end for first bullet of Appendix 3A, under Sensitivity Studies to: "A mesh sensitivity is required to make sure the analysis results are mesh independent".
NEI 61	4349	Delete "plots." Including plots of all results generates an enormous amount of unneeded data in the FSAR.	As stated in the SRP, the SAR or related documents should include all relevant results (including plots).	No change
NEI 62	4411	The guidance stating that the decay heat removal system should operate reliably under off-normal and accident conditions is inappropriate given that some of the abnormal and accident conditions themselves involve impairment or loss of the decay heat removal system (e.g., blocked air ducts).	Agree with comment.	Revised SRP to state: "Evidence must be provided by the applicant that the decay heat removal system will operate reliably under normal and loading conditions."
NEI 63	4551	In item (2), it appears that this is an option addressing when fuel cladding temperature does exceed 400°C (i.e., delete "not"). Please clarify.	Agree with comment. For low burnup fuel the maximum allowable peak cladding temperature may be higher than 400°C as long as the hoop stress is less than 90 MPa, as indicated in the SRP.	Review Revised SRP to state: "(2) the maximum calculated temperatures for normal conditions of storage do exceed 400°C (752°) and…"

Comment	SRP Location	Public Comment	Resolution	Changes to SRP
NEI 64	4469-4471	Clarification should be provided for "address, quantify and report the degree of conservatism associated with the proposed models and the resulting safety margin." This statement is vague. It is unclear what the specific information is requested and to what level of detail.	Agree with comment	Changed last sentence in 2nd paragraph of Section 4.4.4 to read: "The applicant must discuss, quantify, and report in the SAR any conservatism associated with the proposed thermal models. The level of detail of the discussion should be comparable with sections of the SAR that describe the analytical thermal models. A table of results should be provided in the SAR showing how the associated conservatisms affect the safety parameters (e.g. calculated peak cladding temperature, confinement seal temperatures, etc.). The table of results must be supported with fully documented analytical models and calculations"
NEI 65	4580	Editorial: Change "on" to "in."	Agree with comment.	Changed as stated in comment.
NEI 66	4612	Editorial: add a closing parentheses at the end of the sentence	Agree with comment.	Changed to read: "SNF pool's technical specification maximum temperature limit (typically 46°C) (115°F)."
NEI 67	4686-4687	Delete this sentence. It does not appear to add value to the review guidance. Alternatively, clarify why this is only applicable to horizontal basket designs.	Partially agree with comment. Internal natural convection however should be verified through physical experiments or use of validated CFD codes.	Changed last sentence in Section 4.5.4.1 to read: "Traditionally, the staff has maintained that natural convection in enclosed cavities should be validated through robust CFD calculations or physical experiments."

Comment	SRP Location	Public Comment	Resolution	Changes to SRP
NEI 68	4768-4770	The SRP requires test data for each thermal effective conductivity. Are correlations from handbooks which are based on test data acceptable? Is test data still a requirement if a CFD sub-model is used to calculate the effective conductivity as specified in Line 4686 to 4687? It is recommended that "from test data" be changed to "from test data, or CFD sub-models, or other appropriate sources"	Agree with comment.	Changed 2nd sentence, 2nd paragraph of Section 4.5.4.1.2 SRP to read: ""If effective thermal conductivity is used in this manner, the reviewer should verify that the same values have been determined from test data, or CFD submodels, or other appropriate sources that are representative of similar geometry, materials, temperatures, and heat fluxes used in current application."
NEI 69	4678-4681	Limiting convection to the outer surface of the cask contradicts already-approved designs that credit convection inside the fuel canister. This is clearly permissible with appropriate justification.	Agree with comment.	Deleted the following sentence from SRP: "Convection by natural circulation should be limited to that between the external surface of the cask and the ambient environment."
NEI 70	4687	Delete the word "robust." Words like this are vague and subjective, allowing each reviewer to apply his or her personal definition of "robust" in their review and generate RAIs if the model is not "robust" enough.	Partially agree with comment. This guidance is intended to advise the reviewer that CFD calculations are not trivial and many times are subjective to errors if not used adequately.	Replaced the word "robust" with the word "sufficient"
NEI 71	4742	Editorial: Delete misplaced closing parentheses in this line	Agree with comment.	Changed this line to read: "width and height of the air channel." Removed parenthesis after "air channel" and punctuated this phrase with

Comment	SRP Location	Public Comment	Resolution	Changes to SRP
				commas.
NEI 72	5041	Allowance should be made for a properly scaled mock-up instead of an "as-built cask system" to confirm the thermal analysis.	Agree with comment. Design verification testing could be achieved by using as-built cask system or mock-up system	Changed 1st sentence of last paragraph in Section 4.5.4.7 to read: "As an alternative to a confirmatory analysis, the applicant may be required to perform design-verification testing of an as-built cask or properly scaled mock-up system (when applicable) to confirm the thermal analyses presented in the SAR."
NEI 73	5185-5187	Delete or clarify this sentence. No such "periodic surveillance program" has "typically" been required or performed for stainless steel welded canister confinement systems. Periodic surveillance of the confinement boundary, if any, should only be required case-specifically, if the particular design features of the confinement system require it. Inspections of the air vents or temperature monitoring have been accepted as the sole periodic surveillance.	Agree with comment.	Replaced subject sentence with the following: "This practice is consistent with the fact that other welded joints in the confinement system are not monitored since the initial staff review ensures the integrity of the confinement boundary for the licensing period. Typical surveillances include checking for blockage of the air vents or temperature monitoring depending on the specific design."
NEI 74	5347-5348	The statement that the monitoring systems are not important to safety <u>and</u> classified as Category B (an ITS class) does not appear to be consistent.	Monitoring systems are not specifically mentioned in NUREG/CR-6407 "Classification of Transportation Packaging and Dry Spent Fuel Storage System Components According to Importance to Safety". ISG-5 which	Added "It is termed as not important to safety since most of the associated hardware have not met the important to safety programmatic controls, like design, or procurement" to the 3rd paragraph before the

Comment	SRP Location	Public Comment	Resolution	Changes to SRP
			was incorporated into the revised SRP, mentions that monitoring systems are a Classification Category B because as stated in Table 2, a Category B component is one whose failure in conjunction with the failure of an additional item, like the containment boundary seal, could result in an unsafe condition (potential release of radioactive material). It is termed as not important to safety since most of the associated hardware have not met the important to safety programmatic controls, like design, or procurement.	last sentence of Section 5.5.2
NEI 75	5384	Editorial: Change "Review" to "Evaluation."	Agree with comment	Changed as stated in comment
NEI 76	5413-5426	This paragraph does not appear to be consistent with ISG-5 and ISG-18 and would only apply to non-welded-canister type confinement systems. Based on NUREG/CR-6397, damaged fuel would not have a driving force to release fines form from the fuel matrix. What is the technical or safety issue of concern? What factors <u>are</u> suggested for damaged fuel?	If the canister is welded and tested to be leaktight then the size of the source term is immaterial for determining a release.

The staff agrees that fuel rods that are classified as damaged due to a preloading cladding breach may not have a pressurized fuel rod driving force for the release of particulates from the rod. However, under an impact accident, damaged fuel rods might release additional fuel fines due to the fracture of the fuel, especially in the rim region of high burnup fuel. In addition, some canisters may be | Added to end of the 3rd paragraph in Section 5.5.3:

Fuel rods that are classified as damaged due to a preloading cladding breach may not have a pressurized fuel rod driving force for the release of particulate from the rod under off-normal events and design basis events. However, under an impact accident, damaged fuel rods might release additional fuel fines due to the fracture of the fuel, especially the in the rim region of high burnup fuel. In addition, some |

Comment	SRP Location	Public Comment	Resolution	Changes to SRP
			pressurized to several atmospheres and cask blowdown may also affect release fractions. Each applicant should establish release fractions for damaged fuel based on applicable physical data and other analyses appropriate for the specific type of fuel, damaged condition, and accident conditions. This will be clarified in Section 5.5.3.

Alternatively, a leak-tight confinement boundary may be specified to preclude the release analyses of damaged fuel.

Also, see resolution to NAC 5426. | canisters may be pressurized to several atmospheres and cask blowdown could also affect releases fractions. Each applicant should establish release fractions for damaged fuel based on applicable physical data and other analyses appropriate for the specific type of fuel, accident impacts, and damaged condition of DSS. Alternatively, a leak-tight confinement boundary may be specified to preclude the release analyses of damaged fuel. |
| NEI 77 | 5800-5801 | "radionuclide content, and estimated radiation source strength in Becquerel's, ... should be described": This appears to be a new expectation from the NRC. It is not clear what the basis of this request is as radiation source strength in Ci or Bq is not clearly related to gamma/neutron source strength (e.g. beta emitters). | This guidance is provided to evaluate source terms of different types of contents, for both the shielding and confinement analyses. The SRP is revised to clarify this guidance. | Replaced the 2nd sentence in Section 6.4.2 which begins with "The physical and chemical form, ..." with the following:

"For spent nuclear fuels, the source terms in particles/s or MeV/s per energy bin should be described in form of either group structure or a continuous function of energy. For non-fuel hardware, source in Curies or Becquerel is acceptable. For contents other than fuel or non-fuel hardware components, isotopic composition and photon yields for each |

Comment	SRP Location	Public Comment	Resolution	Changes to SRP
				constituent should be specified. For confinement evaluation purposes, the physical and chemical form, source geometry, radionuclide content, and estimated radiation source strength should be described."
NEI 78	5809-5810	"characteristics for each gamma-ray source type should be provided, including isotopic composition, and photon yields": Is a tabulation of spent fuel isotopics requested here? If so, for what purpose? Typically, inputs into depletion analysis are provided, but not isotopics of depleted materials.	The guidance is intended for different types of contents in the shielding and areas of review. Isotopic concentrations are needed because the input file, as described in the SRP, is typically a representative input and not necessarily bounding. For cases in which the source terms are not derived from a depletion calculation, the applicant should provide isotopic concentration and photon yields. The SRP is revised to clarify this guidance.	Replaced the 1st paragraph of Section 6.4.2.1 with the following: statements. "The SAR should specify gamma source terms for both spent fuel and activated materials. For spent nuclear fuels, the source terms should be described in a format that is compatible with shielding calculation input, typically in the form of photons/s or MeV/s per energy bin. For assembly hardware and non-fuel hardware, source terms should be specified by ^{60}Co activity (in Curies or Becquerel). For contents other than fuel or non-fuel hardware components, isotopic composition and photon yields for each constituent should be specified. A tabulated form of the radiological characteristics is acceptable."
NEI 79	5813-5814	Within gamma source	Agree with comment	Replaced with the following

Comment	SRP Location	Public Comment	Resolution	Changes to SRP
		description "describe extent to which radioactivity may be induced by interactions involving neutron originating in the stored materials": If this implies n-gamma reactions, then the current SRP version is clearer. If activation is to be considered for decommissioning, that should be clarified.		statement: "The SAR should include discussion of energetic radiations created by nuclear reactions such as (n, γ) in the packaging materials and the contents."
NEI 80	5868-5870	Shielding analyses do not need to be "bounding analyses." Applicants need only provide representative dose rates to demonstrate reasonable assurance that the system is capable of meeting the offsite dose limits or 72.104 for an entire ISFSI. (See line 5723 and subsequent text.)	Shielding analyses should provide the bounding dose rates and demonstrate that the system is capable of meeting the requirements of 72.104 and 72.106. The bounding doses rates should be based on the design basis loadings that are defined through the applicant by maximum burnups, minimum cooling times, and minimum enrichments.	No Change
NEI 81	5873-5882	High burnup fuel has been licensed for storage on several dockets. There is no indication that high burnup fuel produces substantially high dose rates due to limited validation data. If limited data is available it leaves an open ended question as to how to specify uncertainties. "Conservative assumptions" and "design margins" are not defined, leaving it up to each reviewer	Fuel assembly with higher burnup will produce higher gamma and neutron sources. The gamma source increases linearly proportional to burnup and the neutron source increases proportional to the fourth power of burnup. It is expected that the magnitude of uncertainties in exposure rates and decay heats would propagate proportionally in the same manner. There exist biases and	

Comment	SRP Location	Public Comment	Resolution	Changes to SRP
		when, and how much, in uncertainties to apply. There is no correlation as to how maximum fuel assembly heat load is related to uncertainties - low heat capacity /minimal shield system may be affected by low fuel assembly heat load, and vice versa.	uncertainties in the computer model and input data. These errors, bias, and uncertainties are in general not quantified if the computer code and models are not benchmarked and validated against experimental data. The NRC recognizes that the nuclear industry has not developed experimental data for the high burnup fuel that is proposed for storage. NRC has traditionally allowed applicants to use isotopic codes beyond their validated range. However, some applicants have applied penalties to assure these un-quantified uncertainties are sufficiently considered for high burnup fuel source term and decay heat predictions. The penalty factors should account for reasonable uncertainties in both gamma radiation, neutron radiation, and decay heat source terms. The magnitude of uncertainties in these three source terms may be significantly different at high burnups.

Alternatively, the applicant may propose measurement programs in the technical specifications that directly monitor shielding and thermal performance (e.g., cladding temperatures,, and detect | |

Comment	SRP Location	Public Comment	Resolution	Changes to SRP
			abnormalities that could result from unaccounted uncertainties in the source term predictions. Regarding the uniformity of the practice among the staff on the additional safety margin, the Criticality Safety and Dose Assessment branch of SFST has working groups to share review experience and develop consensus. The staff is in general aware of the common practice.	
NEI 82	5968	Editorial: Incorrect spelling of "Principle."	Agree with comment	Changed as stated in comment
NEI 83	5996	Editorial: Figure 6-2 is missing from the document.	Agree with comment	In Section 6.5.2.1 Replaced the words: "in Figure 6-2 (reproduced from NUREG/CR-6716)" with: The larger neutron fluence generates a larger actinide content which results in larger neutron source term and secondary gamma source term as illustrated in NUREG/CR-6716, Section 3.4.1.2.
NEI 84	6003-6004	"...applicant and the staff should not attempt to establish specific source terms as operating control and limits for cask use.": If this is true, why does the SRP focus in the Section 6.4.2 on curie content and isotopic description of the	The focus of this requirement is different from that of Section 6.4.2. The requirement here is for consideration of the cask operations. Fuel assembly initial enrichment, burnup, and cooling time are the readily usable and inspectable parameters for cask	Replaced last sentence in Section 6.5.2.2 with the words "However, the applicant ... limits for cask use" with: However, the staff should not attempt to use specific source terms as bases for establishing

Comment	SRP Location	Public Comment	Resolution	Changes to SRP
		spent fuel? For Cobalt-60 dominated hardware sources, a source term may be more appropriate than other limits (e.g., mass, exposure, cool time).	operation. Source terms would be additional parameters that are calculated from enrichment, burnup, and cooling time.	operating controls and limits for cask use because these are not readily inspectable parameters. The fuel assembly initial enrichment, burnup, and cooling time are more appropriate for use as loading controls and limits.
NEI 85	6036	Editorial: add a closing parenthesis.	Agree with comment	Changed as stated
NEI 86	6449-6450	"…homogenization should not be used in neutron dose calculation when significant neutron multiplication can result from moderated neutrons…": While not changed from the current SRP, it should be noted that standard, NRC-approved, practice is to homogenize the rod lattice in shielding calculations (not necessarily homogenizing basket structure into the fuel region).	Although this assumption has been acceptable in many applications, there may be instances where homogenization may not be appropriate.	No change
NEI 87	6188	Incorrect spelling of the word "Evaluation"	Agree with comment	Changed as stated in comment
NEI 88	6221-6222	Review staff should recognize that importance functions may also be produced with Monte Carlo, point-kernel and transport codes.	Agree with comment.	Replaced the 1st sentence of the 2nd paragraph in Section 6.5.4.1 with the following: "The reviewer should be aware that the applicants often use transport or point-kernel methods to calculate neutron and/or gamma importance functions (unit of

Comment	SRP Location	Public Comment	Resolution	Changes to SRP
				mrem/hr/particle/s-cm)." Added the following statement to the end of the 2nd paragraph in Section 6.5.4.1 "The reviewer, however, should pay close attention to the applicability of the importance function to the actual cask content and geometry of contents and shielding."
NEI 89	6246-6248	"The applicant should use the latest released computer code version that is valid for the particular computational platform used to perform the analysis.": This item in particular has been discussed with NRC staff as a significant issue. A licensed code for the same type of application should not require a code version change simply because the code developer has issued a new version. Use of different code versions within one or more applications is difficult to reconcile and potentially leads to unnecessary confusion. Such burdens should only be borne by the applicant if a significant safety issue has been identified with the previous code version. Typical	Partially agree with comment. The staff would prefer models to be based on latest released computer code versions because NRC typically upgrades its shielding computer codes on a regular schedule with code vendor upgrades. However, computer codes used for shielding analyses do not necessarily need to be updated to the most recent version. The applicant should demonstrate that use of a code version, that is no longer supported by a vendor, is valid for the specific analysis, and also that the code has been properly maintained in accordance with the requirements contained in 10 CFR Part 72, "Subpart G-Quality Assurance." The SRP is revised to clarify this guidance. The letter, dated July 2, 2009, from Mr. Raymond Lorson to Mr. Steven	Replaced the sentence with: "The applicant should use a computer code version that is demonstrated to be adequate for the analysis and is valid for the particular computational platform used to perform the analysis. The staff should also consider if additional confirmatory assessments and review is needed to validate the shielding predictions by an applicant that uses older or unsupported codes, especially in cases were NRC may have upgraded codes and no longer have the capability to directly examine unsupported code models from the applicant."

Comment	SRP Location	Public Comment	Resolution	Changes to SRP
		new release code versions tend to contain a certain amount of bugs that get resolved through user feedback to code originator. While it could be postulated that newer code provide more "accurate" results, but if the previous version was found to be acceptable for system approval with no safety issues identified, why should applicants be required to change? The goal per draft SRP Section 6.4 is to provide reasonable assurance that system will meet limits. This is also inconsistent with how NRR deals with updated codes (e.g., ASME Code).	P. Kraft (ADAMS: ML0918802633) provides the regulatory basis and detailed explanations for this requirement. The applicant's quality assurance program should also be capable of identifying and addressing "bugs" in cases in which they chose to use new codes for their shielding analyses.	
NEI 90	6302-6309	"by verifying that the following information has been provided in the SAR ... The computer code solutions to a series of test problems ...": The draft SRP revision does not contain the previous SRP statement "that these solutions may be referenced, and need not be submitted in the SAR". This change would add a substantial amount of information to the SAR without any safety benefit as the referenced documents, per	Agree with comment	Added to the second bullet the following words; "Or the specification of publically available references for commonly used and well-established codes (e.g. SCALE and MCNP) that demonstrate validation.

Comment	SRP Location	Public Comment	Resolution	Changes to SRP
		current SRP, should be public information and/or have been previously submitted to NRC.		
NEI 91	6578	This implies that only boron can be employed as a fixed absorber. It is recommended that "boron" be changed to "neutron poison material"	Agree with comment.	Changed to specify "neutron poison material"
NEI 92	6739	Neither Section 8.5.4.3 nor Attachment 8-3 exists in the document.	Agree with comment.	The citations to other parts of the SRP are corrected.
NEI 93	7099-7104	This section requires explicit analyses of atypical control rod insertion while Section 7.5.5.6 (lines 7138-7157) discusses margin to cover higher-than-modeled reactivity due to control rod insertion. These two sections appear to conflict. Please clarify what is required in the design basis calculations.	Agree with comment.	The following statement is inserted in Section 7.5.5.6: "While the applicant should make every effort to identify and appropriately address these potential uncertainties explicitly, data limitations may make it difficult to quantify these uncertainties precisely and assure that they are adequately bounded."
NEI 94	7102-7104	These lines explicitly require the analysis of integral fuel burnable absorbers. However, there are NUREG/CR reports that provide guidance on when these absorbers need to be considered in the analysis. These lines should be revised accordingly.	Agree with comment.	A reference to NUREG/CR-6760 is added to last paragraph of Section 7.5.5.3.
NEI 95	7242 7390	"Foreign standards are not generally acceptable..." What is the basis for this statement? For non-ASME code	Agree with comment. The applicant must provide an analysis that shows the foreign standard is equivalent to a comparable US	Changed wording in Section 8.1, 3rd paragraph and Section 8.4.2.1, 3rd paragraph to state

Comment	SRP Location	Public Comment	Resolution	Changes to SRP
		applications, there are many recognized standards essentially equivalent to ASTM, such as Euronorm, JIS, etc. The applicant should be able to use foreign standards with appropriate justification.	standard, or otherwise sufficient for its intended use. The staff may review foreign standards in greater depth depending on the familiarity and applicability of the standard to the proposed DSS design.	Foreign standards (and codes) may be acceptable on a case-by-case basis. The applicant should provide complete documentation supporting the use of the foreign standard and show that the foreign standard is equivalent to a comparable US standard (e.g. ASME, ASTM, etc.), or otherwise sufficient for its intended use. The staff may need to review foreign standards in greater depth, depending on the familiarity with the standard and applicability of the standard to the proposed DSS design.
NEI 96	7248	The Chapter 8 convention of indicating with an asterisk the items that should be addressed in the Technical Specifications is not used in any other chapter. All of the chapters should be consistent and not use this convention.	Agree with comment. The convention is removed. The specification of review areas that should be considered for Technical Specifications is clarified in the text of Chapters 8 and 13.	Section 8.4.1 and 13.5 were revised to clarify items that should be considered in the technical specifications.
NEI 97	7266 7554-7564	Replace "weathering steel" with "0.20% copper steel" or "carbon steel with a minimum copper content of 0.20%". Also, add "salt water" to "coastal marine sites". The term "weathering steel" applies to a class of low-alloy steels that contain small amounts of such alloying elements as Cr,	Agree with comment	Changed Section 8.4.6 to: To address the increased atmospheric corrosion rates found at coastal marine (salt water) sites, some applicants have specified the use of 0.20%, minimum, copper-bearing steels, or, "weathering steels" such as Cor-Ten. The

Comment	SRP Location	Public Comment	Resolution	Changes to SRP
		Ni, P, Si and Cu. These steels are covered by ASTM A242 and A588. Also "copper bearing steel" should be generalized to allow for other appropriate measures to control corrosion.		Kennedy Space Flight Center has collected data which has demonstrated the benefit of copper-bearing and weathering steels for significantly reducing corrosion at coastal marine sites. Therefore, for coastal marine ISFSI sites, the use of copper-bearing steels (containing a minimum of 0.20 percent copper), or weathering steels, may be necessary. Such steels are covered by ASTM A-242 and A-588, and supplemental requirements to ASTM A-36, and/or other specifications. Other corrosion control measures may be employed, provided adequate documentation is supplied to demonstrate efficacy.
NEI 98	7317-7321	This paragraph should be deleted for several reasons. The portion of the sentence stating that the body of the SAR "is not enforceable" is incorrect. Users must comply with the Part 72 cask SAR unless a change, appropriately reviewed and authorized under the provisions of 10 CFR 72.48, is performed. If not, NRC enforcement action may	Partially agree with the comment. General licensees and cask vendors have authority to change an FSAR under the requirements of 10 CFR 72.48. If 10 CFR 72.48 is not performed correctly, the NRC may take enforcement action. Enforcement action may only be taken for violation of regulatory requirements, license/CoC	Also see response to NEI 18. This paragraph is accurate. However, this paragraph was removed because it did not belong in Section 8.4. "Review Procedures and Acceptance Criteria" of the materials chapter. The statements are more appropriately reflected in Chapter.

Comment	SRP Location	Public Comment	Resolution	Changes to SRP
		be taken. In addition, using this logic as the basis for putting information in the CoC or TS is flawed because it is not risk-informed, is too subjective, and dilutes the CoC holder's and licensee's ability to implement changes that meet the criteria of §72.48. Moreover, this increases the NRC's need to spend resources reviewing changes to the CoC that are not risk- or safety-significant.	(including TS which are appendices to the CoC) conditions, and NRC Orders. Paragraph 72.48(c)(1)(B) recognizes the role of certificate conditions and TS in limiting design changes without NRC approval. These certificate conditions and TS are established at the discretion of NRC for design features, operations, and contents that should not be changed without additional NRC licensing review ..	
NEI 99	7334	a) Amendments are not "completely new designs." New designs are submitted as a new CoCs. This statement should be revised. b) Use of the term "beware" is derogatory in that it implies the applicants are trying to sneak changes through the NRC without them being noticed. Please revise.	a) Although some modified designs have been submitted for NRC review as new certificate applications, other modified designs which represent new, major design components such as canisters, storage overpacks, and transfer casks have been submitted as CoC amendment requests. The staff will revise the phrase to clarify this issue. b) The statement is not intended to be derogatory towards vendors. Given past review experience, each vendor has used unique styles and formats in their amendment requests, including the integration of new analyses into existing FSARs, and the demarcation of textual changes	Section 8.4.1 was changed to begin as follows: "The reviewer should survey the SAR and design drawings (generally SAR Chapters 1 and 2) to identify the various materials issues that may be associated with the specific design proposal in the application. The reviewer should also examine the criticality, shielding, confinement, and thermal chapters to identify cross-cutting issues that should be coordinated among the technical disciplines. Note, not all design and license changes in the amendment will necessarily be neither separately identified by the applicant nor obvious.

Comment	SRP Location	Public Comment	Resolution	Changes to SRP
			and 72.48 changes. It has been challenging for the staff in some cases to understand exactly what information is new and has changed since the last version of the FSAR was formally reviewed by staff in a previous licensing actions. In some cases, new or removed information has not been properly identified, or was ambiguous to the staff, in SAR change pages submitted during the review process. Given this experience, the purpose of the statement is meant to caution the reviewer to not overlook all potential changes in an amendment request, and ensure there is a clear understanding of the changes being requested for approval. However, the statement is revised for clarity and moved to the Introduction of the SRP for generic application to all disciplines.	The reviewer should examine the following Technical Specification (TS) items to verify its proposal by the applicant and understand the specific limits, design requirements, and operating constraints proposed by the applicant."
NEI 100	7338-7345	This paragraph should be deleted for a couple of reasons. It is incorrect to state that things previously approved and outside the scope of the amendment request are subject to review again. This is contrary to good regulatory practice and re-reviewing approved	Amendments such as content and design changes, are founded upon the design and methodologies previously reviewed by NRC. Compliance of a DSS is often based on the performance of the contents, canister, and overpack as a system. As a result, portions of these designs and methodologies in the SAR may be re-examined as	Removed the subject paragraph and added the following to the Introduction as the fifth paragraph under Review Process: Some amendments such as content and design changes, are founded upon the design and methodologies previously

Comment	SRP Location	Public Comment	Resolution	Changes to SRP
		information could create a contradiction with a previous staff SER. In addition, the sentence in lines 7341 and 7342 could be viewed as derogatory towards both the NRC project management and the applicant.	part of good regulatory practice to ensure the new amendment proposal meets Part 72 requirements. It is not the intent of the staff to re-examine designs previously approved in a CoC for re-approval. However, an amendment audit review may from time to time detect deficiencies or errors that were not identified during a previous audit review. It should be noted that it is the primary responsibility of the cask vendor to ensure such errors do not exist. Also, new information regarding operational experience or new phenomena may come to light which requires NRC consideration, in order to assure the design remains safe and compliant with applicable regulations. However, the statement is revised to clarify this issue and is moved to the Introduction of the SRP for generic application to all disciplines. Issues involving the licensing process (line 7341) and interactions are appropriately described in internal operating procedures for NRC staff. Therefore, this discussion is	reviewed by NRC for that system. Evaluation of amendment changes to a DSS are often based on the performance of the contents, canister, and overpacks as an integrated system. As a result, portions of previously approved components, contents, or methodologies in the SAR may be re-examined to ensure that the new system under the amendment proposal meets Part 72 requirements. During the audit review of an amendment, the staff may occasionally find errors or other safety questions that affect part of the previously approved design. The staff may need to review that part of the SAR and ask questions to assure the design remains safe and compliant with applicable regulations. The questions should be limited to understanding and resolving the specific technical issue, and should consider past precedents, regulatory guidance, and risk significance, as appropriate. The staff should also consider other processes (e.g. inspections, enforcement actions, generic

Comment	SRP Location	Public Comment	Resolution	Changes to SRP
			eliminated from the SRP.	issue program, etc..) to resolve these potential type of safety questions with a previously approved design
NEI 101	7362-7263	"copper bearing structural carbon steel" should be generalized to allow for other appropriate measures to control corrosion. Also, it seems inappropriate to single out one DSS design in review guidance.	Agree with comment	Changed this line in Section 8.4.1 to: "Use of copper bearing or weathering steel for structural steel components at coastal marine ISFSI sites (or other corrosion mitigation measures)."
NEI 102	7382	This should read "All ASME materials are a subset of AWS and ASTM materials"	Agree with comment	Changed as stated in comment
NEI 103	7394	The statement that all ITS materials are typically ASME II materials is not correct. That is only true of components subject to ASME Section III jurisdiction, typically confinement boundary and fuel basket. ITS attachments to the confinement boundary, as well as structural components of the overpack, are likely not ASME section II materials; for non-ASME ITS components, ASTM materials can be used.	Agree with comment	Changed 4th paragraph in Section 8.4.2.1 to: ITS components subject to ASME Section III jurisdiction, typically confinement boundary and fuel basket, are normally ASME Section II materials. ITS attachments to the confinement boundary, as well as structural components of the overpack, may be ASME or ASTM materials, depending on the code of record for the component. For non-ASME ITS components, ASTM materials may be used.
NEI 104	7400	Non-ITS materials specified to ASTM. This is not correct. According to Reg Guide 7.10, Appendix A, ITS Category B	Agree with comment	Changed 5th paragraph in Section 8.4.2.1 to: Non-ITS items can be specified

Comment	SRP Location	Public Comment	Resolution	Changes to SRP
		must be used in accordance with rigorous specifications; ITS Category C need not. Therefore, it is correct to state that ITS A and B should be specified to ASTM, ASME, or equivalent standards; ITS Category C, and non-ITS items can be specified by generic names such as "stainless steel", "aluminum," "carbon steel," etc., as appropriate for the application.		by generic names such as "stainless steel", "aluminum," "carbon steel," etc., as appropriate for the application.
NEI 105	7408	Editorial: Delete. This line repeats lines 7391-7392.	Agree with comment	Changed as stated in comment
NEI 106	7411-7412	No changes in neutron absorbers without NRC review. This is not correct; changes should be acceptable with appropriate review or testing by the certificate holder, with only select critical limiting characteristics included in the TS. 72.48 provides adequate change control for these items given the risk of dry cask storage operations.	Agree with comment	Changed 6th paragraph in Section 8.4.2.1 to: Proprietary materials which are ITS (specifically neutron poisons) must be described adequately in SAR Chapter 8, "Materials" to permit the staff to make a safety finding. The governing quality assurance and quality control (QA/QC) documents, key manufacturing procedures, and key testing protocols for proprietary materials should be incorporated by reference into the TS. Limited changes to the materials composition, performance, or manufacturing methods may be allowed if the

Comment	SRP Location	Public Comment	Resolution	Changes to SRP
				changes satisfy the criteria of 10 CFR 72.48.
NEI 107	7420-7425	Editorial: This information repeats prior information.	Agree with comment	Deleted subject lines.
NEI 108	7470-7471	Remove "transportation" as transfer is already listed. Remove "retrieval". In this context it is the same as unloading.	All of the terms in this sentence are applicable to the storage activity except "transportation"	Removed "transportation" from the sentence.
NEI 109	7515-7518	The information pertaining to steel producers is unnecessary for review guidance and should be deleted. If it is retained, at a minimum delete the last sentence regarding "defeating" a steel producer and clarify who is meant by "steel producers."	Agree with comment. The information pertaining to steel producers was meant to reflect lessons learned in past evaluations of steel certification. However, the language is clarified to only reflect the use of ASME Code values and CMTR values.	Replaced the last sentence in 6th paragraph of Section 8.4.5.1 with: "Examine the SAR adopted material properties for ITS component materials and ensure ASME Section II, Part D, properties and stresses are employed. The staff position (developed by NRR) regarding material properties is that ASME Code values must be used. Use of certified material test report (CMTR) values of UTS, yield, etc., is generally not permissible. Use of CMTR values is at risk of being non-conservative because samples may be taken at a portion of the ingot, billet, or forging that have optimum materials properties during certification."
NEI 110	7520-7523	This paragraph appears to be an editorial opinion and serves no value as review guidance. Delete.	Agree with comment.	Paragraph removed.
NEI 111	7554-7557	References to specific dry	Agree with comment.	Revised to remove vendor

Comment	SRP Location	Public Comment	Resolution	Changes to SRP
		storage vendors are typically not appropriate in the SRP. Please consider revising this section. If reference to a vendor is appropriate, the corporate name should be used rather than abbreviations. Therefore, change TN to Transnuclear, Inc.		names
NEI 112	7562-7564	What is the basis for no credit for coatings unless periodically inspected? Thermal spray Al-Zn coatings and hot dip galvanizing are widely used in marine applications, and are much more predictable than paint with respect to adhesion.	Without supporting data to demonstrate predicted coating life, monitoring is needed to assure intended performance. The guidance is revised to clarify this information.	Changed last paragraph in Section 8.4.6 to: Coatings may be specified to alleviate the coastal atmospheric corrosion issue. The coating must be periodically inspected and maintained, unless supporting data is available to demonstrate a predicted coating life.
NEI 113	7577	It is recommended that "AWS D1.6 (current edition), "Structural Welding Code – Stainless Steel" be added to this list of codes.	Agree with comment	Add "AWS D1.6 (current edition) Structural Welding Code-Stainless Steel," to Section 8.4.7.1, Welding Codes.
NEI 114	7608	The full penetration welds should only apply to the confinement boundary of the canister. In some designs the bottom closure weld is not a confinement boundary weld. For non-confinement boundary welds, other design should be acceptable. Please	Agree with comment	Changed first sentence in Section 8.4.7.2 to: Verify that the canister confinement welds are full penetration welds.

Comment	SRP Location	Public Comment	Resolution	Changes to SRP
		clarify		
NEI 115	7621-7622 8465	"helium leakage test is performed of the entire shell" – Please clarify that this testing only applies to the confinement pressure boundary (i.e., not attachment shell welds).	Agree with comment, except that Line 8465 is not applicable to this comment. The NRC has also issued draft ISG-25 "Pressure Testing of Confinement Boundaries" This guidance has been administratively incorporated into the SRP, which addresses this comment.	Revise Section 8.4.7.2 to incorporate the technical review guidance of draft ISG-25.
NEI 116	7621-7622	What is the basis for requiring a helium leakage test? The confinement boundary is designed in accordance with ASME Section NB, NC, or ND. The Code includes pressure tests to confirm pressure boundary integrity. If this is sufficient for high pressure vessels and piping systems in a power plant, it should be acceptable for a confinement boundary given the relative risk and service conditions.	See response to NEI Comment 115. Pressure tests, examinations, and leakage tests serve different functions. The volumetric and surface examinations of welds ensure geometric compatibility with the design requirements, but can only detect flaws down to a certain size. The ASME Code pressure test provides additional assurance that the component has been properly fabricated by stressing the component to a minimum Code required loading. The helium leakage test ensures there are no flaws or leak paths that could result in significant release of the helium and radioactive content to the environment. The weld non-destructive examinations, ASME Code pressure test, and helium leakage test are not considered	See response to NEI Comment 115.

Comment	SRP Location	Public Comment	Resolution	Changes to SRP
			equivalent substitutes for each other. The regulations mentioned in the text (i.e. 72.236 (d), (j), and (l)) provide the regulatory basis for the helium leakage rate test. Designing a component in accordance with ASME Code does not ensure that it is fabricated to prevent small potential gaseous leaks. Section 8.4.7.2 is updated to clarify guidance for leakage tests and to administratively incorporate the guidance of draft ISG-25. Sections 10.5.1.1 and 10.5.1.2 also capture this guidance.	
NEI 117	7624-7625	Not all of these tests (e.g., hydrostatic or pneumatic) are performed in the fabrication shop. Testing is in accordance with the design code. No additional review guidance is necessary. Shop helium testing would be an additional commitment beyond what the design code requires. Please clarify.	The helium leakage test provides assurance there are no flaws or leak paths that could result in significant release of the helium and radioactive contents to the environment. It is required to demonstrate compliance with 10CFR 72.236 (d), (j), & (l).The Code required pressure test ensures fabrication integrity of the component, but it does not ensure prevention of small gas leaks. Meeting Code requirements for pressure testing does not ensure meeting regulatory requirements for helium leakage rate testing. He leakage testing derives from Part 72, not ASME.	Section 8.4.7.2 has been updated to incorporate the guidance of draft ISG-25. Refer to response to NEI Comments 115 and 116.

Comment	SRP Location	Public Comment	Resolution	Changes to SRP
			Also refer to Sections 10.5.1.1 and 10.5.1.2 that capture this guidance for pressure testing and leak testing, respectively.	
NEI 118	7630	Editorial: Add "as" after "or."	Agree with comment	Changed as stated in comment
NEI 119	7641	Editorial: Change "designedto" to "designed to."	Agree with comment	Changed as stated in comment
NEI 120	7646	The N45.2 series has been replaced by NQA-1. Suggest referring to both for older commitments and newer commitments to the QA code.	Agree with comment	Changed the 2nd sentence of the fourth bullet in Section 8.4.7.3 to: Records documenting the lid welds shall comply with the provisions of 10 CFR Part 72.174, "Quality Assurance Records" or with NQA-1, "Quality Assurance Requirements for Nuclear Facility Applications," depending upon the standard in effect at the time of licensing.
NEI 121	7697-7701	For stainless steel canisters and welding, this is too limiting. The J-integral method to evaluation flaw size is used, which limits the size of a single weld pass. In order to be consistent with line 7682, it should explicitly state that the applicant can use J-integral methodology incorporating plasticity for ductile weld materials such as stainless steel.	Agree with Comment	Revised Section 8.4.7.4 to identify the use of either ultrasonic or multi-pass liquid penetrant examination for the structural lid-to-shell weld. Guidance is also provided on determining critical crack size including use of J-integral or net section stress methods.
NEI 122	7700	The canister is designed per ASME Section III, Division 1,	Agree with comment. Division 3 has not yet been endorsed by	Changed Division 3 to read "Division 1," in Section 8.4.7.4

Comment	SRP Location	Public Comment	Resolution	Changes to SRP
		Subsection NB, not Division 3. Has Division 3 been endorsed by NRC? If so, both Division 3 and Division 1 should be discussed. If not, reference to Division 3 should be deleted.	NRC.	
NEI 123	7715	Delete "Pursuant to NRC to Bulletin 96-04 (1996)." This language implies regulatory requirements are contained in the bulletin. An NRC bulletin is a request for information at a particular point in time. It is not something to be referenced as a source of information upon which to base a review of an application. The SRP should stand alone and refer to regulations and approved guidance only.	Partially agree with comment. An NRC Bulletin is not a regulatory requirement. Bulletin 96-04 addressees the potential for chemical, galvanic, or other reactions among the materials of a spent fuel storage cask, to assure no adverse reactions exist. This guidance is still applicable to the certification of DSS in order to meet the regulatory requirements of 10 CFR Part 72. The SRP is revised to clarify that the Bulletin may be used for guidance.	Changed to read: "The reviewer can find operational issues associated with hydrogen generation and guidance for evaluating galvanic or corrosive reactions in NRC Bulletin 96-04 (1996). Also, The reviewer should confirm the DSS will perform adequately under the operating environments expected (e.g., short-term loading/unloading or long-term storage) for the duration of the license period such that no adverse galvanic or corrosive reactions occur between the canister materials, fuel payload, and the operating environments."
NEI 124	7743	The statement that aluminum-based metal matrix composites are employed for all presently utilized neutron poison materials is incorrect. Boral, for example, is used through the industry and is not a metal-matrix composite.	Agree with comment	Replaced with: Aluminum based metal matrix composites and aluminum / boron carbide laminates (e.g. Boral^{TM}) are employed for all presently utilized neutron poison materials.
NEI 125	7750 7763	Analysis of creep for all aluminum based structural	Gap analysis can change drastically if aluminum components	Changed 1st sentence of 2nd paragraph of Section 8.4.9 to:

Comment	SRP Location	Public Comment	Resolution	Changes to SRP
		materials, including those only supporting dead weight – "any kind of loading." There is no sound basis for requiring a creep review of materials that have no structural function except bearing accident loads through their thickness, and supporting their own dead weight during normal storage.	creep and increase basket gap.	Review the design maximum temperatures and stress for any aluminum components and verify a creep analysis has been performed if any structural load bearing aluminum components operate at a design temperature above approximately 200F.
NEI 126	7724 7824 7881	This section is entitled "Exterior Protective Coatings" but lines 7824 and 7881 refer to interior coatings.	Agree with comment. The section is applicable to both interior and exterior coatings.	Title of Section 8.4.11 is revised to the title Protective Coatings.
NEI 127	7772	Exterior coatings. Scope and level of review for this area appears excessive and inconsistent with the "low priority" given. This should be reduced to specifying the generic coating systems that are acceptable, with surface preparation and paint application in accordance with manufacturer's instructions. Specifying the manufacturer and submitting the paint technical data sheets requiring qualification testing (lines 7881) are overly burdensome given the low risk.	Partially agree with comment. With the exception of coating issues that may result in adverse chemical or galvanic reactions described in NRC Bulletin 96-04, coatings are generally a low priority item with low safety significance. In these instances, most of the guidance in this section is not applicable. However, instances may exist in which unique or innovative coatings are specified by the applicant to perform a specific function unique to the cask system. In these instances, the reviewer may use discretion in implementing the detailed guidance in this section.	Section 8.4.11 is revised to include "Coatings generally have a low safety significance with the exception of coating issues that may result in adverse chemical or galvanic reactions. Typically, the detailed guidance in this section is not generally subject to further confirmation as part of the review. However, there may be instances in which unique or innovative coatings are specified by the applicant to perform a specific function unique to the cask system. In these instances, the reviewer may use discretion in implementing the detailed review guidance in this section.
NEI 128	7824-7825	It is not necessary to include	See response to NEI 127. Most	See response to NEI 127.

Comment	SRP Location	Public Comment	Resolution	Changes to SRP
		the coating manufacturer's technical literature in the SAR. The critical characteristics of the coating material are what is important and should be sufficient. The supplier should be free to use whatever coating material and manufacturer that has these characteristics for the service conditions.	coatings have unique properties and application steps. Often the characteristics of the coating and coating performance are dependent on the precise steps that were taken to apply the coating.	Deleted "The coating manufacturer's technical literature for all coatings specified for cask interiors must be submitted in the SAR for staff review". Add "Due to the unique nature of coating properties, and coating application techniques, the manufacturer's literature may be the only source of information on the particular coating.
NEI 129	7832-7942	Delete Sections 8.4.11.4 through 8.4.11.6. Surface preparation coating repairs, and coating qualification testing are all details not necessary for the staff to review. These attributes of the coating system are dictated by the coating manufacturer or the CoC holder for the particular coating material and service conditions. Appropriate surface preparation, repairs and qualification testing are all adequately governed by the CoC holder's or licensee's coating specification and procedures developed under the applicable QA program and the coating manufacturer's requirements. All of the above is subject to NRC inspection	See response to NEI 127	See response to NEI 127

Comment	SRP Location	Public Comment	Resolution	Changes to SRP
		for verification of compliance.		
NEI 130	7882-7884	It appears that this sentence is written for paints and does not account for the possibility of plating as a coating.	The statements in this paragraph are applicable to any coating, including paints or plating.	Phrase "(including paints or plating)" was added to sentence.
NEI 131	7950	The statement that neutron shielding materials are not ITS appears to conflict with NUREG/CR-6407, which specifies that shielding materials are ITS Category B. Please clarify.	Agee with comment	Paragraph 8.4.12.1 "Neutron Shielding Materials" was revised to indicate that shielding materials are ITS.

Staff also noted that the qualification and acceptance testing of neutron shielding materials should not be required in the TS. Only characteristics directly related to performance (e.g., composition and density) of the neutron shielding material should be specified in the TS. |
NEI 132	7963	The first sentence in this line is unnecessary. Delete.	Agree with comment	Changed as stated in comment
NEI 133	8021	Impurity limits may or may not be established as a result of qualification testing; that is not the main purpose of qualification testing.	Agree with comment	Deleted the following: "Qualification tests would be useful in establishing that the impurity concentration limits for borated absorbers are not exceeded. Agreement on these limits can be done by agreement between buyer and seller."
NEI 134	8008	Editorial: "Surrey" should be "Surry."	Agree with comment	Changed as stated in comment
NEI 135	8048	Submittal of manufacturer's data sheet for neutron	Agree with comment	Replaced; "The manufacturer's data sheet should be submitted

Comment	SRP Location	Public Comment	Resolution	Changes to SRP
		absorber is only applicable if the applicant is proposing a trade name product. Add "as applicable" at the end of the sentence.		to supplement the above information" with the following; "If the applicant intends to use an absorber material with a specific trade name, the manufacturer's data sheet should be submitted to supplement the above information."
NEI 136	8103	ZrB2 standard: All standards are a compromise of some kind: homogeneous standards like ZrB2 must be paired with aluminum sheets to simulate the scattering by aluminum in the neutron absorber; scattering by carbon in boron carbide is generally not simulated. Non-homogeneous standards that have a very fine uniform dispersion of the boron-containing phase are only an approximation of the homogeneous material assumed in the criticality safety calculations, but they get the appropriate aluminum and carbon scattering. Therefore, change "a qualified homogeneous standard such as ZrB2" to "a calibrated standard that is either homogeneous, such as ZrB2, or that has a very fine and	Agree with comment	Changed to: Aa = acceptance value of neutron attenuation, based on a qualified homogeneous absorber standard such as ZrB2, or a heterogeneous calibration standard that is traceable to nationally recognized standards, or calibrated with a monoenergetic neutron beam to the known cross section of boron-10. Calibration standards should be evaluated at 111 percent (i.e., 1/0.90) of the poison density assumed in the criticality computational model.

Comment	SRP Location	Public Comment	Resolution	Changes to SRP
		uniform dispersion of boron such that it approximates homogeneity."		
NEI 137	8110	P=0.999: Previously the staff has accepted P=0.95 and should continue to do so considering all the conservatisms involved (e.g. keff ≤ 0.95, the 90% maximum credit for boron 10).	Agree with comment	Changed; "Let P = 0.99 and γ = 0.95." To: "Let P = 0.95 and γ = 0.95.":
NEI 138	8122	Quantitative measures (porosity testing, tensile testing, etc.) are now preferred over qualitative examination (TEM, SEM). Metallic/ceramic systems are generally accepted as not susceptible to radiation damage from gammas or from neutrons at the fluences encountered in dry storage.	Agree with comment .	Replaced first two paragraphs of Section 8.4.13.3 with a detailed description of the qualification testing previously accepted by the staff. Also added qualification tests needed to be performed on structural neutron poisons.
NEI 139	8155	A sample from every other piece is too prescriptive for a standard review plan; according to ASTM C1671, random or systematic sampling should be applied.	Agree with comment	Changed: "Adequate numbers of samples should be taken from every other component....." to "Adequate numbers of samples should be taken from components......"
NEI 140	8156-8157	Lot definition based on billet may not be appropriate for material from small billets; allow alternate definitions that are uniform for sampling purposes.	Agree with comment	Removed the sentence defining 'lot.'
NEI 141	8186	Please delete the following	ACI 349 Section R3.5.3 States:	Changed the sentence to:

Comment	SRP Location	Public Comment	Resolution	Changes to SRP
		sentence "Zinc, zinc rich coatings, zinc-clad materials, and aluminum should not be used for any embedded objects that will be in contact with wet concrete, because of the potential for concrete degradation from an adverse chemical reaction". Zinc galvanized reinforcing steel and zinc plated/galvanized embedded lifting devices are common and widely used in the concrete industry. Even though chemical reaction between the zinc and water in concrete may occur at any age, this reaction is not proven to have any adverse impact on concrete. Note that Section 3.5.3.8 of ACI 318-08 allows the use of galvanized reinforcing steel per ASTM A 767.	"*Deformed reinforcement*—Zinc used in the galvanizing process may negatively react with alkaline materials commonly found in concrete. In addition, potential galvanic corrosion with other embedded metals, as well as hydrogen generation and potential for hydrogen embrittlement, suggest that such coatings may be detrimental. Research conducted by Sergi et al.3.1 concluded that zinc coatings provide little value in providing long-term protection of reinforcing steel, and cautionary statements in ACI 201.2R3.2 support this position. These industry concerns have prompted ACI Committee 349 to prohibit the use of zinc coatings on reinforcing steel in nuclear safety-related structures until adequate data justifying its use can be reviewed."	"Zinc, zinc rich coatings, zinc-clad materials, and aluminum should not be used for any embedded objects in structures designed to ACI 349 or ACI 359 that will be in contact with wet concrete, because of the potential for concrete degradation from an adverse chemical reaction"
NEI 142	8202	Editorial: Change "used" to "use."	Agree with comment	Changed as stated in comment
NEI 143	8228-8229	Delete this sentence. Requirements for water-to-cement ratios and air content (mainly controlled by the use of air entraining admixtures), which are based on the severity of the anticipated exposure of concrete, are provided in ACI 349/318. The	Agree with comment	Deleted as stated in comment

Comment	SRP Location	Public Comment	Resolution	Changes to SRP
		w/c ratio and air content are design requirements and not fabrication details.		
NEI 144	8301-8303	Samples normally taken in HAZ, same weld thickness and materials of construction, etc.: This area needs clarification. Testing is done per ASME Section III and Section IX. Weld thickness relation to the thickness of the design weld is governed by Section IX. Impact testing is required of the base metal (NX-2300 and the weld metal (NX-2400), but not the HAZ. Weld qualifications are performed using materials of the same class (P-number), but not necessarily the same material and grade as that used in construction.	Agree with comment	Replaced sentence with: Metals having a face-centered cubic crystal structure such as austenitic stainless steels, remain tough and ductile to very low temperatures and are not a concern in this regard.

Added as separate following paragraph: Toughness testing (e.g., Charpy impact) of welds is governed by ASME Section III, as supported by Section IX. |
NEI 145	8319-8320	Specifying peak rod burnup is inconsistent with past practice, which has been to specify assembly average burnup.	See response to NEI Comment 9	Changed as noted in NEI Comment 9
NEI 146	8358-8359	The text refers to "the following Part 72 regulations" yet no regulations are discussed in the text that follows.	Agree with Comment.	The 3rd paragraph in Section 8.4.17.1 was removed. The 1st sentence in the following paragraph was modified to read "The acceptance criteria below and review procedures..."
NEI 147	8453	Delete "and retrieval" since this is covered by fuel handling	Agree with Comment	Changed as stated in comment
NEI 148	8567-8568	The text states that this review	Agree with Comment	Revised to clarify coordination

Comment	SRP Location	Public Comment	Resolution	Changes to SRP
		should be coordinated with the materials reviewer. The guidance in this section is specifically for the materials reviewer. Please clarify.		with thermal reviewer.
NEI 149	8593-8595	Delete the last sentence of this paragraph. It is opinion, not review guidance.	Agree with Comment	Changed as stated in comment
NEI 150	8636	Replace the word "dangerous" with "large" or "significant."	Agree with Comment	The term "dangerous" was changed to "large"
NEI 151	8645-8656	Helium testing of the entire confinement boundary is not necessary. Confinement boundary welds are volumetrically tested in the fabrication shop and the entire vessel is pressure tested after loading. Both the inspections and testing are performed per the ASME Section III Code. Additional testing beyond what the ASME Code requires should not be necessary. Please revise.	See response to NEI Comment 116. The ASME Code non destructive examinations and pressure test are performed for different reasons than the helium leak rate test which assures no significant radiological leakage.	No Change
NEI 152	8726	RG 1.183 should be RG 1.193.	Agree with comment	Changed as stated in comment
NEI 153	8914-8956	References to Part 71 regulations do not appear appropriate in these lines. Please revise accordingly.	Agree with comment	Removed sentence that mentions Part 71. Also, under the definition of damaged fuel, transportation was similarly removed.
NEI 154	8990-9013	Editorial: The numerals in the compound names should be subscripts to be consistent with the convention in other portions of the SRP. Please	Agree with comment	All chemical formulas in the paragraph were changed to be written in subscripted form, not U4O9 but U_4O_9

Comment	SRP Location	Public Comment	Resolution	Changes to SRP
		revise.		
NEI 155	9077 9271	Sections 8.7.3 and 8.8.3 should be removed and the references moved to the consolidated references in Appendix A to be consistent with the treatment of references in other chapters and to eliminate duplicate references (e.g. line 9089 and line 12923).	Agree with comment	Delete Sections 8.7.3 and 8.8.3. Integrate references into Appendix A and eliminate redundancy
NEI 156	9090	Editorial: The reference incorrectly lists the upper temperature as 400. The correct value is 360 as listed in line 12923.	Agree with comment	Changed as stated in comment
NEI 157	9231-9232	The limit could be interpreted as the limit in any one cycle is 65°C. It needs to explicitly state that the 65°C range can be exceeded but for less or equal to 10 cycles.	This paragraph provides support for SRP Section 4.4.2, which discusses thermal cycling.	No change
NEI 158	9518 9520	Editorial: Sketches A and B should more appropriately be listed as Figures and the references to the sketches appropriately revised.	Agree with comment	Sketches A and B have been redesignated as Figure 8-3 and Figure 8-4, respectively. The list of Figures has been revised to include them.
NEI 159	9518 9520	Information was removed from the sketches when they were incorporated from ISG-18 Rev. 1 (e.g. identification of cover plate and vent and drain port cover plate). This information should be restored.	Agree with comment	Changed as stated in Comment
NEI 160	9737	Suggest changing "use and	The SRP provides examples of the	No Change

Comment	SRP Location	Public Comment	Resolution	Changes to SRP
		operation" to "function". The cask vendor may not offer all of these specialized tools or require a particular tool to be used to accomplish a task. The user needs to understand the intended function for them to purchase the equipment needed to accomplish the task.	specialized equipment and tools with enough detail for the staff to understand their use and operation (i.e. lifting yokes, transporter equipment, welding and cutting equipment and vacuum equipment). If their use and operation was changed to function, then the prior named examples would be sufficient since their name is self descriptive. The staff should review the description of how this specialized equipment is used and operated with the DSS, as stated in the SRP.	
NEI 161	9752	Delete "receipt inspection activities." Receipt inspection is a separate QA function not related to the operations described in Chapter 9.	Agree with comment	Delete the words: "receipt inspection" from the activities listed
NEI 162	7124-7125 9767-9768	Delete references to performing measurements to confirm assembly burnup values. Reactor records have repeatedly shown to be reliable for performing core reloads and to estimate boron concentration and rod position for reactor startup. They should be equally sufficient to validate assembly burnup for cask loading, a much lower risk activity.	Current guidance on implementing burnup credit recommends a burnup verification measurement. This measurement is recommended to prevent misloading of assemblies that do not meet the loading criteria for burnup credit. The guidance indicates the level of agreement between the measurement and the record burnup values expected in order to confirm that the record is the correct record for the particular assembly being measured. Given the uncertainties in the	Remove the text of Section 7.5.5, "Burnup Credit" and replace it with: "For guidance regarding the use of burnup credit, see the current revision of ISG-8.

Comment	SRP Location	Public Comment	Resolution	Changes to SRP
			measurement and record values, these uncertainties are to be accounted for in assigning the assembly's burnup for loading into the cask. Based upon ACRS comments, as well as the expectation that the guidance on burnup credit will be revised in the near future, the guidance on burnup credit is being retained as a separate ISG (i.e., ISG-8) and the SRP is being modified to refer the reviewer to the current revision of this ISG. Revision to the ISG is expected to expand several aspects of the current guidance, including burnup verification.	
NEI 163	9847-9848	Delete the requirement to re-evacuate and re-backfill. The necessary helium purity can be obtained with a single backfill of high enough purity. More generally, care should be taken in using the PNL document referenced because it is over 20 years old. Cask operations have changed in that time. For example, one current cask vendor dries the canister without the use of vacuum. We realize these are examples, but the reviewer should understand that the	Agree with comment.	Deleted the following sentence from the SRP: "The cask is then re-evacuated and re-backfilled with inert gas before final closure." Section 9.5.1 is clarified to recognize forced helium drying.

Comment	SRP Location	Public Comment	Resolution	Changes to SRP
		reference document is out of date.		
NEI 166	9973-9974	Delete this item. Dose rates do not belong in TS and do not verify proper loading of the cask.	Surface dose rate measurements are a parameter used to verify cask fabrication and operation. It is a measureable parameter during deployment of the cask onto the storage pad. Measurements may not detect all types of fabrication errors, but it provides a means for identifying potentially serious problems with the loaded contents and cask shielding system.	The guidance in Section 9.5.1 was revised to clarify the measurement of surface dose rates.
NEI 167	10343	Editorial: Change "i.e." to "e.g."	Agree with comment	Changed as stated in comment
NEI 168	10345	Editorial: Delete close parentheses after "Program" and move the period inside the close quotation.	Agree with comment	Changed as stated in comment
NEI 169	10366-10367	The "basis of tests deemed acceptable" should be from regulations or something more definitive and stable than prior staff acceptance.	10CFR72.82, Inspections and Tests, Paragraph (d) states: "Each licensee shall perform, or permit the Commission to perform, such tests as the Commission deems appropriate or necessary for the administrator of the regulations in this part". As such the regulations state the basis for performing the test as those deemed appropriate. The guidance also is specified to address unforeseen design proposals and considers the operational experience and precedent from previous licensing of licensing actions of storage casks..	Section 10.5.1 revised to state:

The following guidance is presented on the basis of tests deemed acceptable by the staff in previous SAR reviews. The guidance is based on operational experience and the knowledge from past licensing reviews. Alternative tests and criteria may be used if the SAR provides appropriate explanation and adequate justification. Additional tests and criteria may be needed, depending on the operational experience and uniqueness of |

Comment	SRP Location	Public Comment	Resolution	Changes to SRP
				the design proposal.
NEI 170	10381-10382	Recurring trunnion load tests for transfer casks is not consistent with ANSI N14.6, which permits NDE to be performed periodically rather than load testing.	The guidance for the load tests recognizes the lifting trunnion test provisions in accordance with ANSI N14.6. As such, periodical NDE, in lieu of annual load tests, is acceptable for the trunnion provided that other conditions as specified in ANSI N14.6 are also met	Added the following sentence to the 1st paragraph in Section10.5.1.1: Periodical NDE, in lieu of annual load tests, is acceptable for the trunnion provided that other conditions as specified in ANSI N14.6 are also met.
NEI 171	10418-10433	Please clarify the guidance pertaining to testing. Clarification should include ASME Code concurrence that fracture testing is not required for material with wall thicknesses of less than 5/8 inch.	NUREG/CR-1815 "Recommendations for Protecting Against Failure by Brittle Fracture in Ferritic Steel Shipping Containers Up to Four Inches Thick" RG-7.11, ASME Code for Transport Packages and DOE guidance all establish fracture toughness testing for 3/16-inch and thicker material. Therefore, FT testing is required below 5/8-inch to down to 3/16-inch, unless other justification is provided.	Changed the last sentence in the 4th paragraph of Section 10.5.1.1 to read as follows: NUREG/CR-1815, "Recommendations for Protecting Against Failure by Brittle Fracture in Ferritic Steel Shipping Containers Up to Four Inches Thick," provides staff guidance concerning materials and thickness ranges subject to brittle fracture testing. On the basis of guidance in NUREG/CR-1815, Section 5.1.1, the NRC established two methods for identifying suitable materials.
NEI 172	10476-10479	Delete the sentence pertaining to inspection personnel qualifications. This is something governed by the QA program and outside the scope of a cask design review. At a minimum, delete "the current revision of." The	Partially agree with comment. The reference is appropriate but is changed from "current" to "appropriate" version.	Deleted "current revision" and replace it with "appropriate revision" in referring to SNT-TC-1A

Comment	SRP Location	Public Comment	Resolution	Changes to SRP
		fabricator should not be forced to adopt the most recent revision of SNT-TC-1A to qualify personnel if a different code or older version of SNT-TC-1A is acceptable within their QA program. If and when to adopt a later Code should be at their discretion.		
NEI 173	10513-10516	Why specify the particular NDE method if the Code does that? Suggest deleting this detail. Also, AWS should be offered as an acceptable weld code for non-confinement boundary welds.	Agree with comment regarding AWS. Specific guidance on the Code requirements is provided to avoid misunderstanding and possible conflicting interpretations. In addition, the specific guidance assists reviewers in focusing on important elements of the NDE methods with respect to the associated review objectives.	Added the following after the last non-confinement weld paragraph in Section 10.5.1.3: "(LOW Priority)Non-confinement welds may also be welded, repaired and examined in accordance with AWS D1.1, Structural Welding Code – Steel, D1.3, Structural Welding Code – Sheet Steel and D1.6, Structural Welding Code – Stainless Steel. Use of these standards shall be called out on the licensing drawings."
NEI 174	10576-10577	Delete these lines. Dose rate measurements of every cask after SNF is loaded are of little value in determining whether the design criteria have been satisfied because the shielding analyses are extremely conservative. Users will perform appropriate dose rate measurements on the loaded casks as a part of their	Partially agree with comment. The guidance is revised to indicate that dose measurements of loaded SNF, in lieu of an auxiliary source, may be used to verify shielding effectiveness with appropriate scanning of the shield and appropriate consideration of the actual source strength of the loaded contents.	Revise Section 10.5.1.4 to include: Dose measurements of loaded SNF, in lieu of an auxiliary source, may be used to verify shielding effectiveness with appropriate scanning of the shield and appropriate testing program that considers the actual source strength of the

Comment	SRP Location	Public Comment	Resolution	Changes to SRP
		Radiation Protection Program and ALARA procedures.		loaded contents.
NEI 175	10588-10597 and 10620-10629	Duplicated paragraphs.	Agree with comment	Deleted 5th paragraph in Section 10.5.1.5.
NEI 176	10613	Editorial: "bench marked" should be "benchmarked" (one word).	Agree with comment	Changed as stated in comment
NEI 177	10741	Clarify "periodic tests to verify shielding and thermal capabilities." Such tests are usually not necessary for passively cooled systems beyond periodic checks of the air vents. Also, there are no credible age-related means to degrade shielding. Such tests should only be required if the particular cask design has unique features or active components requiring such tests.	Aging and degradation of shielding materials may be a credible phenomenon. Degradations of components, such as cracks on concrete over-pack, corrosions of steel components, are examples that may impair their shielding capabilities. The applicant should otherwise justify that aging of materials related to the shielding, confinement, and thermal designs are not credible during the licensed period of the DSS.	Section 10.5.2.2 is revised to clarify that justification is required to eliminate shielding, confinement, and thermal tests.
NEI 178	10955	Delete "including minors." Minors are not part of the working staff at power plants subject to occupational exposure.	Disagree with comment. In the event, that a minor is present in an occupational capacity at a licensee's facility, that licensee is responsible for ensuring that the requirements of 10 CFR 20 are met.	No Change
NEI 179	10956	Delete "retrieval and".	The regulations (in 10 CFR 72.236(h)) require that the spent fuel storage cask be compatible with both wet and dry loading and unloading facilities. It is reasonable to expect that fuel cannot be unloaded without first	No Change

Comment	SRP Location	Public Comment	Resolution	Changes to SRP
			retrieving the storage cask from the storage location. The potential exists for there to be differences in the radiological conditions encountered in retrieving and unloading fuel from an ISFSI as compared to loading and emplacing the casks. The applicant should consider the possible differences in radiological conditions and provide dose estimates if any differences are expected to be significant.	
NEI 180	11003-11005	The value of applicants calculating and NRC approving dose versus distance from a hypothetical ISFSI is of questionable value in the application because of the arbitrary nature of: the number of casks, the arrangement of the casks on the ISFSI, the distance to the site boundary, and the cask contents. Licensees are required to perform a 72.104 dose analysis for their particular ISFSI by 72.212.	The application should demonstrate that there is reasonable assurance that the requirements of 10 CFR 72.234(a) will be met by the proposed system. One of these requirements is presented in 10 CFR 72.236(d) which indicate that shielding and confinement features must be sufficient to meet the requirements of sections 72.104 and 72.106. The applicant must demonstrate the proposed system is capable of meeting these requirements. Past review experience has shown that the dose rate versus distance calculation for a standardized array has been beneficial in confirming these requirements.	Replaced 2nd paragraph in Section 11.5.3.1 with the following: The reviewer should verify that the applicant includes a dose rate versus distance curve in its evaluation of offsite dose for a hypothetical cask array. The theoretical cask array should consist of at least 20 storage casks (2x10 array), and the analysis may include the effect of shielding among casks in the array. The reviewer should examine predicted dose rates and compare them to the dose rates from previously approved casks, and any associated annual doses that have been observed for the casks at existing ISFSIs.

Comment	SRP Location	Public Comment	Resolution	Changes to SRP
NEI 181	11007-11018	As only hypothetical array and single cask are evaluated, it is not clear when features would be required to show compliance with regulations and should be included in the conditions of cask use. Specific distance and shielding options and inclusion of such limitations in the CoC are not consistent with the 72.212 evaluation that a site would do to establish compliance with the requirements.	As indicated in response to the previous comment, staff must be able to have reasonable assurance that shielding and confinement features for a proposed dry storage system are sufficient for users of the system to design ISFSIs that can meet the requirements of Sections 72.104 and 72.106. If the dose requirements of Section 72.104 can be met at the minimum distance specified in the regulations (100 meters, specified in Section 72.106(b)) for a single cask and the hypothetical array described in this section of the SRP, the staff considers there is reasonable assurance that the regulatory requirements will be met. If additional distance or shielding is needed to meet the dose limits beyond the controlled area boundary for either the single cask or the proposed hypothetical array, then staff needs some basis on which to make its determination. The applicant should provide a justification for how a general licensee could reasonably meet the requirements of Section 72.104. Including a shielding or distance requirement in the CoC conditions of use would only be needed if the applicant chose to use either or both of these as a basis for its SAR	Moved the sentence "In addition, the SAR should determine the degree to which the normal condition dose rates could change for the identified off-normal conditions" to the end of the first paragraph in Section 11.5.3.1.

Replaced the text from the 3rd and 4th paragraphs in Section 11.5.3.1 with the following three paragraphs:

It is important to note that the general ISFSI licensee is permitted to use either distance between the ISFSI and the controlled area boundary or engineered features (supplemental shielding) such as berms to mitigate doses to real individuals near the site. The SAR needs to provide sufficient information to support informed choices on the part of the general licensee. If the SAR analyses were performed for the minimum 100 meter distance and did not use any additional shielding, and the projected dose at 100 meters exceeded the regulatory limits, the reviewer should verify that the application contains a |

Comment	SRP Location	Public Comment	Resolution	Changes to SRP
			evaluations and did not provide a SAR analysis without the added distance or shielding. Site-specific features, or extra distance or additional shielding that a general licensee chooses to evaluate and/or implement to further reduce doses outside the controlled area boundary are not included as limitations in the CoC, but are included in the site's 72.212 evaluation. Analogous to the evaluations for 72.104, if evaluations for accidents show that a distance greater than 100 m or additional shielding is needed to meet the dose limits at or beyond the controlled area boundary, the preceding discussion would also apply for meeting the requirements of Section 72.106. Thus, in responding to this comment, staff also modified the guidance regarding accidents in Section 11.5.3.2 of the SRP to include similar guidance as provided for normal conditions in Section 11.5.3.1 of the SRP.	justification for how a general licensee could reasonably meet the requirements of Section 72.104. If the dose versus distance curves for the single cask and hypothetical array in the SAR were only evaluated at distances greater than 100 m, or assumed some engineered feature, then the CoC should contain a condition of use to that effect. An example of such a condition may be similar to the following: "The use of this system may require more than the minimum 100-meter distance between the ISFSI and the controlled area boundary, or engineered features (i.e., berms or shield walls), or both to ensure the dose limits in 10 CFR 72.104 can be met. In cases where engineered features are used to ensure that the requirements of 10 CFR 72.104(a) are met, such features are to be considered important to safety [ITS] and must be evaluated to determine the applicable [QA] category." If an engineered feature is used in the SAR evaluations, then

Comment	SRP Location	Public Comment	Resolution	Changes to SRP
				that feature is to be considered to be part of the system. As such, it should be described in the CoC. Inserted the following sentence at the end of the 2nd paragraph of Section 11.5.3.2: For those systems for which these curves indicate the need for greater distance and/or engineered features, such as berms, a condition similar to that described in the preceding section of this SRP may be needed, with the requirement being 72.106(b) in this instance.
NEI 182	11265-11266	Clarify this statement. Not all DSS monitoring equipment is ITS. It is only ITS if it meets the definition of ITS in the NUREG based on its design function. Suggest: "DSS monitoring equipment is classified in accordance with NUREG/CR-6407..." This also conflicts with lines 1678 and 5347.	Agree with comment and changed to be consistent with Lines 5347 and 1678. Pressure monitoring systems are not ITS because they cannot withstand the design basis loadings, nor are they required to be procured in accordance with ITS practices. Their failure does not result in an unsafe condition, but their failure in combination with another failure (e.g. confinement seal) could result in an unsafe condition which makes it a Category B item under the guidelines of NUREG/CR-6407.	Changed this sentence to: "DSS monitoring equipment (such as a pressure monitoring system) are classified as not important to safety, but are classified as Category B under the guidelines of NUREG/CR-6407, 'Classification of Transportation Packaging and Dry Spent Fuel Storage System Components According to Importance to Safety (INEL-95/0551)' since they aren't designed nor procured under the same requirements as the confinement boundary, but

Comment	SRP Location	Public Comment	Resolution	Changes to SRP
				whose failure in combination with another failure could result in an unsafe condition."
NEI 183	11364-11368	What is the purpose of capitalizing this text?	The statement is meant to caution the reviewer about terms and conditions of the CoC and technical specifications. The capitalization is removed from the text.	Revised the paragraph as follows (no capitalization): If a reviewer determines that a design feature, content feature, analytical assumption, operating assumption, control, limiting condition of operation, program or other SAR item is important and should not be changed without NRC staff approval, then it should be further evaluated and considered as a potential CoC condition or technical specification. The reviewer should further consider the guidance in this chapter for establishing conditions and technical specifications in the CoC. Only the terms and conditions of the CoC, including the attached technical specifications and drawings, are legally enforceable. If a reviewer deems an item so important that it should not be changed without NRC staff approval, the item should either be included directly in the CoC terms, conditions or technical specifications.

Comment	SRP Location	Public Comment	Resolution	Changes to SRP
NEI 184	11440-11452	Most of the text about the Code in this paragraph is of limited value. Suggest replacing this with simpler guidance that states the applicant should state the applicable design codes, sections, subsections, as appropriate, and any alternatives to the code being implemented.	The referenced paragraph presents the current and historical basis for the use of the ASME Code for DSS as guidance for the staff.	No Change
NEI 185	11460	Editorial: Add "s" to the end of "specification."	Agree with comment	Changed as stated in comment
NEI 186	11588	Editorial: Change "12" to "13."	Agree with comment	Changed as stated in comment
NEI 187	12723-12724	ISG-15 should not be listed in the reference section since it has been incorporated into this document. Other ISGs are not listed in the reference section.	Agree with comment	Changed as stated in comment
NEI 188	13025	Editorial: Change "to" to "10."	Agree with comment	Changed as stated in comment
NEI 189	13158	Editorial: Insert a close parenthesis at the end of this line.	Agree with comment	Changed as stated in comment
NEI 190	13237	Editorial: Change 'uncorrectd" to "uncorrected."	Agree with comment	Changed as stated in comment
NEI 191	13475	Editorial: Change "austentic" to "austenitic."	Agree with comment	Changed as stated in comment
NEI 192	13475	With regard to ISG 12, the status block states that a new revision is pending. This is inappropriate information for the SRP. In addition, a pending revision to this ISG has not been announced by NRC, yet draft revisions to	Agree with comment	Deleted "new revision pending"

Comment	SRP Location	Public Comment	Resolution	Changes to SRP
		ISG-2 and ISG-23 have been issued by NRC and are not noted in this appendix.		
NAC	General	The Draft NUREG-1536 does not appear to reflect NRC's position on risk based regulations. It appears to be too prescriptive in areas that have little to no impact on safety

Reconsider detailed prescription of requirements that are covered by other regulations, measurements and controls, e.g., shielding design, related computer verification, measurements required during loading operations, measurements on loaded casks for site operations to manage site boundary dose. Technical Specification material should be limited to system operational limits that the licensee must meet and not repeat regulatory requirements or include material property and test requirements addressed by Quality Assurance requirements. Technical specifications should not be used as a control on the licensee use of | NRC staff is not revising the Part 72 regulatory requirements as part of the update to this SRP. Some Part 72 regulations are prescriptive and others are performance based. In addition, the NRC does not endorse a risk-based approach, but rather a risk-informed approach as delineated in SECY-98-144. The SRP is being revised to risk-inform, or risk prioritize the review guidance used to verify that the established Part 72 regulations are met. The areas of review in the SRP have been prioritized considering the potential relative risk impact of not meeting the requirements. It provides guidance that reduces the intensity of the review for low risk areas. It is in this sense of focusing staff attention on areas important to safety that the SRP is risk-informed. In fact, this is the NRC definition of risk-informed.

Also see response to NEI 18 regarding Technical Specifications. | To clarify the approach that the staff is using to prioritize the review procedures sections of this SRP and eliminate any confusion with a more classical quantitative PRA approach, the text has been modified to generally substitute "prioritized" for "risking informing" when referring to the review procedures. Attachment B also has been similarly modified. |

Comment	SRP Location	Public Comment	Resolution	Changes to SRP
		72.48 revisions.		
NAC	1914-1917	"Nevertheless, for assessment purposes and to demonstrate … the DSS should be evaluated for effects of a confinement boundary failure." This is not duplicated in confinement SRP discussion. Evaluation of the effect of a confinement boundary failure is not a standard evaluation set for current licensed systems (ISG-5). Nonmechanistic failure should not be a system analysis requirement. This imposed analysis is beyond regulation requirements.	See NEI Comment 28	See NEI Comment 28
NAC	3106, 12537	Imposes excessive conservatism for seismic evaluations. RG 1.60 should be replaced by NUREG/CR-6728 and also NUREG/CR-6865.	See NEI Comment 46.	See NEI Comment 46
NAC	3139-3140	"Confinement casks" is poor terminology. It should read: "for the confinement boundary of the cask."	Agree with comment	Changed as stated in comment
NAC	3153	In the previous paragraph, Subsection NB is used to define stress qualification for the confinement boundary, which is a pressure retaining	See NEI Comment 48	See NEI Comment 48

Comment	SRP Location	Public Comment	Resolution	Changes to SRP
		boundary. In this paragraph, it does not clearly state that the basket is a nonpressure-retaining boundary and that the applicant should use Subsection NG. Need to state that Subsection NG is acceptable or else the reader is left to believe that Subsection NB applies to nonpressure boundary baskets. It should confirm that Appendix F is applicable for use with Subsection NG.		
NAC	3168	Includes excessive conservatism that is not consistent physical testing. It should state that Subsection NB and Subsection NG permit the use of Appendix F, which does permit the use of inelastic properties for components that serve as the pressure boundary or also non-pressure boundary applications, such as baskets. It should also state that strain base criteria can be employed for energy limited accident conditions, provided the applicant provides such basis for its use."	The strain-based criteria are not recognized by the ASME Code or other applicable standards. The NRC staff may consider use of other acceptance criteria on a case-by-case basis.	No change
NAC	3171	In many applications for drop	The SRP does not preclude use of	No change

Comment	SRP Location	Public Comment	Resolution	Changes to SRP
		conditions, it should be acceptable to strain rate sensitive properties. Appendix F permits its use. Need to include "strain rate properties, which need the appropriate references."	strain-rate-sensitive material properties for design analysis of cask drop conditions	
NAC	4302	Annotation of input files. It is important to be able to use the applicant's files. It is not necessary to understand all aspects of the input files. Some of these files come from Journal files or Log files which are generated by the program. It is not feasible to add comments to these files. Open-ended statements such as adding "annotation" lead to overstatement by the reviewer for the need of such documentation.	See NEI Comment 57	See NEI Comment 57
NAC	4313	Annotation of the load steps. This would lead to excessive documentation in the computer solutions. 4311-4315 should be removed. It is the responsibility of the applicant's QA program to ensure that the analyses are performed correctly.	See NEI Comment 58	See NEI Comment 58
NAC	4332	Sensitivity study on mesh type. Lack of clarity. "Mesh type" should be removed. It is not	See NEI Comment 59	See NEI Comment 59

Comment	SRP Location	Public Comment	Resolution	Changes to SRP
		clear.		
NAC	4335-4336	Mesh study. Not required when stress linearization is being used for primary loading. Such detailed studies should be restricted to fatigue evaluations at stress discontinuities. Remove these lines. Too subjective, allowing the reviewer to specify detailed mesh studies for any part of the model he so desires.	See NEI Comment	See NEI Comment 60
NAC	4349	Including plots of the results. Generates extra data to be included in the SAR, while it is not needed. Remove "plots" from line 4349.	See NEI Comment 61	See NEI Comment 61
NAC	4680	Exclusion of natural convection internal to the canister. Too restrictive for convection designs. It states: "…should be limited to…the external surface…" This is an unacceptable statement that will be taken by the reviewer that internal convection cannot be used without some excessive burden of proof provided by the applicant. Remove line 4680. There is sufficient test data to confirm that convection internal to the canister is acceptable.	See NEI Comment 69	See NEI Comment 69
NAC	4687/JS	Convection. What does "robust" mean? This allows the	See NEI Comment 71	See NEI Comment 71

Comment	SRP Location	Public Comment	Resolution	Changes to SRP
		reviewer to apply his personal definition of "robustness" to the applicant's analyses. Remove "robust" from Line 4687.		
NAC	5185	Confinement Monitoring Capability. Welded closure seal. "However, the lack of a closure monitoring system has typically been coupled with a periodic surveillance program that would enable the licensee to take timely and appropriate corrective actions ..." Dry cask storage systems have been approved without a closure weld seal monitoring system, as within the storage cask, surveillance of the closure weld is not feasible. Temperature monitoring and/or visual surveillance of the air cooling vents is a standard part of concrete cask (welded canister) licensing.	See NEI Comment 73	See NEI Comment 73

Comment	SRP Location	Public Comment	Resolution	Changes to SRP
NAC	5426	Table 5-2, Release Fractions. "…should not be used for spent fuel described as damaged." Based on NUREG/CR-6497, damaged fuel would not have a driving force to release fines from the matrix. What is the postulated issue here? Is there data available to NRC that indicates a safety concern? Provide additional guidance and describe what factors are suggested for damaged fuel.	There is lack of data regarding the release fractions from damaged fuel, which makes this a safety issue. The data available does not apply to damaged fuel but rather to a single breach of one fuel rod. Compounding the issue is that many of the storage canisters are pressurized with helium to aid in the heat removed of larger thermal payloads. Without compelling factual information and data regarding the release fractions associated with damaged fuel, the staff does not feel there is not adequate evidence to generically make assumptions regarding release fractions associated with potential types of damaged fuel. Therefore, a leaktight confinement boundary is the recommended accepted practice to ensure radiological safety for damaged fuel, without additional data and analyses from the applicant.	See NEI Comment 76.
NAC	5799-5801	Radiation Source Definition. "radionuclide content, and estimated radiation source strength in Becquerels, …. should be described…." New requirement. Provide clarification as to what the basis of this request is, as radiation source strength in Ci	See NEI Comment 77.	See NEI Comment 77.

Comment	SRP Location	Public Comment	Resolution	Changes to SRP
		or Bq is not clearly related to gamma/neutron source strength (e.g. beta emitters)		
NAC	5809-5810	Radiation Source Definition (Gamma Sources) "characteristics for each gamma-ray source type should be provided, including isotopic composition, and photon yields" Is a tabulation of spent fuel isotopics requested here? If so, to what purpose? Typically, inputs into depletion analysis are provided, but not isotopics of depleted materials. Clarify requirement if a tabulation of spent fuel isotopics is requested and describe purpose	See NEI Comment 78	See NEI Comment 78
NAC	5813-5814	Radiation Source Definition (Gamma Sources) Within gamma source description, "describe the extent to which radioactivity may be induced by interactions involving neutrons originating in the stored materials" If this implies n-gamma reactions, then the current SRP version is clearer If activation is to be considered for decommissioning, that should be clarified.	See NEI Comment 79	See NEI Comment 79

Comment	SRP Location	Public Comment	Resolution	Changes to SRP
NAC	5868-5870	Shielding Analyses (Computer Codes) "The applicant should defend any simplifications and assumptions by showing that the approach used will result in conservative (bounding) estimates." Clarify if results need to be bounding or "provide reasonable assurance" as stated in Section 6.4, Line 5723: "Reasonable assurance that the proposed design fulfills the acceptance criteria "	See NEI Comment 80.	See NEI Comment 80
NAC	5873-5874	Shielding Analyses (Computer Codes) "...SAR should numerically specify source term uncertainties for high burnup fuels" in combination with "...validation data is relatively limited for burnup above 45 GWd/MTU." High burnup fuel is licensed and in storage. No indication that substantial dose effects occurred. If limited data is available it leaves an open ended question as to how to specify uncertainties. Conservative assumption and desired design margins are not defined, leaving it up to each reviewer when, and how much, in uncertainties to	See NEI Comment 81.	See NEI Comment 81.

Comment	SRP Location	Public Comment	Resolution	Changes to SRP
		apply. Provide correlation why maximum fuel assembly heat load is related to uncertainties. Low heat capacity/minimal shield system may be affected by low fuel assembly heat load and vice versa		
NAC	6003-6004	Radiation Source Definition (Initial Enrichment) "Applicant and the staff should not attempt to establish specific source terms as operating control and limits for cask use." If that is the case, why does the SRP focus in the Section 6.4.2 on curie content and isotopic description of the spent fuel? For Cobalt-60 dominated hardware sources, a source term may be more appropriate than other limits (e.g., mass, exposure, cool time).	See NEI Comment 84	See NEI Comment 84
NAC	6149-6150	Shielding Model Specification (Configuration of the Shielding and Source) "...homogenization should not be used in neutron dose calculation when significant neutron multiplication can result from moderated neutrons..." While not changed from current SRP statement, it	See NEI Comment 84	See NEI Comment 84

Comment	SRP Location	Public Comment	Resolution	Changes to SRP
		should be noted that standard practice is to homogenize the rod lattice in shielding calculations (not necessarily homogenizing basket structure into the fuel region). Provide additional guidance and/or justification why the standard practice of homogenizing the rod lattice in shielding calculations should not be used.		
NAC	6221-6222	Shielding Analyses (Computer Codes) "The reviewer should be aware that often adjoint calculations are performed by the applicant ... importance functions…" Review staff should recognize that importance functions may also be produced with Monte Carlo, point-kernel and transport codes. Include importance functions produced with Monte Carlo, point-kernel and transport codes	See NEI Comment 88	See NEI Comment 88
NAC	6246-6248	Shielding Analyses (Computer Codes) "The applicant should use the latest released computer code version that is valid for the particular computational platform used to perform the analysis." This item in particular has been discussed	See NEI Comment 89	See NEI Comment 89

Comment	SRP Location	Public Comment	Resolution	Changes to SRP
		with NRC staff as a significant issue. Licensed code for same type of application should not require code version change unless safety issue has been identified. Continual use of different code version within an application is difficult to reconcile and potentially leads to unnecessary confusion. Typical new release code versions tend to contain a certain amount of bugs that get resolved through user feedback to code originator. Could be interpreted that a newer code provides more "accurate" result; but as previous version was found to be acceptable for system approval, there should be no requirement for change. The goal per draft SRP Section 6.4 is to provide reasonable assurance that system will meet limits.		
NAC	6302-6309	Shielding Analyses (Computer Codes) "by verifying that the following information has been provided in the SAR ... The computer code solutions to a series of	See NEI Comment 90	See NEI Comment 90

Comment	SRP Location	Public Comment	Resolution	Changes to SRP
		test problems ..." Draft SRP does not contain the previous SRP statement "that these solutions may be referenced, and need not be submitted in the SAR". This change would add a substantial amount of information to the SAR without any safety benefit, as the referenced documents, per current SRP, should be public information and/or have been previously submitted to NRC. Adopt current SRP verbiage and add: "These solutions may be referenced but need not be submitted in the SAR."		
NAC	7697	Methodology to Evaluation Flaw Size For stainless steel casks and welding, this is too limiting. NAC uses the J-integral method to evaluate flaw size which limits the size of a single weld pass. In order to be consistent with 7682, it should explicitly state that the applicant can use J integral methodology based incorporating plasticity for ductile weld materials such as stainless steel.	See NEI Comment 121	See NEI Comment 121

Comment	SRP Location	Public Comment	Resolution	Changes to SRP
NAC	9131-9232	Fuel Temperature Range Limits This could be interpreted as the limit in any one cycle of fuel temperature is limited to 65°C. It needs to explicitly state that the 65°C range can be exceeded, but for less or equal to 10 cycles.	See NEI Comment 157	See NEI Comment 157
NAC	10418-10433	Charpy Test Requirements. Use of carbon steel less than 5/8 inch thickness. NRC's position/guidance should be stated. Clarification should include ASME Code concurrence that fracture testing is not required for material with wall thicknesses of less than 5/8 inch	See NEI Comment 171	See NEI Comment 171
NAC	11007	Exposures at or Beyond the Controlled Area Boundary (Normal Conditions) Focus added on "additional engineering features and distance from array." As only hypothetical array and single cask are evaluated, it is not clear when features would be required to show compliance with regulations and should be included in the conditions of cask use. Specific distance and shielding options and inclusion of such	See NEI Comment 181	.See NEI Comment 181

Comment	SRP Location	Public Comment	Resolution	Changes to SRP
		limitations in the CoC do not seem to be consistent with the 72.212 evaluation that a site would do to establish compliance with the requirements. Further guidance is required		
NRC Clarification	1445, 1507, 1659, 1682, 1701, 2501, 4533, 4590, 5094, 5249, 6066, 6648, 9993, 11242,	Corrected the typo for the word "Principal"	Correct typo.	Change SRP as indicated
NRC Clarification	5765	Deleted the word "rate". Deleted the words "for occupational exposure and".	Revise the statement to make it more accurate.	Changed SRP as indicated
NRC Clarification	5793	Deleted the words "placed"	Revise the statement to make it more accurate.	Changed SRP as indicated.
NRC Clarification	5873	Changed the verb "is" to "are"	Correct typo..	Changed SRP as indicated
NRC Clarification	5880	Deleted the word "each"	Correct grammar error	Changed SRP as indicated
NRC Clarification	5908	Deleted the word "operations"	Delete redundant word	Changed SRP as indicated
NRC Clarification	5915	Replaced the word "functions" with "operations"	Revise the statement to make the meaning clearer	Changed SRP as indicated
NRC Clarification	5960	Changed the words "5" to "6"	Correct typo.	Changed SRP as indicated
NRC Clarification	5977	Corrected the typo for the word "Principal"	Correct typo.	Changed SRP as indicated
NRC Clarification	5999	Corrected the typo for the word "Principal"	Correct typo.	Changed SRP as indicated
NRC Clarification	6018	Changed "fall" to "falls"	Correct typo	Changed SRP as indicated

Comment	SRP Location	Public Comment	Resolution	Changes to SRP
NRC Clarification	6103	Added verb "are" between "there" and "specific"	Correct typo.	Changed SRP as indicated
NRC Clarification	6230	Changed "ORNL" to "EPRI"	Correct typo.	Changed SRP as indicated
NRC Clarification	6373-6379	Modified line 6373 to add the following words between "to use" and "additional ...": "distance or" Modified line 6374 to add the following words between "berms," and "to mitigate": "or both," Modify line 6375 to replace the words "to show compliance with the regulations" with "evaluations" Modify line 6376 to replace the words "cask conditions for use" with "system and described in the CoC"	Revised the statement to make them more accurate and consistent with staff's response to NEI's Comment 181	Changed SRP as indicated
NRC Clarification	6405	Changed "1m (3.3ft)" to "100m (328 ft)"	Corrected typo.	Changed SRP as indicated
NRC Clarification	6408	Delete the word "phenomenon"	Revised the statement to make it consistent with 10 CFR 72.92.	See NEI Comment 84. Change SRP as indicated
NRC Clarification	8643	Changed priority of Section 8.4.20 to Medium from Low	Change was result of ongoing industry practice of omitting leak testing at fabrication facility	Added "(MEDIUM Priority)" after Section 8.4.20 heading.
NRC Clarification	10976-10979	This sentence mis-states what is required by the regulation.	Reword the sentence so that it correctly reflects the regulatory requirement.	Replaced the first sentence of the first paragraph in Section 11.5.3 with the following text: As required by 10 CFR 72.236(d), the application must

Comment	SRP Location	Public Comment	Resolution	Changes to SRP
				demonstrate that the shielding and confinement features of the cask are sufficient to meet the requirements for real individuals in 10 CFR 72.104, and for DBA conditions in 10 CFR 72.106. These demonstrations in the application facilitate future site-specific evaluations for each general ISFSI licensee.
NRC Clarification	10980	Added the word "boundary" after "controlled area"	Revised the statement to make the meaning clearer	Changed SRP as indicated
NRC Clarification	6863,6864,6993, 6994,7005, 7098,7100	Editorial corrections are needed to make the citation for references compatible with the format of the Consolidated References in Appendix A.	Comment implemented. See NEI Comment 86.Correct typo.	Changes to the reference citations were made in the lines indicated to be compatible with the format in Appendix A.
NRC clarification		The meaning of the first statement in Section 6.1 is unclear regarding what is evaluated for the review.	The statement was clarified to indicate that the evaluation is of the ability of the shielding features to provide adequate shielding.	Modified the first sentence in Section 6.1 to read as follows:

The shielding review evaluates the ability of the proposed shielding design features to provide adequate protection against direct radiation from the dry storage system (DSS) contents. |
| NRC clarification | | The statement in the third paragraph of Section 6.2 refers to dose rate limits; this is not correct. The regulatory limits are dose limits | The statement was corrected to refer to dose limits. | Removed the word "rate" from the sentence in the third paragraph of Section 6.2 (see bottom page 6-1) so that the sentence will read "...meet the dose limits at the controlled |

Comment	SRP Location	Public Comment	Resolution	Changes to SRP
				area boundary …"
NRC clarification		The shielding chapter guidance covers all of 72.104, not just 72.104(a). Table 6-1 should indicate this.	Table 6-1 was corrected to refer to 72.104.	Deleted "(a)" next to 72.104 in Table 6-1.
NRC clarification		The terms "general licensee" and "DSS user" refer to the same entity; so, using them together, as is done in the first paragraph of Section 6.4, is redundant.	The sixth sentence of the paragraph was modified to delete the redundant reference.	Deleted "DSS user" from the sixth sentence of the first paragraph in Section 6.4.
NRC clarification		The guidance described as part of this item 1. in Section 6.4 (second paragraph) includes all of 72.104 and not just 72.104(a).	The statement was corrected to refer to 72.104.	Deleted "72.104(a)" in the first sentence of item 1. of Section 6.4, second paragraph, and replaced it with "72.104".
NRC clarification		Part 72 establishes dose requirements for ISFSI operations (first paragraph in Section 6.5.1.1, second sentence).	The sentence was clarified to indicate that ISFSI operations are included in the dose requirements.	Modified the second sentence of the first paragraph in Section 6.5.1.1 to read as follows: While 10 CFR Part 72 establishes dose requirements for the ISFSI and its operation, it does not impose specific dose rate limits on the individual casks.
NRC clarification		The box for Chapter 9 in Figure 6-1 shows an item for pre-shipment surveys. This language seems to be inappropriate since there is no shipment of the cask under Part 72.	The figure was corrected to remove references to pre-shipment surveys. Surveys are still a part of operations; so, the surveys are linked with the item for radiation levels.	Deleted "Pre-shipment Surveys" from the Chapter 9 box in Figure 6-1 and modify the first bullet to read as: "Radiation Level Surveys".
NRC clarification		The limits in 10 CFR 72.104 are for doses and not dose	The reference to the regulation was corrected to state dose limits. The	Deleted the word "rates" from the third sentence of the fourth

Comment	SRP Location	Public Comment	Resolution	Changes to SRP
		rates. Also, applicants may also use dose v. distance curves for evaluations for this requirement instead of dose rate v. distance curves. (See fourth paragraph of Section 6.5.4.3.)	text is also clarified to indicate that dose v. distance or dose rate v. distance curves may be used for evaluations for this requirement.	paragraph in Section 6.5.4.3. Add the words "dose or" to the fifth sentence of the same paragraph so that the sentence begins with "The applicant should include a dose or dose rate versus ..."
NRC clarification		Section 72.104(c) also applies to operations. Section 72.104 applies to all operations; there is nothing in the regulations to indicate that Section 72.104 doesn't apply to cask unloading. (See Table 9-1.)	Table 9-1 was corrected to include reference to 72.104(c) and show that 72.104(b) and (c) apply to cask unloading.	Included "72.104(c)" in the list of applicable regulations in Table 9-1 and added a bullet in the table entry for 72.104(b) and (c) and cask unloading.
NRC clarification		The text in Section 10.5.1.4 should be clear that the testing is for materials used for neutron shielding; guidance for testing neutron absorbers for criticality control is described in the following subsection of the SRP	The text in the second paragraph of the section was clarified to focus on neutron shielding materials.	Deleted the word "absorption" and replaced it with "shielding" in the first sentence of the second paragraph in Section 10.5.1.4.
NRC clarification		The guidance in Sections 10.5.1.6 through 10.6 applies to the DSS. The text should be clear and use this terminology.	The text of these sections was corrected to refer to the DSS.	Replaced the word "cask" with the word "DSS" near the beginning of the first sentence in Section 10.5.1.6 and deleted the word "cask" where it appeared again at the end of the sentence. Replaced the word "cask" with "DSS" in the title of Section 10.5.1.7 and in the first sentence of Section 10.5.1.7. Replaced the words "Storage casks" with "DSS" in the

Comment	SRP Location	Public Comment	Resolution	Changes to SRP
				sentence in Section 10.5.2. Replaced the words "storage cask" with "DSS" in the first sentence of both paragraphs in Section 10.5.2.1. Replaced the word "cask" with "DSS" in the first sentence in Section 10.5.2.3. Replaced the word "cask" with "DSS" in F10.1, F10.3, F10.4, and the first and second sentences of the last, or summary, paragraph of Section 10.6.
NRC clarification		The second sentence of the summary paragraph in Section 10.6 is missing the object of what is being stored (i.e., the words "spent fuel" are missing).	The sentence was corrected to include the words "spent fuel". In conjunction with the previous clarification comment, the word "cask" was replaced with "DSS."	Corrected the second sentence of the summary paragraph of Section 10.6 to read: "...the DSS will allow safe storage of spent fuel throughout its licensed or certified term."
NRC clarification		Table 11-2 is missing 72.126(d), which is also applicable to radiation protection. 72.126(a) and (d) apply to all areas of review listed in the table. Also, there is no 72.1064(b); it should be 72.106(b) and ALARA does not apply to 72.106(b).	Table 11-2 was corrected to include the stated items.	Include "72.126(d)" and added bullets to show the applicability of 72.126(a) and (d) to occupational exposures and exposures at or beyond the controlled area boundary. Also, "72.1064(b)" was corrected to read "72.106(b)" and the reference to ALARA was removed..
NRC clarification		The last sentence in Section 11.4 lists reviews that should be coordinated with the Chapter 11 review; "the chapters" should read "these chapters".	This editorial correction was implemented.	Edited the last sentence of Section 11.4 to read: "...therefore, the reviews of these chapters should be coordinated."